Combined Proceedings, International Plant Propagators' Society 2018

Abstracts from papers presented at the conferences held annually in each of the IPPS regions around the world.

IPPS Volume 68

Editors:

Robert L. Geneve
Clive Larkman
Charles W. Heuser Jr.
Spencer Gunn
Hadyen Foulds
Elsa du Toit
Fred T. Davies Jr.
Gene K. Blythe

ISBN: 978-1-5108-9398-6

Printed from e-media with permission by:

Curran Associates, Inc.
57 Morehouse Lane
Red Hook, NY 12571

International Plant Propagators' Society (IPPS)
174 Crestview Drive
Bellefonte, PA 16823-8516
USA

Phone: +01 814-355-2467

secretary@ipps.org

Additional copies of this publication are available from:

Curran Associates, Inc.
57 Morehouse Lane
Red Hook, NY 12571 USA
Phone: 845-758-0400
Fax: 845-758-2633
Email: curran@proceedings.com
Web: www.proceedings.com

THE INTERNATIONAL PLANT PROPAGATORS' SOCIETY

Combined Proceedings

Volume 68

2018

IPPS

Sharing Plant Production Knowledge Globally

Published 2018

Combined Proceedings International Plant Propagators' Society (ISSN 0538-9143) is a yearly publication of the International Plant Propagators' Society (IPPS) headquartered in Pennsylvania, U.S.A.
Email: Secretary@ipps.org
Website: www.ipps.org

Business Office. Memberships are processed at the regional level, with dues being allocated to both international and regional expenses. Contact the regional secretary / treasurer in your area of the world for membership information and applications.

The International Plant Propagators' Society is not responsible for statements and opinions published in the *Combined Proceedings International Plant Propagators' Society*; they represent the views of the authors or persons to whom they are credited and are not necessarily those of the Society. The publication of research or other types of information by the Society does not constitute a recommendation or endorsement of products or procedures involved; for general use, label recommendations and directions must be followed.

Table of Contents

Eastern Region – North America

Western Region – North America

Southern Region – North America

PROCEEDING'S PAPERS

SOUTHERN AFRICA REGION

Dr. Elsa du Toit, Regional Editor

Twenty-first Annual Meeting - 2018

Port Elizabeth, Eastern Cape Province, South Africa

Combined Proceedings IPPS 68:1-3. 2018.

Where Propagating Can Take You

Dermot Molloy
Birdwood Avenue, Melbourne Victoria 3004, Australia,

Dermot.Molloy@rbg.vic.gov.au

Keywords: Botanic gardens, wetlands, Melbourne, IPPS tours.

INTRODUCTION

Royal Botanic Gardens Victoria

The gardens have two sites one in Melbourne and one in Cranbourne. The Melbourne site sits on 38 Ha and is 171 years old, it also contains the National Herbarium of Victoria. The gardens have 1.6 million visitors a year and this is expected to double by 2070. The Royal Botanic Gardens Victoria has over 50,000 plants in their collection representing over 8,000 taxa. The gardens have 30 living collections including; Southern Africa, Southern China and Californian collection. The Cranbourne site is 363 Ha and was built in 1970. The Australian Garden takes up 10 Ha and the remaining 353 Ha is remnant bushland.

Royal Botanic Gardens Nursery Melbourne

The Royal Botanic Gardens Nursery has 3 staff and was built in the 1850's. The nursery has 3 main functions – propagation and production; research; and collections.

Propagation and Production

The nursery produces approximately 50% natives and 50% exotics by various methods including; cuttings, seed, grafting, layering, spore and tissue culture.

Plants are grown to order for horticulturists, botanists, research students, Government Department of Environment, Regional Botanic Gardens and local government.

The garden has contracts to grow rare and threatened species for reintroduction to the wild as well as for seed orcharding for the Victorian Conservation Seedbank.

Research

Research at the garden includes seed collecting trips throughout Australia. Australian native plants are grown for taxonomic assessment at National Herbarium of Victoria as well as plants grown to produce herbarium voucher specimens. In the garden, plants are placed in areas divided into families, Rhamnaceae, Asteraceae, etc.

Collections

The gardens currently house a Terrestrial Orchid and a Tropical Plant collection located in the nursery. Victorian terrestrial orchids are grown as an ex-situ collection as well as for reintroduction in the wild.

The tropical glasshouse grows plants of economic importance to assist our education department with their programs. It also houses plants collected from all over the world such as *Amorphophalus titanum.*

Recent Projects at RBGV

Working Wetlands Project

Rainfall harvested from the surrounding streets and stored in our lake and treatment tanks. The lake holds 30 Megalitres which saves 40% of potable water use now with a plan to be 100% self-sufficient. Floating treatment wetlands will filter the water as it circulates the three main lakes. This project also gave the nursery experience in germinating and growing wetland species.

Figure 1. Fern Gully area.

Fern Gully Boardwalk Project

Opportunity to learn how to grow ferns from wild collected spore. The boardwalk is now completed, and the second stage is underway (Figures 1 and 2). The second stage will contain health and wellbeing spaces as well as a sensory garden.

Figure 2. Fern growing in the Fern Gully.

Landscape Succession Plan

The RBGV landscape to remain the same just with different species able to adapt to a drier climate. The whole garden has had every tree and shrub audited for its suitability for the future climate. All garden beds have been placed into high, medium and low water requirements. Climate modelling suggests that Melbourne's future climate will be similar to Dubbo in New South Wales. The RBGV is projecting climate change to 2090 and unsuitable species will be sent to other gardens with a better climate.

Benefits of IPPS

I joined IPPS in 2006 and was elected to the board in 2016. The benefits of being a part of IPPS are many and include:

- Attending Local and International IPPS Conferences and tours.
- Broadening my plant knowledge.
- Encouraging young propagators.
- Sharing plant production knowledge.

Tours

I have attended many tours and workshops around the world and gained valuable knowledge and many friends. These tours include:

- Seedbank, Herbarium and Nursery workshop in United Arab Emirates.
- Oman Botanic Gardens, David Hancock and I ran field trips and propagation workshops for staff.
- IPPS Europe preconference tour through England, Belgium, Netherlands and Germany (Figure 3).
- IPPS Conference and post conference tour in Port Elizabeth, South Africa.
- IPPS New Zealand post conference tour of Southern New Zealand.

Figure 3. Solitair Nursery, Belgium

Combined Proceedings IPPS 68:4-7. 2018.

The Role of Soil Surveys in Land Use Evaluation and Planning

Gavin N. Schafer

terramap@net4all.co.za

Keywords: Soil productivity, soil classification, erosion, plantation productivity.

Abstract

Soil Surveys play an important role in both agriculture and forestry to apply best land practice, for farm planning, suitable crop choices, yield predictions and for control of erosion and compaction. Land evaluation includes assessing terrain, climate, geology, soil and vegetation cover. Soil information provides an essential facet for land assessment and suitability for a particular use or crop. A soil survey can be conducted at various levels from a detailed survey for irrigation at a scale of 1:5,000 to a reconnaissance level at scales of 1:20,000 to 1:50,000.

INTRODUCTION

Soil surveys are conducted on a portion of land, farm or estate for the purposes of gathering soil and terrain information that will enable the creation of a soil map. This map in turn should provide a spatial reference to different soil types each with their specific soil characteristics that can be used to make decisions about that unit of land in terms of its suitability for a specific use or crop, and its productivity potential.

Soil survey interpretation comprises the organization and presentation of knowledge about characteristics, qualities and behavior of soils as they are classified and outlined in maps. The purpose being to provide people with the best possible information for appropriate land use planning and decisions.

An overview deals with important soil characteristics, how they are described and the use of the South African soil classification system. The use of soil surveys for various agricultural crops, forestry and even estate landscaping projects is discussed.

SOIL DESCRIPTION

Field work entails the description of soil profiles using either soil pits or the use of a handheld auger. For the purposes of subsequent mapping these observations are conducted either in grid form or free mapping. The grid method provides better data, but is more expensive, whereas free form mapping conducted with the aid of aerial satellite photo imagery and contours can provide a satisfactory soil map especially for reconnaissance mapping. Table 1 lists the important soil characteristics to be described for assessing land use options.

Table 1. Important soil characteristics and related land qualities used in land evaluation (after McRae and Burnham 1981).

Soil characteristic	**Related land qualities**
Soil texture and stoniness	Ease of cultivation, moisture availability, drainage and aeration, fertility, water and wind erosion, soil permeability, irrigateability, rootability.
Visible boulders, rock	Ease of cultivation, moisture availability.
Soil depth	Moisture availability, ease of cultivation, rootability.
Soil structure	Wind and water erosion hazard, rootability, moisture availability.
Organic matter	Moisture availability, nutrient retention and availability, wind and water erosion.
Soil colour	Drainage and aeration.

Influence of soil structure on rooting

Soil structure refers to the degree, size and shape of peds or soil particles. An apedal soil is one without structure and provides a good medium for rooting but is more prone to compaction. A fine crumb structure, especially in the topsoil, provides a good medium for rooting due to greater surface area of small sized peds. In contrast a course firm blocky or prismatic structure reduces rooting surface area, as roots will only grow between the ped surfaces. Figures 1 and 2 illustrate two different structure types.

Figure 1. Strong blocky soil structure.

Figure 2. Crumb soil structure.

Figure 3. Eroded barren land.

The inset on figure 3 illustrates a duplex soil with a light textured topsoil overlying a strongly structured clay subsoil. This soil is also sodic and disperses easily once the cover crop has been removed. This has resulted in the extent of erosion on such a soil, which has been incorrectly managed and not suitable for irrigation. This land is now useless.

SOIL CLASSIFICATION

The classification and naming of soil types is very useful for communication, besides aiding with appropriate land preparation and crop choices. There are several soil classifications worldwide viz. USA, Australia, New Zealand, FAO (Rome) and South Africa to name but a few. The South African Classification has two levels, soil form and family (Soil Classification Working Group, 1991).

Each soil horizon has specific characteristics such as structure, colour, organic carbon content which is used to define what is referred to as a diagnostic horizon. The soil profile is classified depending on the vertical sequence of these horizons. In figure 3 there are three diagnostic horizons. An orthic A overlies an E or eluviated layer which abruptly overlies a firm gleyed clay. A Kroonstad form soil is poor in respect of the abrupt transition to a clay that perches water in the rainy season and has an E horizon that sets hard in the dry season.

Figure 3. Kroonstad soil form

SOIL MAPPING

Soil mapping requires adequate soil information gathered at each soil pit or auger drilling. This will be used as a data base for the delineation of like soil types. These take into account factors mentioned earlier such as soil depth, soil form, drainage, texture and structure.

Delineation of the units is normally done with the aid of aerial photography or satellite imagery and the use of contours, as soil normally changes with the slope shape and gradient. Vegetation and soil patterns are often quite distinct unless the vegetation is dense forest or plantation and provide a means to delineate these soil units.

The soil descriptions are now used to build a map legend, which can provide useful data for planning on each soil type.

SOIL SUITABILTY AND PRODUCTIVITY

Suitability is largely a matter of producing high yields with relatively low inputs. In order to establish this much research has gone into long term trial, and measurement of yield of numerous crops on various soil types and soil conditions.

Productivity ratings or yield predictions will be based on this research. Productivity ratings were developed for the main commercial tree species in the Cape regions. The end result was that the plantation soils were mapped on 150x150m and the productivity models were applied, using soil, terrain and climate data. The development of these models made use of long-term tree growth yield from all plantation compartments - see Table 2.

Table 2. Site Productivity models for the Cape regions (MAI20 is mean annual increment in m3/ha/annum at age of 20yrs)

MANAGEMENT PRACTICE & RECOMMENDED SPECIES	P.C.	PREDICTED MAI_{20}
Veneer or sawtimber:-*P.radiata, P.taeda* or furniture timber species	I	>20
Sawtimber:-*P.radiata* or *P.elliottii*	II	16-20
Sawtimber: *P.radiata*	III	10-15
Poles/short rotation: *P. radiata*	IV	5-9
P. pinaster	V	<5

Literature Cited

McRae, S.G. and Burnham, C.P. (1981). Land Evaluation. Monographs on soil survey. Clarendon Press. Oxford. 239pp.

Soil Classification Working Group (1991). Soil Classification; A Taxonomic System for South Africa. Memoirs on the Agricultural Natural Resources of South Africa No.15. Department of Agricultural Development, Pretoria.

Combined Proceedings IPPS 68:8-9. 2018.

The Role of Nurseries in Preserving Citrus Biosecurity in Southern Africa

Paul Fourie and Vaughan Hattingh
Citrus Research International, Nelspruit, South Africa

phf@cri.co.za

Keywords: Disease, Huanglongbing, HLB, Asian Citrus Psyllid.

INTRODUCTION

South Africa is the 2nd largest exporter of fresh citrus in the world, shipping citrus further and to more export destinations than any other country. Sustainable citrus production, as well as the access to markets, are continuously threatened by exotic pests and diseases. When new pests or diseases are found in a production region, these might impact significantly on tree health and production and/or have significant trade implications, ranging from regulated treatments to quarantine measures.

As a worst-case scenario, incursion of the Huanglongbing (Asian Citrus Greening) bacterium and its insect vector the Asian Citrus Psyllid into Florida (USA) is effectively crippling this once thriving citrus industry. Huanglongbing (HLB) is near-impossible to control and countries still free from this dreaded disease, such as South Africa, are taking pre-emptive measures to prevent or delay its incursion and establishment.

Many of the dreaded citrus diseases, including HLB, will not show symptoms on propagation material or young nursery trees, and infected propagation material, especially rooted plants, are the most effective means of (inadvertently) spreading pests and diseases. Moreover, these pests and pathogens are often not restricted to *Citrus* spp. alone and can also be spread on related species.

Role of Nurseries

The roles of nurseries growing citrus and related (mainly Rutaceae) plants are most important in preventing incursion and spread of exotic citrus pests and diseases. Control measures applicable to nurseries include the use of pest- and disease-free propagation material, effective pest and disease control measures during the production and retail stages, and quarantine measures. Nurseries furthermore play a very important role in creating awareness and educating the general public.

To this end, nurseries must be aware of legal restrictions on movement of plant material in South Africa and ensure compliance with all import requirements. In the interest of the nursery sector, potentially affected agriculture and industry sectors and the country as a whole, nurseries must be vigilant and are urged to take action when receiving information that indicates a biosecurity risk or legal transgression.

PROCEEDING'S PAPERS

NEW ZEALAND REGION

Hayden Foulds, Regional Editor

Forty-seventh Annual Meeting - 2018

Palmerston North, New Zealand

Combined Proceedings IPPS 68:10-12. 2018.

Technology Innovation for Plant Propagation in New Zealand

Paul Fisher
University of Florida, PO Box 110670, 2549 Fifield Hall, 2550 Hull Rd., Gainesville FL USA

pfisher@ufl.edu

Keywords: Labour, costs, media, water, air, automation.

INTRODUCTION

With a high labour cost in New Zealand (NZ) and increasing competition between nursery firms, efficiency in the use of labour and other resources is critical for future growth of this industry. Technologies already being used in North America, Europe, and other overseas locations provide opportunities for increased productivity in NZ, but an "appropriate technology" approach is important because automation is only one aspect of improved efficiency.

GROWING SUBSTRATE

Many NZ growers root plant cuttings in a highly porous mix, often based on pumice or bark, in a community or "hygiene" tray that lacks individual cells. This practice has two main downsides, which include the labour requirement to prick out rooted cuttings to transplant to larger liner containers, and the disturbance and potential damage of root systems as large substrate particles fall from roots during the pricking out process. Pumice substrate has large air spaces between particles, allowing for some air diffusion even when using coarse and frequent sprinkler irrigation. However, the low matric potential in a pumice substrate results in an uneven vertical distribution of air and water meaning that the zone of ideal water/air balance for root initiation and growth is limited in the soil profile.

There are several potential improvements. Trays with individualized cells facilitate grading for quality, and faster transplanting. Several innovations in tray design would benefit NZ growers, such as vent holes in the top of the tray to increase air movement and drying, a more solid tray structure rather than thin plastic to assist handling, and non-circular cell cross-section to reduce root girdling. Substrate in a cell holds together with fibrous components, such as peat or coconut coir, compared with coarse pumice. This means that root systems are less stressed during transplant than if substrate falls off roots during handling.

Crop time can be further reduced, and automatic transplant can be facilitated, by using a stabilized substrate such as foam, peat/polymer, paper-wrapped pots, which do not need a large and mature root ball to hold the cell together. Substrates with higher matric potential, such as peat-based blends, can have a large zone of ideal water/air balance for rooting if irrigation is managed carefully. Roots grown in a substrate (such as peat) that has a similar physical and chemical environment to that of the final container are more likely to grow rapidly after transplant.

WATER AND AIR MANAGEMENT

Time clocks or artificial leaves are commonly used in NZ for controlling mist, with stationary sprinklers that have coarse irrigation droplets to help provide even coverage in semi-open and windy structures. Controlling mist using time clocks means that an experienced (and highly paid) grower is needed to adjust mist depending on weather, crop stage, and cutting type, thereby increasing labour cost and challenges when the lead grower is absent. Artificial leaves are mechanical devices prone to error and inconsistency, and do not function well with fine droplet sizes or fog.

Advanced environmental control computers control mist based on vapor pressure deficit (the combination of temperature and relative humidity), sometimes in combination with wind speed and radiation, and ideally with an in-built crop aging program to reduce mist frequency as cuttings go through callus and root development stages. Simple digital controllers are available with a light meter (sometimes called a calory counter in NZ) that triggers misting during high light times, plus an incorporated time clock to control mist during low light times, and separate day and night settings to avoid over-misting at night.

Most large-scale propagators internationally have moved to boom irrigation to increase mist uniformity, with both a fine mist nozzle for initial hydration and callusing of cuttings and a coarser nozzle for wet-dry substrate cycles during the final root development and hardening off stage. Greenhouse structures that avoid high air flow during the root initiation stage, thereby maintaining high relative humidity, can result in a lower mist frequency and less potential overwatering compared with the open-walled structures often used in NZ. Most NZ propagation structures have fixed shade, however movable shade and automated ventilation allow the environment to adapt to changing weather (which is a feature of NZ climate) and other factors such as crop stage.

The growing substrate must be matched to the irrigation system. Irrigation improvements are essential if NZ growers move to a peat-based substrate with a high matric potential, rather than pumice. Otherwise, the substrate will be over-watered, root diseases and algae growth will increase, and productivity gains from faster rooting and reduced losses will not occur.

LABOUR AND AUTOMATION

Labour costs are high in NZ relative to many countries, however many NZ nurseries are small to middle-scale with high plant diversity. Some NZ growers have moved to automated transplanting, in combination with use of individualized-cell trays and peat-based substrates, but this is primarily for bedding plants rather than natives and other perennials that have widely varying plant forms and sizes. For some larger NZ growers, transplanting robots being used in North America and Europe would have a positive return on investment. However, regardless of size all growers should evaluate simple and low-cost automation of tasks such as dibbling of planting holes in trays.

In addition, hiring lean consultants is standard practice in large international propagators to review the transplant manufacturing process and identify opportunities to reduce labour and other waste even without automation. There are many online articles and training materials available on lean concepts applicable for growers of any scale.

Periodically taking a step back from the day-to-day hard work of running a nursery is essential for business owners in order to identify opportunities to reduce waste and increase efficiency.

In some cases, this may mean automation, however nursery pro-duction will always be labour-intensive and reliant on managing an efficient team of people.

Therefore, focusing on staff training has a rapid return on investment, along with participating in nursery tours with IPPS, and bringing in experienced consultants who can look at processes with a fresh perspective.

Combined Proceedings IPPS 68:13-16. 2018.

Grow Superfoods in Containers: Ginger and Turmeric

Rosanna Freyre, Sofia Flores and Paul R. Fisher
Dept. Environmental Horticulture, University of Florida, PO Box 110670, Gainesville, FL 32607, USA.

rfreyre@ufl.edu

Keywords: Edibles, *Zingiber, Curcuma*.

INTRODUCTION

Ginger and turmeric are traditionally used as spices in powder form, especially in curries and masalas. Ginger is also used as an ingredient in beverages such as ginger ale or ginger beer, and to make candies, preserves or tea. Fresh or dried ginger rhizomes can be easily found in most supermarkets and ethnic grocery stores. However, fresh turmeric rhizomes are less widely available in New Zealand.

These plants have also been used in Indian and Chinese medicine for centuries. Ginger is mostly taken to improve digestion and to counter nausea, and turmeric as an anti-inflammatory and for skin health. In recent years these two crops have been included in the category of superfoods in Western markets – i.e., foods that are rich in antioxidants, nutrients or vitamins, or that have health benefits. There are claims that both ginger and turmeric relieve inflammation and are good for joint health, for reducing blood sugar, and even fight cancer. Nowadays ginger and turmeric can be commonly found in capsules, drinks or tonics, and their teas are sometimes labeled as herbal supplements. They are also used as ingredients in healthy smoothies, therefore fresh rhizomes are sought-after items and can obtain high prices in farmers' markets or grocery stores. There is also a potential market for nursery owners to grow and sell edible ginger and turmeric container plants in the spring to consumers that want to grow and harvest their own product.

Both ginger and turmeric (Figure 1) are in the Zingiberaceae family. They are ancient crops domesticated centuries ago, and there are no live wild relatives. Edible ginger is *Zingiber officinale*, and there are about 150 species of *Zingiber* – 34 of them from India and 24 from China. *Zingiber zerumbet* and *Z. spectabilis* have spikes with colorful bracts and are used as ornamentals, as well as *Z. mioga*, which is a dwarf variegated plant with edible buds. Another ornamental ginger is *Hedychium gardeniarum*, considered an in-vasive plant in New Zealand and Hawaii. However, *Z. officinale* is sterile and does not produce seeds. Edible turmeric is *Curcuma longa*, which is also mostly sterile, and *C. amada* is also edible and tastes like green mango. There are about 110 species of *Curcuma*, and the greatest diversity is found in India, Myanmar, and Thailand. There are more than 10 ornamental *Curcuma* species, among them *C. alismatifolia, C. elata* and *C. roscoeana*, with beautiful and colorful spikes. *Curcuma caesia* or black turmeric has attractive leaves with a dark central vein, rhizomes with dark blue centers, and is used as a traditional medicine.

Fig. 1. Rhizomes of ginger (left) and turmeric (right).

CULTIVATION

To plant ginger or turmeric, rhizome fingers should be cut to 5 to 8 cm (15 to 75 g) with two to four buds. The cut areas should be surfaced sterilized with a 10% bleach solution, and then the seed pieces should be dried (cured). Because sprouting of buds can be uneven, it is recommended that seed pieces are maintained on humid potting mix under 75 or 80% humidity for sprouting before planting. Alternatively, tissue culture plantlets may be available for planting. These provide uniform and pathogen-free planting material, however the yield and quality of the first-year harvest is usually lower than when planting rhizome seed pieces.

When planting containers, a well-aerated potting mix should be used, with components such as coarse coconut coir, peat or bark. Tissue culture plantlets should be planted at the crown, and seed pieces should be planted about 5 cm below the surface. Enough empty space should be left at the top of the containers to allow mounding of the plants twice, around 45 and 90 days after planting, which will help increase the rhizome size. Bigger containers work best – we harvested only 350 g of rhizomes per plant of Hawaiian Red turmeric in 7 L pots, whereas yields were 1000 g per plant in 60 L pots. Plastic bags can also be used as containers, with the tops of the bags being unrolled as needed after mounding the plants. Ginger plants can show tip burn if the substrate is not kept sufficiently moist, or if the fertilizer levels are either too low or too high.

In the field, ginger and turmeric plants prefer a pH of 5.5 to 6.5. They are usually planted in early spring in raised beds, spaced 20 x 20 cm apart, and covered with 5 to 10 cm of soil. As with container production, plants should be mounded twice during the growing season, around 45 and 90 days after planting. Plants yield the most in tropical to subtropical climates of 25 to 35°C, but in colder climates they can be grown in high tunnels or greenhouses. Plants also tend to be more vigorous under 30% shade rather than full sun in high light environments.

Ginger and turmeric are quantitative short-day plants for flowering and rhizome swelling. As the temperature gets colder and the days are shorter in the autumn, plants enter into dormancy and leaves yellow. At this point the irrigation can be stopped and after three weeks the wilted plant tops can be cut off. Three additional weeks may be allowed for rhizome drying before harvest. The harvest time after planting depends on the end use. Five months after planting is enough for rhizomes that will be sold as fresh vegetables, with low fiber content, and with segments of green leaves attached. Rhizomes harvested between five and seven months after planting are suitable for curing and selling in retail, and for making preserves. Rhizomes with longer growing periods of eight to nine months are more suitable to be dried or used to extract essential oils.

In a greenhouse, the temperature and day length can be manipulated to extend the growing season. Providing a night interruption period with incandescent lights strung above the plants between 10 pm and 2 am with a minimum of 2 micromol$\cdot m^{-2}\cdot s^{-1}$ of photosynthetically active radiation provides a similar effect to naturally long day length, and plants remain vegetative. This allows for off-season harvest of fresh rhizomes, which may result in higher sale prices.

After harvest, it is recommended to wash the rhizomes by soaking or with pressure hoses to remove soil, and then treat them with a sanitizing 10% bleach solution to disinfect them. Rhizomes that will be used for seed can also be treated with fungicides. Storage should be at low temperature and high relative humidity to prevent drying (13°C and 70%), and any moldy rhizomes should be discarded.

At the University of Florida, we are evaluating different species and varieties of ginger and turmeric under different growing conditions. In our first-year harvest (February 2018) we found significant differences in yield between turmeric varieties grown in 60 L containers under natural days, with Hawaiian Red having the highest yield (average of 1 kg per plant) compared to the lowest yielding tissue culture plantlets (0.28 kg per plant). For ginger there were no significant differences in yield between plants started from seed pieces or tissue culture plantlets started at two different dates, and plants yielded on average between 0.8 – 0.98 kg per plant (Figure 2).

We are now starting our second year of research and evaluation of species and varieties for rhizome yield as well as ornamental value as landscape or greenhouse plants. We will also use different treatments on seed pieces aiming to obtain uniform sprouting as compared to the control, such as a water soak for 24 hours, a water soak at 50°C for ten minutes and ten days before planting, and hormone treatments with different concentrations of ethephon or benzyladenine. Additionally, we plan to do chemical analyses on rhizomes, and research the profitability and marketing of live transplants and rhizome use as local fresh food or for use in the beverage industry.

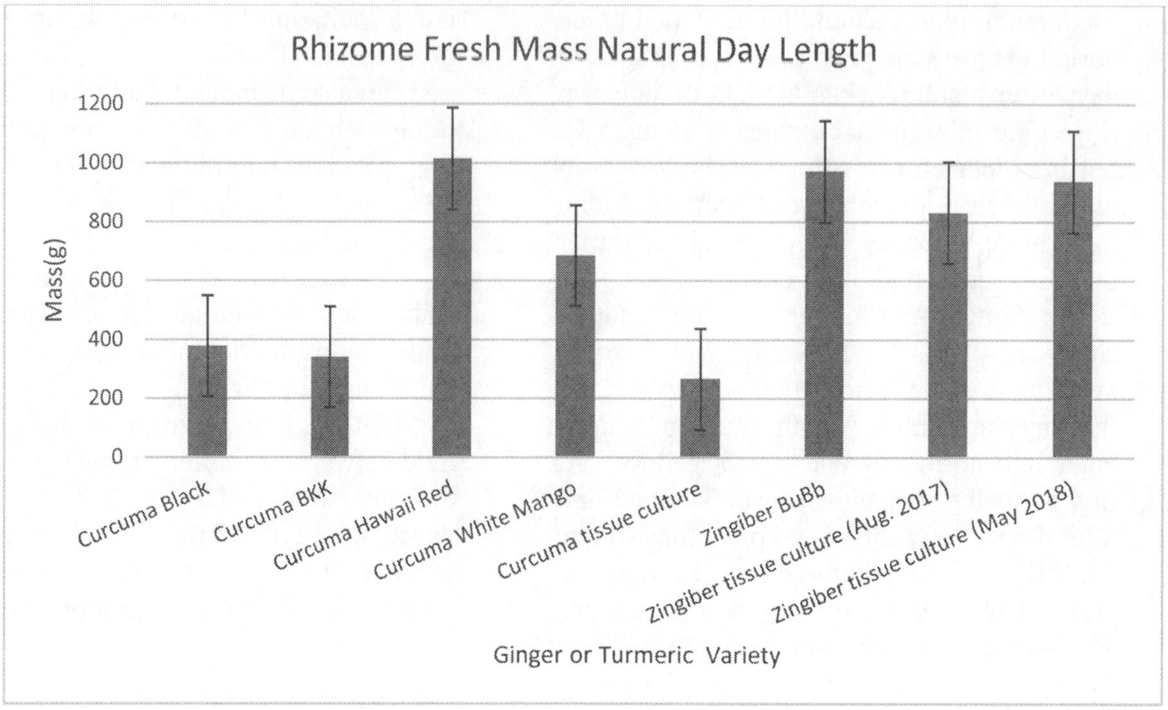

Figure 2. Rhizome production in ginger and turmeric.

Combined Proceedings IPPS 68:17-20. 2018.

International Plant Propagators' Society New Zealand – Western Region Exchange 2017

Dharini Marinkovich[1] and Dr. Keith Hammett[2]
[1]11 Barrett Road, RD3, Albany, Auckland 0793; [2]Plant Breeder at 488C Don Buck Road, Massey, Auckland 0614

Email: dharini@xtra.co.nz

Keywords: Canada, United States, scholarship.

IPPS Western Region exchange 2017

When I applied for the International Plant Propagators' Society New Zealand - Western Region Exchange in 2017 I knew that I could have the opportunity to attend the Western Region annual conference in Wilsonville, Oregon. Little did I know that IPPS would send me on a journey through two countries and over 3900km total I will try to summarize some highlights of the exchange.

I landed in Vancouver, Canada, on the 7th of October. Valerie Sikkema picked me up from the airport and although I had told myself not to do this, I immediately went to get in the driver's side of the car! It was the day of the cranberry festival, so we visited Fort Langley to see some flooded fields with floating cranberries and sample some berry wines.

Valerie and Arnold Sikkema took me to Lynn Canyon for a hike (not a 'tramp'). I tried my best to learn some of the local flora, feeling quite disoriented without the usual New Zealand natives around. The canopy was filled with Douglas fir (*Pseudotsuga menziesii*), Canadian hemlock (*Tsuga canadensis*), and Pacific red cedar (*Thuja plicata*). Huckleberry (bear food!), Salal (*Gaultheria shallon*) and swordferns dominated the undergrowth.

Douglas Justice is the Associate Director of the Horticulture and Collections at the UBC Botanic Garden. After attending his lecture on Woody Plant Identification at the University of British Columbia (UBC), I was treated to a tour of the UBC Botanical Gardens. They have a focus on rhododendrons, magnolias and maples to name a few, many of which are collected from seed from all over the world. Every plant that comes in and goes out is tracked, identified and meticulously recorded.

To finish the day, I was taken on a tour of the UBC Nitobe Memorial Garden (Figure 1), one of the top five Japanese gardens outside of Japan. Each stage of the garden is carefully yet subtly designed to take you through a symbolic journey of life; from birth, through rebellious teenage years, family time, and death.

Figure 1. Nitobe Memorial Garden.

Before leaving Canada, we visited Valerie's work, Van Belle Nursery. They were big users of the Ellepot system (biodegradable paper based 'pots' in customizable soil mixes and volumes) and recycled as much as possible (Figure 2). They are also working on putting up cross-poles for hawks to land on and help with pest control, as well as a robotic crop scanner to monitor plant health via detection of chemical volatiles.

Figure 2. Ellepot system for biodegradable pots.

Crossing the border, I made it to Washington to stay with the IPPS Western Region's exchange coordinator Todd Jones, owner of Fourth Corner Nurseries, in their beautiful farmhouse surrounded by mountains. Todd arranged a visit to Sakata Seed in Skagit Valley, a breeding station for cabbage, spinach, broccoli and beets (Figure 3). It really emphasized the hand labour involved in breeding work, with thousands of cabbages all being hand pollinated and spinach plants grown in special isolation bags. I could really appreciate and relate to all the labour involved in the pollinations, seed cleaning and record keeping!

Figure 3. Sakata seed production area.

My next stop was Olympia with Sarah and Jim Brackman (Sarah is a technical sales representative for BioSafe Systems) but first I got to view the famous city of Seattle in slow-motion as we drove past in rush hour traffic. Jim took me to two Weyerhaeuser nurseries, one bareroot and one container. Weyerhaeuser owns thousands of acres of forestry land and supply their own seedlings. Douglas Fir is the main crop and takes 40 years to mature. Weyerhaeuser had also experimented with the Ellepot system, but they found that some plants just do not like the paper sleeves.

A five-hour journey on the Amtrak train flying past fields of bright orange pumpkins at harvest took me to Eugene, Oregon. Tony Shireman, Laboratory Manager from Fall Creek Nursery showed me their blueberry tissue culture lab and nursery grounds where they carry out their own breeding program. Blueberry season sees the lead breeder eating blueberries from the start of the day until about 2 pm! All part of sensory evaluation of course.

The USDA National Clonal Germplasm Repository was next on the list; they hold and maintain a large quantity of plant cultivars and species, including strawberries, hops and pears. It was surreal for me to be there in Corvallis learning how the germplasm repository functions because I had written about it for an assignment on pear breeding at Massey University the year beforehand.

Figure 4. Strawberries at the USDA National Clonal Germplasm Repository.

Finally, the conference itself; a jam-packed three days of nursery tours and stimulating lectures. Themes that stood out from tours and talks were the use of automation technology and beneficial insects, labour shortages and increased labour costs, climate change, and the importance of critical thinking/planning. With beneficial insects, the key was prevention rather than treatment. For example, Little Prince nursery in Oregon housed beneficial insects long term on host plants kept in mesh boxes in each plastic-house.

Aroma Cannabis was one of the highlights being somewhat a novelty seeing large scale, high-tech glasshouse production of cannabis. All watering, lights, fans and heaters were automated and could be controlled remotely by mobile phone. There are very strict growing regulations, every plant has a Radio-frequency identification (RFID) chip tag and every scrap of plant material must be weighed and recorded as plants are trimmed and harvested.

Figure 5. Aroma Cannabis greenhouse production.

Another highlight was Monrovia nursery and North American Plants tissue culture facility for sheer scale of production. With 900 acres at Monrovia, we only saw a small corner of it, they produce approximately 10 million plants per year, 2/3 container production and 1/3 stock plants. With 10 crew 'canning' they could produce 25,000 plants in a day. North American Plants boasted 50 million plants produced per year; with 23,000 square feet of laboratory space they could produce 67,000 jars per day.

Heirloom Roses was also a big Ellepot user for their rose propagation (Figure 6). An interesting group discussion ended with an emphasis on the importance of social media. They avidly follow and post on Instagram, Pinterest and Facebook to both advertise and monitor trends. Some of these were surprising such as the trendy "brown rose bouquets".

Figure 6. Rose cuttings at Heirloom Roses in Ellepots with peat-bark-pumice mix.

The final two weeks of my extended trip were also a whirlwind. After a day's driving lesson from Laurie Rogers-Roach from Eshraghi nursery in Hillsboro, Oregon (I am eternally grateful!) I drove over 2600km from Portland to LAX. I definitely took the scenic route, visiting the snow-capped Mount Hood, crystal blue Crater Lake, ominously dark and misty Crescent City and surrounding Redwood forests, San Francisco (a terrifying drive), Yosemite National Park, Santa Barbara and Lompoc.

Takao Nursery was a bonus to see because of their focus on technology, hygiene and quality, all fresh in my mind from the conference talks (Figure 7).

Figure 7. Cutting sanitation at Takao Nursery.

Watering, misting and potting were all automated and there were organized workstations throughout the nursery set up for cuttings/maintenance of plants with sanitation sprays at hand. Many of the staff had been there for decades and emphasised the importance of passing on plant propagation knowledge to the younger newcomers.

Annie's Annuals and Perennials was a treasure trove of interesting and colourful plants. I enjoyed seeing and feeling Annie's passion for plants coming through their nursery garden bed displays. Having mature plants in a garden setting is important for customers to see the potential of what they are buying and might encourage them to buy things that don't look showy at point of purchase but will look great down the track.

Figure 8. Keith Hammett sweet pea cultivar seedlings at Annie's Annuals and Perennials.

The PlantHaven team gave an in-depth presentation of the work they do with emphasis on representing plant breeders' portfolios. It enabled me to better understand the process between the development of a new cultivar and getting it to the market. We also drove to GroLink where the PlantHaven plant trials are grown.

Thank you to IPPS New Zealand and IPPS Western Region for making this incredible trip possible. Thank you to all the hosts, IPPS members and nursery staff for their generous hospitality and organizing such an enriching and life changing experience.

Combined Proceedings IPPS 68:21-29. 2018.

Seed Banking (Conservation) in New Zealand: Supporting in Situ Conservation

Craig McGill[a1], Kate Nolan[2], Jessica Schnell[1], Marion MacKay[1], Cristina Winkworth[1], Georgina Aubia[1] and Vivienne McGlynn[1]
[1]School of Agriculture and Environment, College of Sciences, Massey University, Palmerston North, New Zealand. [2]Institute of Education, College of Humanities and Social Sciences, Massey University, Palmerston North, New Zealand.

C.R.McGill@massey.ac.nz

Keywords: Seed storage, germination, Myrtaceae, Fabaceae, accessions.

ABSTRACT

The conservation status of New Zealand's indigenous flora is continuing to deteriorate with the number of species classified as Threatened or At Risk increasing. The in situ conservation efforts to preserve the flora need to be supported by ex situ conservation approaches. One such approach is seed banking where seed is collected and stored at low moisture, usually in equilibrium with 15-25% relative humidity and low temperature, usually -20°C as insurance against loss in situ. New Zealand has a long history of banking seed important to its primary industries, but, a much shorter history of banking seed of New Zealand's indigenous flora. A project to collect, study and conserve seed of New Zealand's flora began in 2013. Following specific protocols this project banked 327 seed accessions of 195 species from October 2013 to October 2017. The arrival of the disease Myrtle rust in New Zealand in 2017 has meant that since late-2017 the focus has been on banking seed of Myrtaceae collected by the Department of Conservation. To date 500 Myrtaceae collections have been collected and banked within the Myrtle rust response programme. The arrival of Myrtle rust in New Zealand and the potential threat it possesses to the New Zealand Myrtaceae has led to wider questions on what seed banking in New Zealand with the integration of Mātauranga Māori alongside seed bank activities should look like.

INTRODUCTION

New Zealand is listed as one of the world's biodiversity hot spots (Myer3 et al., 2000) with 80% of New Zealand's trees, ferns and flowering plants found nowhere else (https://www.doc.govt.nz/nature/native-plants/). However, New Zealand's unique plant biodiversity is under threat from diseases such as Kauri dieback (*Phytophthora agathidicida*) and Myrtle rust (*Austropuccinia psidii*) (de Lange et al., 2018) as well as climate change and habitat loss.

The status of New Zealand's vascular flora is continuing to decline. There are 403 species classified as Threatened (nationally critical, endangered or vulnerable). This is an increase from 289 in 2012. Similarly, the number of species classified as At Risk (Declining) has increased from 102 in 2012 to 158 in 2017. In contrast, the number At Risk (Recovering) has only increased from 7 to 8 (de Lange et al., 2013; de Lange et al., 2018). The worsening status of flora in situ means that conservation efforts in situ need to be supported by a range of other approaches. One such approach is seed banking where seed is stored at low temperature, usually around -20°C and at low moisture, usually in equilibrium with 15-25% relative humidity. Seed banks are a very efficient and relatively cost-effective way of holding a wide range of biodiversity within a relatively small area (Li and Pritchard, 2009). Seed banks provide an insurance policy against future changes that may threaten plant populations in the wild.

New Zealand has a long history of collecting and banking seed important to New Zealand's primary industries. One such seed bank is the Margot Forde Germplasm Centre, based at the AgResearch Grasslands Research Centre in Palmerston North. The Margot Forde Germplasm Centre was established in the 1930s to hold collections of the world's grassland genetic diversity.

These collections are a key contributor to the resilience of New Zealand's multi-billion dollar pastoral sector providing genetic diversity that can be used in breeding programmes to develop new cultivars in response to climate change or changing pest and disease pressures (Williams, 2010).

ESTABLISHMENT OF THE NEW ZEALAND INDIGENOUS FLORA SEED BANK

In contrast to the species utilized in primary production, banking of seed of New Zealand's indigenous flora is relatively recent. The New Zealand Indigenous Flora Seed Bank began in 2013 as a project to collect, study and conserve seed of New Zealand's flora. Prior to this banking of some seed of the New Zealand flora was achieved through funding from MWH New Zealand Limited. In 2013 funding from the Massey University Strategic Innovation Fund and the New Zealand Lottery Grants Board enabled the expansion of seed banking of New Zealand's indigenous flora through the creation of a partnership to bring together organizations with the expertise and / or facilities to achieve the collection, study and banking of seed of the New Zealand flora:

1. Massey University is coordinating the project, providing research expertise and resources for collecting and banking the seed collected as well as holding herbarium voucher specimens from the population from where seed is collected.

2. The Department of Conservation provides access to public conservation land, threatened species advice and connections for engagement with iwi / hapu. For the seed banking in response to Myrtle rust the Department of Conservation is undertaking the seed collection.

3. AgResearch provides drying facilities and storage for the banked seed, as well as seed collections' data at the Margot Forde Germplasm Centre (https://www.agresearch.co.nz/about/our-subsidiaries-and-joint-ventures/margot-forde-forage-germplasm-centre/).

4. Landcare Research provides taxonomic advice and duplicate storage for herbarium voucher specimens.

5. The New Zealand Plant Conservation Network gives access to its flora database and a network of members to contribute to seed collecting.

The New Zealand Indigenous Flora Seed Bank is part of the Millennium Seed Bank Partnership led from the Royal Botanic Gardens, Kew, UK. This is the largest ex situ conservation project in the world with partners in over 95 countries and with seed of 13% of the world's plant species banked (https://www.kew.org/science/collections/seed-collection/about-millennium-seed-bank). The Royal Botanic Gardens, Kew have provided scientific and technical advice for the New Zealand project.

NEW ZEALAND INDIGENOUS FLORA SEED BANK PROGRAMME

Target Species

At the establishment of the New Zealand Indigenous Flora Seed Bank four target species groups were identified. This was to give a collecting focus to the seed bank collection programme. Non-target species are still collected but the collection programme is built around the target species.

The four target groups were the Myrtaceae, the alpine flora, the Fabaceae and the Podocarps and other trees of the forest. The groups were targeted for a number of different reasons:

1. The Myrtaceae includes 28 species in New Zealand including pōhutakawa (*Metrosideros excelsa*), northern rātā (*Metrosideros robusta*), southern rātā (*Metrosideros umbellata*), mānuka (*Leptospermum scoparium*) and kānuka (*Kunzea* spp.). Myrtaceae are widespread throughout New Zealand and mānuka is a species of economic importance. At the time of the seed bank's establishment in 2013 the potential for the arrival of Myrtle rust in New Zealand made collection of the Myrtaceae a high collecting priority. As a result of the arrival of Myrtle rust on Raoul Island in April 2017 and on the New Zealand mainland in May 2017, all 28 of New Zealand's indigenous Myrtaceae are now classified as threatened (de Lange et al., 2018). One New Zealand Myrtaceae, swamp maire (*Syzygium maire*) is known to be desiccation-sensitive and therefore the seed cannot not be dried to the low moisture needed for banking under standard conditions. Alternative protocols will need to be developed for swamp maire.

2. The alpine flora, one-third of New Zealand's flora is found in the alpine zone. This includes around 500 species that are found exclusively in the alpine zone. Of the species in the alpine zone around 83% are endemic and about one-third are At Risk or Threatened. Those At Risk or Threatened include the alpine forget-me-nots (*Myosotis* spp.), alpine daphne (*Pimelea* spp.), hebe (*Veronica* spp.), the alpine daisies (*Celmisia* spp.), buttercups (*Ranunculus* spp.), speargrass (*Aciphylla* spp.) and gentians (*Gentianella* spp.).

Alpine flora has some protection in National Parks, but at the same time it is threatened by human pressure and browsing animals. The alpine flora is especially vulnerable to climate change, as the opportunity to migrate to new environments is limited. If lost, the alpine biodiversity can never be replaced, so there is some urgency to conserve these species. Seed banking of this group will safeguard the diversity of these species.

3. The Fabaceae which include some of the best-known flowering plants in the New Zealand flora, such as kowhai (*Sophora* spp.), the native brooms (*Carmichaelia* spp.) and two kākā beak species (*Clianthus puniceus* and *Clianthus maximus*). Of the 36 species in this group 26 are At Risk or Threatened (de Lange et al., 2018). There are eight species of kowhai and although kowhai is found in forest and scrub vegetation throughout New Zealand some species are limited in their distribution (Heenan et al. 2001).

The native brooms (*Carmichaelia* species), are predominantly South Island species, many with limited distribution and kākā beak (two *Clianthus* species) are restricted to a very few places in the North Island. Twenty species of *Carmichaelia* and both *Clianthus* are At Risk or Threatened, due to threats in their native habitats such as browsing pests and / or habitat loss. Four *Carmichaelia* (*C. corrugata*, *C. nana*, *C. petriei* and *C. torulosa*) have moved into a worse category since 2013 (de Lange et al., 2018). Seed banking of this group will safeguard these species until the threats in their habitats can be overcome.

4. The broadleaf and coniferous trees and shrubs are the backbone of the New Zealand's forests. Conifers from the family Podocarpaceae include rimu (*Dacrydium cupressinum*), tōtara (*Podocarpus totara* var. *totara*), miro (*Prumnopitys ferruginea*) and matai (*P. taxifolia*), while other indigenous conifers are kauri (*Agathis australis*), and the native cypresses kawaka (*Libocedrus plumosa*) and kaikawaka (*L. bidwillii*). Broadleaf trees include tawa (*Beilschmiedia tawa*), tītoki (*Alectryon excelsus* subsp. *excelsus*), kohekohe (*Dysoxylum spectabile*), and rewarewa (*Knightia excelsa*).

The shrub layer of the forest includes species like māhoe (*Melicytus* sp.), *Pittosporum* sp. and hebe. Nearly 300 tree and shrub species (on-shore and off-shore) are At Risk or Threatened, including kauri and the Myrtaceae which are threatened by disease. A number of species in this group are thought to be desiccation sensitive (for example, rimu, miro matati, tawa and kohekohe) and therefore cannot be banked under standard conditions. New protocols will need to be developed for these species.

The initial project duration was four years, from October 2013 through to October 2017. There were two key aspects to the initial project: 1. to develop a network of volunteer seed collectors to along with seed bank staff collect the seed and 2. to process and bank seeds.

Seed Collection

Prior to collecting, discussion is undertaken with relevant groups, including with iwi/hapu to ensure that appropriate permissions to collect are obtained. Once permission is obtained, strict protocols are followed when collecting seed to ensure that seed is collected with minimum disruption to the environment, including careful control of the numbers of seed collected so sufficient seed remains for natural recruitment, and to ensure that high quality collections that capture the maximum genetic diversity in the target population of the species or sub-species are collected. The higher the quality of the seed at collection the longer the potential storage life of the seed.

A network of volunteers has been trained through a series of regional workshops. At the conclusion of the four-year project 162 volunteers covering most areas of New Zealand had been trained. Collectors have come from a range of backgrounds including Department of Conservation staff, regional and city council staff (including staff from Botanic Gardens), iwi / hapu groups, botanical / conservation groups, university / museum-based researchers and individuals with an interest in plant conservation.

Training includes a one-day practical collecting expedition and covers:

1. Assessing if a seed population was suitable for collection. This includes assessing both seed number and quality.

2. Sampling and collecting strategies. This is to ensure that the seed collected is representative of the population and captures as much of the genetic diversity in the population as possible.

3. Planning and risk assessment (including biosecurity).

4. Collecting of associated data and herbarium voucher specimens. This data is as critical as the seed itself. Without this data the value of the seed collected is greatly reduced. If the seed is used to re-establish a population this data is used to maximize the chances that the plants grown from the seed will re-establish in the wild.

5. Post-harvest handling of seed to ensure that the seed is received by the seed bank in as good condition as possible.

6. Obtaining permissions / permits to collect.

In addition to collecting by volunteers, the seed bank also undertook collecting expeditions. Over the four-year project time 30 collecting expeditions ranging from 1 to 14 days were undertaken. Most collecting expeditions were 1-3 days. The 14-day expedition was focused on collecting seed of the Alpine flora. In the four year of the project, 327 seed accessions of 195 species were banked.

Processing and Banking of Seed

Again, strict protocols are followed to ensure that during the processing and banking of the seed, loss of seed quality is minimized. The processing of the seed involves:

1. Drying the seed as quickly as possible to a moisture in equilibrium with 15% relative humidity. For the majority of seed there is a direct relationship between seed moisture and seed storage life with lower relative humidity / moisture meaning seed will remain viable (alive) in storage longer (Justice and Bass, 1978).

2. Removing the seeds from the fruits. This can be a dry fruit such as a pod or capsule, for example as in kowhai, or a fleshy fruit, for example as in *Coprosma* spp. As much of the fruit debris and any other vegetative material is then cleaned from seed. Extraction and cleaning is largely done by hand to minimize any damage to the seed; damage that can shorten the storage life of the seed. Once cleaned the number of seeds in the collection is determined by weight. Seed is then returned to 15% relative humidity to remove any moisture that may have re-entered the seed during extraction and cleaning.

3. Banking. Once re-dried the seed is split into two unequal size collections and sealed in water impermeable packaging for storage. The larger (main) collection remains in the seed bank and the other, smaller collection, is sent for safety duplicate storage. Safety duplicate storage will be at one or more of New Zealand's botanic gardens. This does not however exclude the possibility of safety duplicate storage at other locations within New Zealand. An accession number on the packages links the collections in the seed bank and safety duplicate storage to the data gathered during seed collection and processing / banking. This data is held in the seed bank database. After one month in storage a sub-sample of the seed from the main collection is assessed for quality and to confirm that the seed has survived the banking process and first month of storage.

4. Assessing the quality of the seed. Seed, if large enough, is x-rayed to determine if the seed is full i.e., contains an embryo and food reserves and / or whether the seed has been predated by insects. Without an embryo and food reserves the seed will not develop into a plant. Insects can also deposit eggs in the developing seed, these hatch into larvae that feed on the seed. Larvae or the damage caused by them can be seen on the x-ray. The ultimate aim is to determine the percentage of healthy (full) seed in the seed lot that can be converted into a plant.

5. To determine the amount of full seed that is viable (alive not dead) a sub-sample of seed is germinated. Seed of many wild species, including species in the New Zealand flora have complex dormancy mechanisms within them. In the natural environment these are designed to disperse the seed in time and space. Nonetheless in the seed bank seed dormancy is problematic. To determine the full normal germination potential of the seed the dormancy in the seed needs to be alleviated. Assessing normal germination will indicate if the seed is viable and whether the resultant seedling has all the structures needed to develop into a healthy plant. For many species in the New Zealand flora the dormancy-breaking techniques are not known. As a result, at the end of the germination test any seed that has not germinated is stained with Triphenyl tetrazolium chloride. This chemical will stain viable tissue red. The test, the "Tz test", will indicate which seed is viable but not necessarily whether the viable seed will produce a healthy seedling.

6. From the seed weight, x-ray, germination and Tz test results the number of viable seed available for restoration if needed can be determined.

7. The viability of collections are monitored over time to determine how quickly the seed is losing viability and whether the population needs to be recollected. Most collections in the New Zealand Indigenous Flora Seed Bank were collected in 2014 or later so no re-checking of the collections has occurred as yet.

Missing Knowledge

Relatively little is known about the seed biology and reproductive phenology of many of New Zealand's indigenous flowering species. There is a need to understand New Zealand's indigenous plant ecologies in specific locations and regions across New Zealand to enable effective seed collecting. Data is missing on:

1. Sensitivity to drying. A number of species in the New Zealand flora are known to produce seed that will not tolerate drying. Species producing seed known to be desiccation sensitive (or recalcitrant) include nīkau (*Rhopalostylis sapida*), swamp maire (*Syzygium maire*) and kohekohe (*Dysoxylum spectabile*). For others such as miro (*Prumnopitys ferruginea*) and mātai (*P. taxifolia*) the storage behaviour of is uncertain (http://data.kew.org/sid/). The extent of desiccation sensitivity in the New Zealand flora is not known. Of the approximately 3000 named and unnamed taxa in the New Zealand flora less than 200 have had their storage behaviour reported (http://data.kew.org/sid/). For those sensitive to drying different storage approaches will be needed. Two research projects are underway at Massey University to develop storage protocols using cryopreservation (freezing the seed or the seed embryo in liquid nitrogen (-196°C)) for swamp maire (PhD research) and kohekohe (Masters research).

2. Optimum time for seed collection. For maximum storage life there is an optimum time for seed collection. The phenological information needed to enable the optimum time for collection to be determined is not

available for most species. The optimum date for seed collection will vary across regions and between seasons. Collecting of the phenological data provides an opportunity for citizen science projects locally, regionally and nationally with the potential to grow scientific capability to ensure that we know at what time to collect indigenous seed for maximum storage potential.

3. Seed storage life. The time for which seed of a given species will retain viability in storage will vary between and within families. For example, seed of Orchidaceae are short-lived in storage (Hay et al., 2010). In contrast most species within the Malvaceae are long-lived in storage whereas within the Fabaceae longevity can be variable across species (Long *et al*., 2015). The length of time most species can be stored before viability declines is not known (collections are monitored to determine when viability begins to decline) nor why seed of some species retains viability in storage longer than other species. A research project (PhD) is underway to determine why orchid seed is short-lived in storage.

4. Dormancy and germination. To assess the quality of the seed coming into the seed bank and to be able to convert the seed into plants the ability to germinate the seed is essential. Seed of many species in the New Zealand flora have complex dormancy mechanisms, which need to be alleviated to allow germination to begin. For many species seed dormancy-breaking protocols still need to be developed. Equally for other species protocols may have already been developed within the wider plant conservation community. A key role for the New Zealand Indigenous Flora Seed Bank may be as a data warehouse where knowledge around germination and dormancy-breaking requirements is held not only for use in banking but also in local and national restoration projects where plant supply will be greatly enhanced if every seed can be converted into a plant.

Impact of Myrtle Rust on the New Zealand Indigenous Flora Seed Bank

The arrival of Myrtle rust in 2017 considerably changed the operational activities of the New Zealand Indigenous Flora Seed Bank. The New Zealand Indigenous Flora Seed Bank is supporting the Myrtle rust response being led by the Ministry of Primary Industries and the Department of Conservation. The seed bank is processing and banking seed of New Zealand's indigenous Myrtaceae being collected by the Department of Conservation. This programme is the largest targeted seed collection programme ever undertaken in New Zealand (https://www.doc.govt.nz/nature/pests-and-threats/diseases/myrtle-rust/). As part of this programme a further 50 collectors, predominantly DOC staff have been trained in sustainable seed collecting in 2018. To date 500 Myrtaceae collections, representing 14-and-a-half million full seeds, have been collected and banked. This adds to the 24 Myrtaceae collections made prior to the arrival of Myrtle rust.

Although collections of seed of non-Myrtaceae are still being sent in by trained collectors for banking, processing and banking Myrtaceae seed is now the primary focus of the seed bank.

SEED BANKING IN NEW ZEALAND: THE FUTURE

The New Zealand Indigenous Flora Seed Bank began as a four-year project to collect, study and conserve seed of New Zealand's flora. The arrival of Myrtle rust in New Zealand and the potential threat it possesses to the New Zealand Myrtaceae, including iconic species such as the two New Zealand species of pōhutukawa (*Metrosideros excelsa* and *M. kermadecensis*), has raised the profile seed banking in New Zealand and the role it can play within New Zealand's internal biosecurity.

This has led to wider questions on what seed banking in New Zealand with the integration of Mātauranga Māori alongside seed bank activities should look like. Should New Zealand follow the traditional centralized seed banking model, or should the New Zealand model be built around a decentralized seed bank. Whichever model is used, at the centre of any seed bank programme would be coordinated research activity, seed storage, information / data sharing and project management. This hub will serve a range of connected groups (scientists, educators, iwi / hapu, community leaders, conservationists) to support and resource local and regional seed conservation activities. A hub and spoke model will facilitate community engagement in the project through education outreach, possibly with the New Zealand botanic gardens who already undertake considerable education outreach, and citizen science.

The long-term vision should be to facilitate the development of networked collaborations that result in sustainable conservation and preservation activities for New Zealand, locally, regionally and nationally. The aim of the seed bank is to partner with other ex situ and in situ conservation activities at the individual, iwi / hapu, community, regional, national and international levels to help ensure the continued existence of indigenous plant life and the biodiversity it contains.

Literature Cited

de Lange, P.J., Rolfe, J.R., Champion, P.D., Courtney, S.P., Heenan, P.B.: Barkla, J.W., Cameron, E.K., Norton, D.A., and Hitchmough, R.A. 2013. Conservation status of New Zealand indigenous vascular plants, 2012. New Zealand Threat Classification Series 3. Department of Conservation, Wellington. 70 p.

de Lange, P.J., Rolfe, J.R., Barkla, J.W., Courtney, S.P.; Champion, P.D., Perrie, L.R.; Beadel, S.M., Ford, K.A., Breitwieser,I., Schonberger, I., Hindmarsh-Walls, R.; Heenan, P.B., and Ladley, K. 2018. Conservation status of New Zealand indigenous vascular plants, 2017. New Zealand Threat Classification Series 22. Department of Conservation, Wellington. 82 p.

Hay, F.R.; Merritt, D.J.; Soanes, J.A. and Dixon, K.W. 2010. Comparative longevity of Australian orchid (Orchidaceae) seeds under experimental and low temperature storage conditions. Botanical Journal of the Linnean Society *164*:26–41.

Heenan, P.B., de Lange, P.J. and Wilton, A.D. 2001. *Sophora* (Fabaceae) in New Zealand: Taxonomy, distribution, and biogeography. New Zealand Journal of Botany *39*:17-53.

Justice, O.L. and Bass, L.N. 1978. Principles and practices of seed storage. Agricultural Handbook No 506. United States Department of Agriculture, Washington.

Li D.Z. and Pritchard H.W. 2009. The science and economics of ex situ plant conservation. Trends in Plant Science *14*: 614–621.

Long, R.L.; Gorecki, M.J., Renton, M., Scott, J.K., Colville, L, Goggin, D.E., Commander, L.E., Westcott, D.A., Cherry, H. and Finch-Savage, W.E. 2015. The ecophysiology of seed persistence: a mechanistic view of the journey to germination or demise. Biological Reviews, *90*:31–59.

Myers, N., Mittermeier, R.A., Mittermeier, C.G., da Fonseca, G.A.B. and Kent, J. 2000. Biodiversity hotspots for conservation priorities. Nature *403*:853–858.

Williams W.M. 2010. The key roles of seed banks in plant biodiversity management in New Zealand. Agronomy Society of New Zealand Special Publication No. 13 and Grassland Research and Practice Series No. 14. pp 5-11.

Combined Proceedings IPPS 68:30-33. 2018.

Rescuing Hawke's Bay's Wild Kakabeak

Marie Taylor
Plant Hawke's Bay Ltd, PO Box 137, Bay View 4149, Hawke's Bay

planthawkesbay@xtra.co.nz

Keywords: *Clianthus maximus,* conservation, East Cape, threatened, Forest Lifeforce Restoration.

INTRODUCTION

This paper addresses the conservation work being done in the Hawke's Bay region to rescue and protect one of the most threatened species in New Zealand, *Clianthus maximus*, ngutu-kaka or kakabeak.

The natural range of the species includes Gisborne, but I am not so familiar with the work Graeme Atkins is doing up there with the Department of Conservation in the Ruatoria area, so I'm keeping to my local patch.

Kakabeak are in a pretty dodgy state, with only 150 plants left in the wild in New Zealand, and they are found from Shine Falls, in Boundary Stream northwest of Tutira, north to East Cape.

The Department of Conservation has categorised it as nationally critical. When the plants were much more widespread (and I've had QEII National Trust covenantors at Putorino tell me that their parents reported rim rock on their farm used to be red with kakabeak) it probably didn't matter if 30% of the population was lost every 10 years.

But now, with so few known plants left, it really doesn't take much to lose a big proportion of the population. The species has disappeared from 98% of its former habitat. Now the only places they can survive are cliff faces where goats can't get to.

Forest Lifeforce Restoration Trust

Pete Shaw, who works for the Forest Lifeforce Restoration Trust, is one of the most important people looking out for and looking after kakabeak in New Zealand.

While he describes himself as a crusty bushman, he is sadly underselling himself with that description. I interviewed him for this talk, and he supplied most of the photos. Pete's brother is Willie Shaw from Wildlands. Willie and Sarah Beadle had found a wild plant on the Waiau Bluffs, south of Panekiri at Lake Waikaremoana, in 1983.

Pete now manages the largest privately-owned forest blocks in Hawke's Bay for Simon Hall and his Forest Lifeforce Restoration Trust. As part of this role, he went back in 2008 with Willie to find the plant. Twenty-five years after first finding the plant, Willie went straight back to the site, and there it was, still remaining. Unfortunately, the plant has since died.

In addition to work on kiwi and whio, Simon has financed a great deal of kakabeak restoration work, including enabling Pete to hunt for new plants, and fencing enclosures with selections of plants grown from cuttings of those in the wild.

The patron of the Forest Lifeforce Restoration Trust is Rachel Hunter, and one of the plants found in the wild is named after her. Pete Shaw found the plant in the Mohaka River gorge, collecting seed and cutting material.

Figure 1: Patron of the Forest Lifeforce Trust, New Zealand Supermodel Rachel Hunter. Photo courtesy - Forest Lifeforce Restoration Trust.

I have to say; I've never seen such a fertile plant in my life. All the seeds germinated, and most of the plants we have grown in the nursery in the past few years have come from Rachel's plant.

The Trust has built five enclosures in Maungataniwha Forest as a way of protecting populations of plants to enable production and collection of large amounts of seed. The first were learning ones, while the last two are in perfect spots.

"We have figured out what they need. You see kakabeak growing on these bluffs in the wilderness, hanging on with a few tenuous roots, but you plant them, and they want fertile soil and dappled light. We have 400 plants in enclosures now."

Figure 2: Brent Gilmore collecting material from the Mohaka Valley. Photo courtesy - Forest Lifeforce Restoration Trust.

The last enclosure was built specifically for all the known current Hawke's Bay plants, of which there are 30, some of which have since died in the wild. Pete says in 2008 when they began searching for plants in the wild was that he had no perception of where they would be living. Willie's plant was the first he had seen in the wild. He searched the Mohaka River by walking and he twice kayaked the river from Willowflat to Raupunga by himself.

There have been several aerial searches with a range of helicopters and a Piper Cub. An Airforce Iroquois was also used to search and winch people down to collect material, and two DOC staff collected material from on a long line under a Hughes 500 helicopter.

Wild kakabeak face many problems:

- Deer
- Goats
- Hares and rabbits
- Snails and slugs
- Storms and droughts
- Weed sprays
- Lack of awareness
- Lack of funds

Figure 3: Hare eating young Kakabeak plants. Photo courtesy Department of Conservation

Goats are the major problem kakabeak have to cope with. Goats can smell kakabeak from miles away and are much more aggressive about getting to the plants growing on the cliff systems, than, say deer. The whole habitat of kakabeak through its Hawke's Bay and Gisborne range is infested with goats. It is very difficult to do any effective restoration and conservation work from Napier north to Gisborne and beyond because of the goat densities.

All the wild kakabeak plants have been DNA tested to see how they relate to each other. Gary Houliston at Landcare Research was contracted by DOC to do this work, and he found the DNA of plants from different areas appears quite distinct.

Interestingly, the DNA from *Clianthus puniceus* is plotted inside the DNA spread of the East Coast plants. His work would indicate that genetically there is no difference between the two kakabeak species, but a taxonomist has identified *Clianthus puniceus* as a different species.

I asked Pete what he thought, and he gave me a lovely quote: "I wouldn't like to say myself. I've got in the crap before saying what I think. I have no opinion myself." A new tool in the race to save the species is shooting seed onto cliff systems – where goats cannot reach - in a technique called aerial propagation using helicopters.

The idea is to collect the seed, fill shotgun cartridges with regular pellets, a pulp medium like peat, and seed. The shells are shot from about 20 metres away, which is about the distance a helicopter could hover from likely sites in the wild.

They have also tried paintball guns, but shotguns have better direction, force, accuracy and penetration.

Recently some historical records made by Dr John Findlay, a Dannevirke GP and keen botanist, were discovered by his daughter Jeana Trent, as she went through his botanical diaries.

She found notes and photos from plants he recorded at Lake Waikaremoana in the early 1980s, and previously unknown. His photographs record a plant about five metres wide and three or four metres tall.

Figure 4: Kakabeak plant photographed by Dr John Findlay at Lake Waikaremoana early 1980's. Photo courtesy of Jeana Trent.

Nowadays we usually tell people that kakabeak grow two by two metres in size, but the one he found was much larger.

In conclusion, if we had vast amounts of money, the ideal solution to saving the species is that all plants would be in safe refuges free of deer and goats so that they could expand in the wild.

But that is unlikely, so the real-world solution is to try and fence some of these areas where it is practical, and to conserve as much genetic diversity as we can.

What I would like to see is a royalty on the sale of plants so people would know what they are buying and which plant it came from in the wild, and then the money could go back to conserving the plants in the wild.

PROCEEDING'S PAPERS

JAPAN REGION

Dr. Masanori Tomita, Regional Editor

Twenty-fifth Annual Meeting - 2018

Wakayama, Japan

Combined Proceedings IPPS 68:34-44. 2018.

Comparison of growth, yield, and fruit quality between trees grafted onto rootstocks propagated by air layering and trees grafted onto seedling in mango 'Aikou' pot culture

Masahiko Fumuro
Experimental Farm, Kindai University, Arida-gun, Yuasa, Wakayama 643-0004

fumuro@nara.kindai.ac.jp

Keywords: Layered, rootstock, grafting, fruit tree, *Mangifera*, nursery production.

Abstract

Growth, yield, and fruit quality were compared at 8 years after planting between Aikou mango trees grafted onto rootstock propagated by air layering and those grafted onto seedling rootstock under pot culture. The rootstock used in this study was a Taiwanese native strain. The two types of trees were planted in 25-L pots constructed of non-woven fabric. The trunk diameter of the air-layered rootstock trees was significantly smaller than that of the seedling rootstock trees until the trees were 5 years old, but there was no difference between the two tree types after 8 years old. Moreover, no significant difference was observed in total length of green branches and numbers of leaves per tree except at 8, or 6 and 8 years of age. The fresh and dry weights of the leaves, green branches, thick branches, and fine roots of the air-layered rootstock trees were significantly greater than those of the seedling rootstock trees. The total weight of the aboveground parts of the air-layered rootstock trees was significantly greater than that of the seedling rootstock trees, but no significant difference was observed in the total weight of the underground parts between the tree types. The fresh weight of an entire air-layered rootstock tree was significantly greater than that of a seedling rootstock tree. The top/root (T/R) fresh weight ratio of the air-layered rootstock trees was significantly greater than that of the seedling rootstock trees. No significant differences in yield or fruit quality were observed between the two tree types. These results indicate that the use of rootstock propagated by air layering during mango nursery tree production is practical, as the growth, yield, and fruit quality of the air-layered rootstock trees and seedling rootstock trees were similar.

IPPS Vol. 68 – 2018

INTRODUCTION

Mango (*Mangifera indica* L.) nursery trees are normally propagated from seeding cultivars for rootstock. The germinated seedlings are grown for approximately 2 years, and the scion is grafted onto the rootstock. Most mango growers purchase nursery trees from dealers, but some farmers produce nursery trees on their own to reduce costs.
In Japan, the seeds of a vigorous Taiwanese native strain are used. The seeds are relatively easy to obtain and are used to grow the mango rootstock.

If trees of the Taiwan native strain are planted, the rootstock can be produced without purchasing seeds, and nursery trees can be propagated by grafting. However, the rootstock cannot be produced if these trees do not flower and bear fruit. In such a case, the rooted shoots can be propagated for rootstock by air layering using the shoots of the Taiwanese native strain, and the nursery trees can be propagated by grafting.

Efficient air-layering techniques for mango have been reported by Fumuro (2011), and a 100% rooting rate was obtained for shoots of the Taiwanese native strain using the same method (Fumuro, unpublished). Fumuro (2017) found no significant differences in growth, yield, or fruit quality of mango cv. Aikou (Sasaki et al., 2005) own-rooted trees propagated by air layering and grafted trees propagated by conventional methods in pots. A prolonged life span of pot-planted trees may also be possible by using own-rooted trees. However, there is no information regarding the yield, fruit quality, or life span of trees produced using a rootstock propagated by air layering.

In this study, at 8 years after planting, the growth, yield, and fruit quality of Aikou mango trees grafted onto a rootstock propagated by air layering were compared with those of trees grafted onto seedlings of the native Taiwanese strain to assess the practicality of using rootstocks propagated by air layering.

MATERIALS AND METHODS

Nursery tree production

Production of nursery trees grafted onto air-layered rootstocks

Taiwanese native strain trees planted in a greenhouse (width: 7.6 m, length: 22 m) at Kindai University's experimental farm (Yuasa, Wakayama Prefecture, Japan) were used in this study. Air layer propagation was performed on 9 September 2008 according to the method of Fumuro (2011). On 12 November 2008, rooted branches (Figure 1) were removed and planted in small pots (diameter: 13.5 cm, height: 11 cm). Aikou scions were grafted onto nursery trees of the Taiwanese native strain on 9 June 2009. On 20 September 2009, the air-layered rootstock trees were transferred to 25-L pots constructed of a non-woven fabric (diameter: 32 cm, height: 35 cm) that were filled with a mixture of mountain soil, perlite, compost, and vermiculite (volume ratio: 1:1:1:1).

Figure 1. Rooting of air-layered Taiwanese native mango strain.

Production of nursery trees grafted onto seedling rootstocks

Aikou scions were grafted onto 2-year-old rootstocks (Taiwanese native strain seedlings) planted in 8.5-L poly pots (diameter: 24 cm, height: 24 cm) on 15 June 2009. The grafting position was about 24 cm above the ground. On 22 October 2009, the plants were transferred to pots constructed of non-woven fabric, as described above. Both the air-layered rootstock trees and seedling rootstock trees were 2 years old.

Cultivation management

Until October 2012, each pot was managed within a large plastic film greenhouse (width: 9 m, length: 54 m, height: 4.3, two buildings, 972 m^2). The pots were arranged 1.4 m apart in the greenhouse in rows separated by 1.5 m. In November 2012, all pots were transferred to a smaller plastic greenhouse (width: 6 m, length: 18 m, height: 4 m, 108 m^2) with the same space between pots, and growth was continued.

The greenhouse was heated from early December to ensure a minimum temperature of 6°C. This minimum temperature was increased gradually beginning in mid-February and maintained at 18–20°C from the middle of March until late April during the flowering period. A fan was used for ventilation to ensure that the internal air temperature remained < 30°C until flowering and < 35°C after flowering. Irrigation was controlled using an automatic timer. Approximately 50 g of slow-release fertilizer (N: P_2O_5: K = 10: 10: 10%) was supplied to each tree in February, March, April, May, June, July, September, and November. Assuming 476 pots per 1,000 m^2, approximately 19 kg each of nitrogen, phosphoric acid, and potassium were supplied annually. Pruning began when the harvest was almost complete and ended in late September. As part of the training method, two to three scaffold branches per tree and an appropriate number of bearing shoots were set within a crown diameter of 1.3–1.4 m. No pruning was carried out in 2016, as a dissecting survey occurred in October. Diseases and pests were controlled according to conventional procedures.

Fruit management and harvesting

The flowering period during each year in both tree types was generally the same. The full bloom period fluctuated with the year but occurred between late March and late April.

In June of every year, the fruits were thinned out so that the leaf-fruit ratio (ratio of the number of leaves to fruits) was 60. Each fruit was covered with a bag-shaped net before harvesting, and tree-ripe fruits that naturally dropped in the net were harvested from late July until early November every year.

Measurements

Tree growth

Four of the air-layered rootstock trees and five of the seedling rootstock trees were used for the following measurements.

The trunk diameter, number of leaves per tree, and length of the green branches per tree were measured in late December every year from 2009 until 2015, and in October 2016 (before the dissecting survey). The trunk diameters were measured using calipers, and the measurements were made 10 cm above the ground. The lengths of green branches with less than 10% lignification were measured, and the total length of the green branches was calculated. Tree heights were measured in October 2016 at the time of the dissecting survey.

Fresh and dry weights of each organ

The dissecting survey was performed between 11 and 24 October 2016. The trees were 9 years old at the time of dissection. The different parts of the tree were categorized as follows: leaves, green branches, thick branches, trunk, thick roots (≥ 1 mm in diameter, including the shoot-derived wood), and fine roots (< 1 mm in diameter), and the fresh weight of each organ was measured. Each organ sample was dried, and the percentage of dry matter was determined. The total dry weight of each organ was calculated by multiplying the percentage of dry matter by the total fresh weight of each organ. The top/root (T/R) ratio was calculated by dividing the weight of the aboveground parts of trees excluding the leaves by the weight of the underground part of the trees.

Forty leaves were randomly sampled from each tree, and leaf area was measured using an automatic leaf area meter (AAM-9; Hayashi Denko Co. Ltd., Tokyo, Japan). The total leaf area per tree was calculated by multiplying the average leaf area and the number of leaves per tree. Tree trunks were cut using a saw at the position used to measure their diameters. The contours were copied onto paper, and the area of each trunk's cross-section was measured using the automatic leaf area meter.

Yield and fruit quality

The harvested fruits were weighed, and the yield, number of fruits per tree, and average fruit weight were calculated. About 10–20 fruits from both tree types harvested from late August to mid-September at the peak of harvest every year were used to measure fruit quality.

Peel color (Hunter's L-, a-, and b-values) was assessed using a color-difference meter (CR-400; Konica-Minolta, Tokyo, Japan) positioned centrally on the side of each fruit. Flesh firmness was determined using a Magness-Taylor-type fruit penetrometer with an 11.3-mm-diameter plunger (FT011; Effegi, Alfonsine, Italy) by removing a piece of peel 3 cm in diameter with a sharp knife. The maximum force generated when the plunger penetrated 7 mm into the flesh through the cut surface was recorded. Measurements were made on both sides of the fruit, and the average value was calculated. In addition, flesh was collected from a central point on both sides of the fruit. Juice from the fruit was squeezed and filtered through gauze, and the total soluble solids (TSS) and titratable acidity were determined. The TSS value was determined using a refractometer (PAL-1; Atago Co. Ltd., Tokyo, Japan), and the titratable acidity was determined by the titration method with 0.1 N NaOH to a phenolphthalein endpoint and converted to citric acid content.

Statistical analysis

The data obtained in this study were subjected to the *t*-test. A p-value < 0.05 was considered significant.

RESULTS

Tree growth

Trunk diameters increased with age in both tree types (Figure 2). The average trunk diameter of the air-layered rootstock trees was significantly smaller than that of the seedling rootstock trees until the trees were 5 years old, but there was no difference between the two tree types after 8 years old.

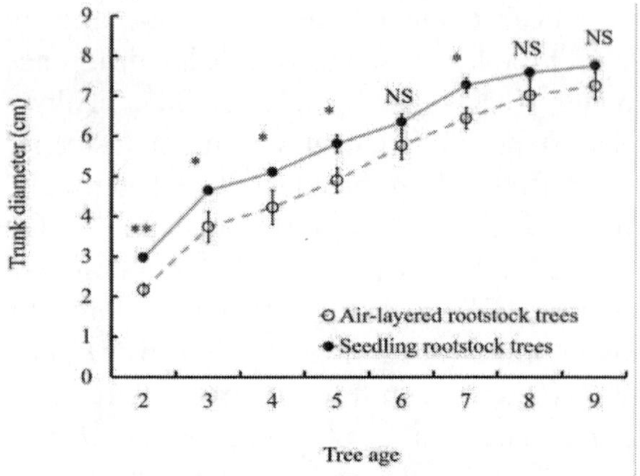

Figure 2. Annual changes in trunk diameters of the air-layered rootstock trees and the seedling rootstock trees in 'Aikou' mango grown in pots. Vertical bars represent ± standard error. NS and * indicate not significant and significant at $P = 0.05$, using *t*-tests.

The trunk cross-sectional areas of the air-layered rootstock trees and the seedling rootstock trees were 38.2 and 44.6 cm^2, respectively (Table 1).

Table 1. Comparison of the total leaf area per tree, trunk cross-sectional area, and tree height of the air-layered rootstock trees and the seedling rootstock trees in 'Aikou' mango.

Propagation method of rootstock	Total leaf area (m^2/tree)	Trunk cross-sectional area [z] (cm^2)	Tree height (m)
Air-layered	9.4	38.2	2.39
Seedling	7.1	44.6	2.08
Significance [y]	*	NS	**

[z] Trunk cross-sectional areas were measured at 10 cm above the ground.
[y] NS, *, and **, a non-significant difference at $P = 0.05$ and significant differences at $P = 0.05$ or 0.01, respectively (*t*-test).

No significant difference in the total length of the green branches per tree was observed between the tree types except at 8 years old (Figure 3).

The numbers of leaves per tree increased with age in both tree types, but no significant difference was detected except at 6 and 8 years old (Figure 4). However, the lead areas of the air-layered rootstock trees were larger in 9-year-old trees than in the seedling rootstock trees (9.4 and 7.1 m^2, respectively; Table 1).

The air-layered rootstock trees were taller than the seedling rootstock trees (2.39 and 2.08 m, respectively).

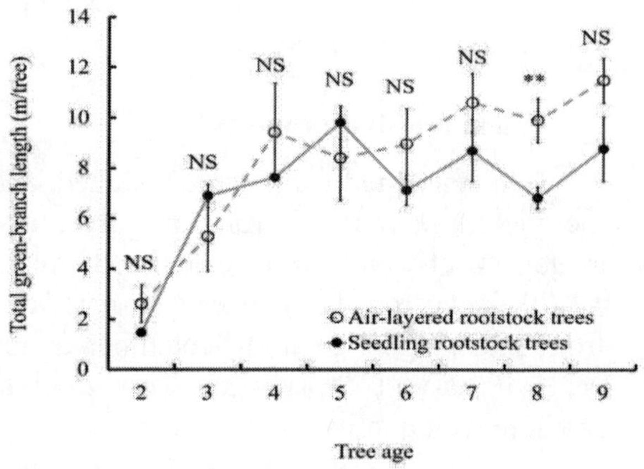

Figure 3. Annual changes in total green-branch length per tree of the air-layered rootstock trees and the seedling rootstock trees in 'Aikou' mango. Vertical bars represent ± standard error. NS and ** indicate not significant and significant at $P = 0.05$, and 0.01, respectively, using *t*-tests.

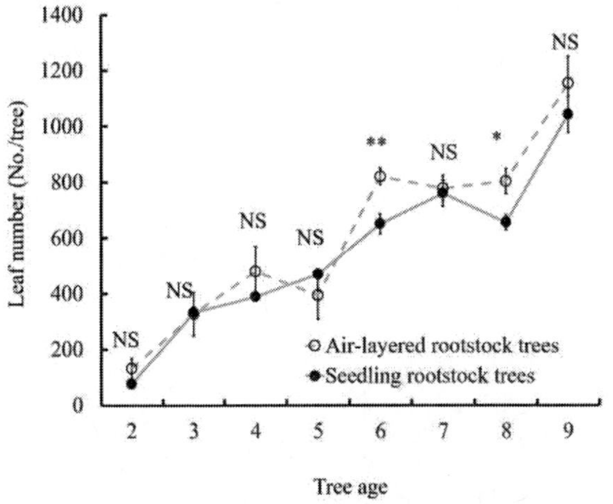

Figure 4. Annual changes in leaf number per tree of the air-layered rootstock trees and the seedling rootstock trees in 'Aikou' mango grown in pots. Vertical bars represent ± standard error. NS, *, and ** indicate not significant and significant at P = 0.05, and 0.01, respectively, using t-tests.

Fresh and dry weights of each organ

The fresh and dry weight values of the tree types tended to be similar (Table 2 and 3).

The fresh and dry weights of the leaves, green branches, thick branches, and fine roots were significantly greater in the air-layered rootstock trees than in the seedling rootstock trees. The total fresh and dry weights of the aboveground parts were significantly greater in the air-layered rootstock trees than in the seedling rootstock trees, but there was no significant difference in the total weight of the underground part between the tree types.

The average fresh weight of a whole air-layered rootstock tree was 13.6 kg and significantly greater than that of a seedling rootstock tree (12.1 kg). However, no significant difference in the dry weights of the whole trees was observed between the tree types.

The fresh weight T/R ratio of the air-layered rootstock trees, which was calculated as the weight of the aboveground parts minus the leaves divided by the weight of the underground parts, was 2.79, and significantly higher than that of the seedling rootstock trees (2.31). However, the dry weight T/R ratios did not differ significantly between the two tree types.

Table 2. Comparison of the fresh and dry weights of the air-layered rootstock trees and the seedling rootstock trees in 'Aikou' mango.

	Propagation method of rootstock	Above-ground part (kg)				
		Leaf	Green branch[z]	Thick branch	Trunk	Total
Fresh weight	Air-layered	3.14	0.64	6.05	1.00	10.83
	Seedling	2.38	0.4	4.9	1.47	9.15
	Significance[w]	*	*	*	NS	**
Dry weight	Air-layered	1.33	0.18	2.1	0.37	3.98
	Seedling	1.07	0.11	1.71	0.52	3.41
	Significance	*	*	*	NS	*

[z] Branches which a ratio of lignification < 10%.
[w] NS, *, **, and ***, a non-significant difference at P = 0.05 and significant differences at P = 0.05, 0.01, or 0.001, respectively (t-test).

Table 3. Comparison of the fresh and dry weights of the air-layered rootstock trees and the seedling rootstock trees in in 'Aikou' mango.

	Propagation method of rootstock	Under-ground part (kg)			Whole tree (kg)	T-R ratio
		Thick root[y]	Fine root[x]	Total		
Fresh weight	Air-layered	2.1	0.66	13.59	13.59	2.79
	Seedling	2.5	0.43	12.08	12.08	2.31
	Significance[w]	NS	***	*	*	*
Dry weight	Air-layered	0.76	0.16	4.90	4.90	2.88
	Seedling	0.91	0.11	4.43	4.43	2.29
	Significance	NS	**	NS	NS	NS

[y] Roots ≥ 1 mm in diameter, including the shoot-derived wood.
[x] Roots < 1 mm in diameter.
[w] NS, *, **, and ***, a non-significant difference at $P = 0.05$ and significant differences at $P = 0.05$, 0.01, or 0.001, respectively (t-test).

The states of the tree before dissecting and the underground part after dissecting of the tree types are shown in Figure 5 and 6, respectively.

Figure 5. The air-layered rootstock trees (left two) and the seedling rootstock trees (right two) before dissection.

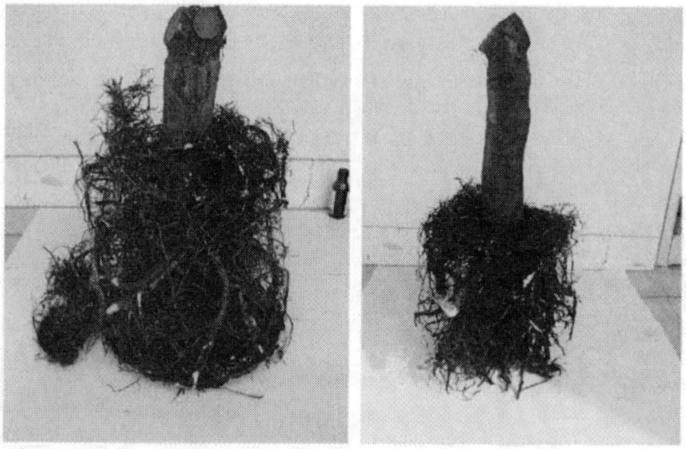

Figure 6. The under-ground parts of the air-layered rootstock tree (left) and the seedling rootstock tree (right).

Yield and fruit quality

The yield per tree was 6.3–7.9 kg in the 6–9-year-old trees, and no significant difference was detected between the two tree types (Figure 7).

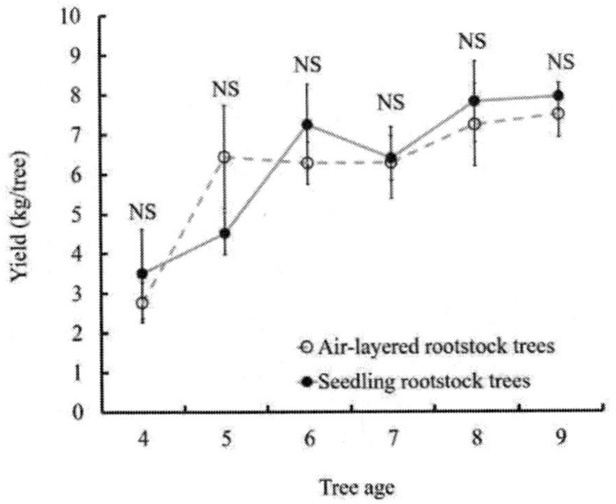

Figure 7. Annual changes in yield per tree of the air-layered rootstock trees and the seedling rootstock trees in 'Aikou' mango grown in pots. Vertical bars represent ± standard error. NS indicate not significant at $P = 0.05$, using t-tests

There were also no significant differences between the two tree types in numbers of fruits per tree or average fruit weight, which were 10.3–13.6 and 539–653g, respectively (Figures 8 and 9).

In addition, no significant differences were observed in peel color, flesh firmness, soluble solid content, or citric acid content (Table 4).

The L-values of both tree types were 35.6–38.3, the a-values were 9.3–18.8, the b-values were 8.1–15.6, flesh firmness was 5.0–9.1 N·cm^{-2}, soluble solid contents were 15.2–17.4%, and citric acid contents were 0.13–0.20, with only small annual variations.

The state of fruiting in the air-layered rootstock trees and the seedling rootstock trees are shown in Figure 10.

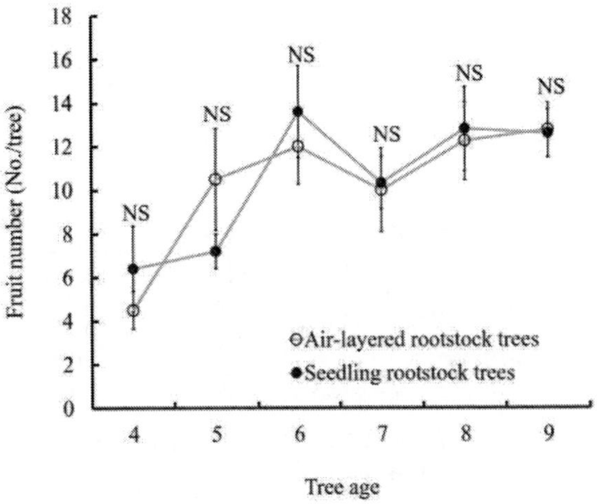

Figure 8. Annual changes in fruit number of the air-layered rootstock trees and the seedling rootstock trees in 'Aikou' mango grown in pots. Vertical bars represent ± standard error. NS indicate not significant at $P = 0.05$, using t-tests.

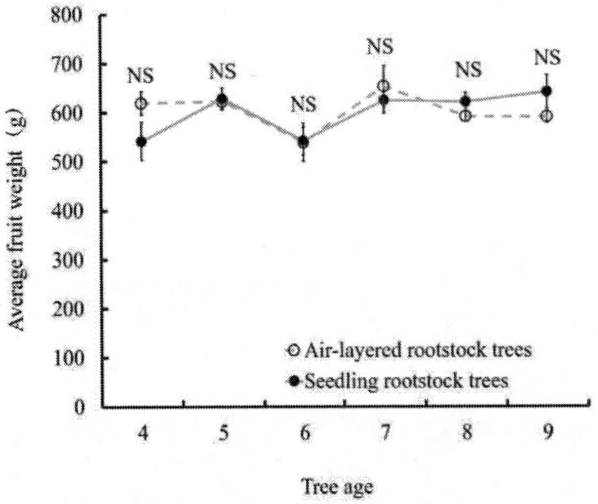

Figure 9. Annual changes in average fruit weight of the air-layered rootstock trees and the seedling rootstock trees in 'Aikou' mango grown in pots. Vertical bars represent ± standard error. NS indicate not significant at $P = 0.05$, using t-tests.

Table 4. Comparison of fruit quality in the air-layered rootstock trees and the seedling rootstock trees in 'Aikou' mango.

Tree age	Propagation method of rootstock	Peel color			Flesh firmness	Total soluble solids	Citric acid
		L-value	a-value	b-value	(N·cm^{-2})	(%)	(%)
4	Air-layered	37.1	10.6	8.1	8.6	16.4	0.20
	Seedling	36.6	10.3	10.0	9.1	15.8	0.18
	Significance	NS[z]	NS	NS	NS	NS	NS
5	Air-layered	37.2	16.2	10.9	5.5	16.5	0.14
	Seedling	35.9	14.2	10.3	5.1	16.6	0.15
	Significance	NS	NS	NS	NS	NS	NS
6	Air-layered	38.3	12.2	13.8	7.6	16.2	0.18
	Seedling	38.1	18.8	12.8	7.4	16.3	0.18
	Significance	NS	NS	NS	NS	NS	NS
7	Air-layered	35.7	15.4	11.8	7.9	15.8	0.14
	Seedling	35.6	16.6	11.4	7.2	16.2	0.17
	Significance	NS	NS	NS	NS	NS	NS
8	Air-layered	37.1	14.4	15.6	5.0	15.5	0.13
	Seedling	35.6	11.4	14.3	6.5	15.2	0.17
	Significance	NS	NS	NS	NS	NS	NS
9	Air-layered	36.2	14.7	12.9	8.0	17.4	0.18
	Seedling	37.9	9.2	16.3	7.3	17.0	0.16
	Significance	NS	NS	NS	NS	NS	NS

[z] NS, non-significant difference at $P = 0.05$ (*t*-test).

Figure 10. State of fruiting in the air-layered rootstock trees (left) and the seedling rootstock trees (right) in 2016.

Discussion

Tree growth

No large differences in trunk diameter, green branch length, or number of leaves per tree were observed (Figure 2, 3, and 4). Therefore, the growth of the air-layered rootstock trees was not inferior to that of the seedling rootstock trees.

In contrast, the air-layered rootstock trees were about 0.3 m taller than the seedling rootstock trees (Table 1). The Aikou trees used in this study are vigorous, and it is easier to increase their tree height than that of 'Irwin', which is the leading cultivar in Japan.

Mango is cultivated in greenhouses in Japan, but the flower clusters and fruits are lifted above the tree crown to promote fruit color. The fruit is covered with a net or bag for tree-ripe fruit production, so the lower the tree height the better. As tree height can be reduced by pruning and training, these treatments make it possible to lower the height of air-layered rootstock trees.

Fresh and dry weights of each organ

The growth of the aboveground parts of trees is closely related to the growth of the underground parts (Fumuro, 1999; Fumuro et al., 1999). Thus, both the aboveground and underground parts must be measured to assess growth. The dissecting survey revealed no significant difference in the weight of the underground parts between the two tree types, but the aboveground parts of the air-layered rootstock trees were significantly heavier than those of the seedling rootstock trees (Table 2).

Fumuro (2017) found that own-rooted Aikou trees propagated by air layering had significantly higher T/R ratios compared to grafted trees propagated using conventional methods. In this study, the air-layered rootstock trees had significantly higher fresh weight T/R ratios compared to the seedling rootstock trees, indicating that the air-layered rootstock trees allowed growth of the aboveground parts with fewer underground parts compared to the seedling rootstock trees.

Although no significant difference was observed in the weight of the thick root between the tree types, the weight of the fine roots of the air-layered trees was significantly greater than that of the seedling rootstock trees, suggesting that the productivity of air-layered rootstock trees is higher than that of seedling rootstock trees. More sustainable fruit production can be realized with a higher T/R ratio and more fine roots.

There is a risk of a decrease in tree vigor caused by root clogging in pot culture. Root clogging in pot culture is caused mainly by enlargement of thick roots and an increased number of fine roots within a restricted root zone. This is accompanied by suppression of new root growth and deterioration of water permeability. However, in this study, no decrease in tree vigor was observed as a result of root clogging after 8 years of culture.

Yield and fruit quality

No study has compared yields and fruit quality between air-layered rootstock mango trees and seedling rootstock trees in culture. In this study, yields were nearly constant for 6 to 8 years in both tree types (Figure 7). The yield per 1,000 m^2 calculated for the average of 6 years, assuming 476 pots per 1,000 m^2, was estimated to be approximately 2.8 t for the air-layered rootstock trees and 2.9 t for the seedling rootstock trees, and no difference between the two tree types was detected.

Fumuro (2011b) reported that the yields of 3- and 4-year-old Aikou trees managed with a leaf-fruit ratio of 60 were 1.2 and 2.2 $kg \cdot m^{-2}$, respectively (5- and 6-year-old roots), which was higher than the values in this study. The average fruit weight of Aikou is greater with a higher leaf-fruit ratio (Fumuro, 2011b). The average leaf-fruit ratio value for 6 years was nearly 60, but differences were detected between years. The average fruit weight in both tree types, excluding the 4- and 6-year-old trees (Figure 9), roughly reached the standard fruit weight (600–700 g) of the Aikou cultivar (Fumuro, 2011b).

Some fluctuations in fruit quality were observed between years, but no significant difference was detected between the two tree types, and the fruit quality roughly reached the standard for the Aikou cultivar (Fumuro, 2011b), which is 16–17% in TSS, 0.15–0.20% in citric acid content, and 7–8 $N \cdot cm^{-2}$ in flesh firmness (Table 3).

Pots with a soil capacity of 40–80 L are typically used for mango pot culture (Yonemoto, 2008). This study implies that cultivation of mango can be continued for at least 8 years even in a small pot with a soil capacity of about 25 L. The pots used in this study were made of non-woven fabric with excellent water permeability and breathability, but the pots that are usually used are constructed of plastic, with inferior water permeability and breathability. In the future, it will be necessary to measure the effects of pot material and size on growth, yield, and fruit quality.

In conclusion, using mango rootstocks propagated by air layering of nursery trees is practical, as the growth, yield, and fruit quality of the air-layered rootstock trees and seedling rootstock trees were found to be similar.

Literature Cited

Fumuro, M. 1999. Interrelationship among tree growth parameters and dry weights of each organ in different aged Japanese persimmon (*Diospyros kaki* L. cv. Fuyu) trees. J. Japan. Soc. Hor. Sci. *68* (*2*), 355–363. https://doi.org/10.2503/jjshs.68.355

Fumuro, M., Ueda, K., and Okisima, H. 1999. Seasonal changes in dry matter production and assimilate partitioning in Japanese pear trees (*Pyrus pyrifolia* Nakai) cv. Kousui and Housui grown under film. J. Japan. Soc. Hor. Sci. *68* (*2*), 364–372. https://doi.org/10.2503/jjshs.68.364

Fumuro, M. 2011a. Effect of several factors on rooting and cultivar differences in rooting abilities of air-layered mango. Engeigaku Kenkyuu *10* (*4*), 451–459. https://doi.org/10.2503/hrj.10.451

Fumuro, M. 2011b. Effects of leaf-fruit ratio on yield and fruit quality in mango cv. Aikou under pot culture. Engeigaku Kenkyuu *10* (3), 383–388. https://doi.org/10.2503/hrj.10.383

Fumuro, M. 2017. Tree growth, yields and fruit qualities of own-rooted and grafted trees in mango cv. Aikou under pot culture. Engeigaku Kenkyuu 16 (Suppl. 1), *52*. 162.

Sasaki, K., Nakazima, A., Simizu, K., Kanzaki, S., and Utsunomiya, N. 2005. Aikou. Cultivar registration. 16162 MAFF Japan.

Yonemoto, Y. 2008. New Special Product Series: Mango. Actuality of Ripe Fruit Cultivation (Tokyo, Japan: Nobunkyo), pp. 1–190. (in Japanese).

Combined Proceedings IPPS 68:45-47. 2018.

Dormancy Period of Pawpaw (*Asimina triloba* (L.) Dunal) Trees in Miyazaki Prefecture, Japan

Shota Miyahara[1], Takaaki Maeda[1], and Katsuya Ozaki[2]

[1]Graduate School of Horticultural and Food Sciences, Minami Kyushu University 3764-1, Tatenocho, Miyakonojo-shi, Miyazaki, 885-0035, Japan.
[2]Field Center, Faculty of Environmental Horticulture, Minami Kyushu University 3764-1, Tatenocho, Miyakonojo-shi, Miyazaki, 885-0035, Japan.

tmaeda@nankyudai.ac.jp

Keywords: Fruit tree, bud dormancy, overwintering.

INTRODUCTION

Pawpaw is a small, deciduous fruit tree species in the family Annonaceae that is native to eastern North America. Among the members of the family Annonaceae, only the genus *Asimina* is found in the temperate zone. Temperate fruit tree species enter a dormant state during the winter. This endogenous dormancy breaks after exposure to low temperatures for a specific period of time. Since the dormancy period of Pawpaw in southern Kyushu of japan is unknown, we collected Pawpaw branches over time and estimated the dormancy period using a stem cutting method.

MATERIALS AND METHODS

Three grafted 10-year-old 'Rebecca Gold' Pawpaw trees being cultivated in open fields in Nichinan, Miyazaki Prefecture, Japan were sampled. To investigate Pawpaw dormancy period, same year living branches were collected approximately every 2 weeks from October 20, 2016 to March 10, 2017 and from August 7, 2017 to March 8, 2018 and subjected to stem cutting. At each sampling, a single medium-length branch (ca. 15 cm) growing at a nearly horizontal angle was collected from each tree. Each branch was stripped, leaving approximately five buds, and cut under water before being immediately inserted into floral foam. The floral foam was placed in a 100-mL beaker and soaked in tap water to ensure sufficient moisture content. The branch and beaker were sealed in a 0.04-mm-thick plastic bag to maintain a humidity of 90% or greater. These beakers were then placed in growth chambers maintained at 25°C under a regime of 10 h light (7 am to 5 pm) and 14 h dark (5 pm to 7 am), and the number of days required for budding, defined as the number of days required for any of the buds on each branch to reach 3 mm in length, was observed. To estimate the temperature and cumulative time required to break dormancy, the soil temperature of the orchard was monitored hourly using a thermo recorder equipped with a data logger (Ondotori Jr. TR-52, T&D Corporation, Japan). For air temperature data, weather data from the Japan Meteorological Agency station at Aburatsu in Nichinan, Miyazaki Prefecture, were used. The cumulative temperature (°C・days) required to break dormancy was defined as the sum of the difference between the reference temperature (2, 5, 7.2, or 10°C) and the daily mean air temperature during the period of deepest dormancy each year.

RESULTS AND DISCUSSION

The results of our 2-year investigation of the depth of dormancy using a stem-cutting method indicate that Pawpaw dormancy is relatively shallow from August to September, with 20 days or less being required for budding (Figure 1). Starting in early October, the number of days required for budding increases gradually, reaching from 46 to 60 days or more by early January when dormancy is deepest. The number of days required for budding began to decrease gradually in early February 2016 and in mid-January 2017. In both years, dormancy was broken completely by early March, at which point the mean air and soil temperatures were both 10°C.

The cumulative number of times not greater than each reference temperature (*i.e.*, 2, 5, 7.2, and 10°C) for the period between entering deep dormancy and breaking dormancy (October 5, 2016 to March 10, 2017 and October 5, 2017 to March 8, 2018) was 58, 282, 604, and 1214 h, respectively, in 2016 and 131, 492, 902, and 1563 h in 2017 (Table 1). Although cumulative temperature decreased with decreasing reference temperature, further investigation is warranted.

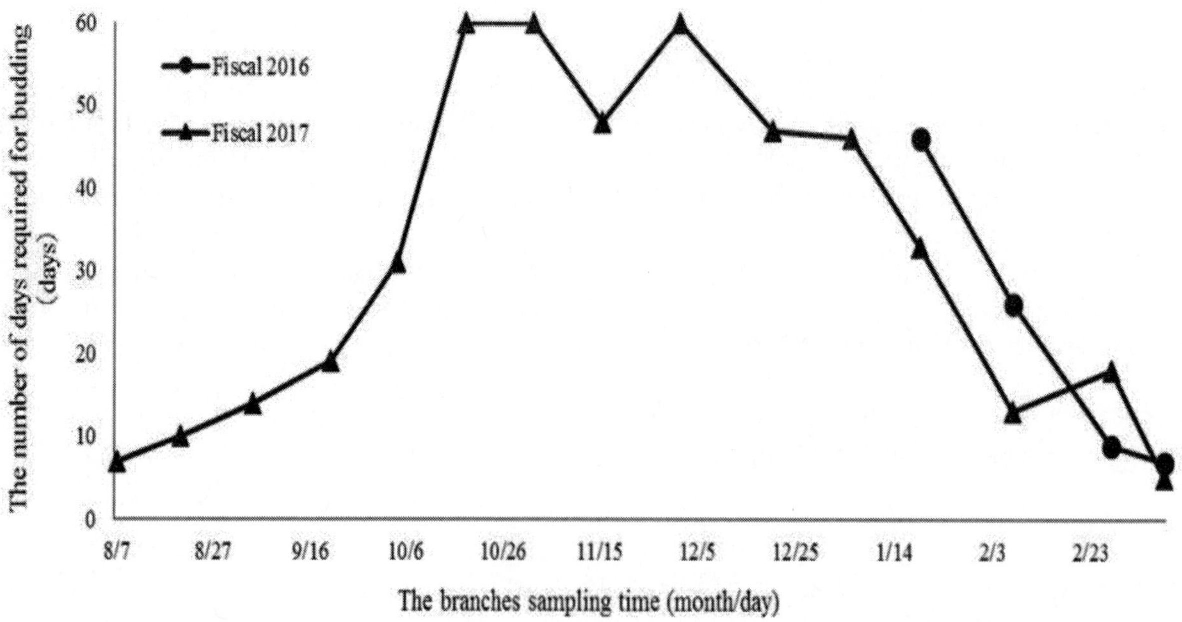

Figure 1. Seasonal change in the degree of Pawpaw bud dormancy.

Table 1. Cumulative time encountered less the reference temperature until the dormancy break.

Fiscal	Cumulative time	Reference temperature			
		2°C	5°C	7.2°C	10°C
2016	16.10.5～17.3.10	58[Z]	282	604	1214
2017	17.10.5～18.3.5	131	492	902	1563

[Z]The cumulative time is the time when it encountered less the reference temperature until the time when bud break started to take place within 10 days from the time when it took more than 20 days to bud.

Combined Proceedings IPPS 68:48-52. 2018.

Effect of Soil Moisture on The Yield and Quality of Rhizome In Turmeric (*Curcuma longa*)

Arisa Noguchi, Yuta Ito and Hiroko Nakatsuka
Department of Agriculture, Tokyo University of Agriculture
1737 Funako, Atsugi, Kanagawa 246-0034, Japan

a3noguch@nodai.ac.jp

Keywords: Moisture level, irrigation, rhizome.

Abstract

Turmeric (*Curucuma longa* L.) includes many functional components in the rhizome, and the most important ingredient is curcumin, a yellow pigment, which has anti-oxidant and anti-bacterial properties. Plant growth, rhizome yield and curcumin content change depending on various environmental factors. We examined to evaluate the effects of relative soil moisture on the yield and curcumin content of turmeric rhizome. The average soil moisture of the control plot during the cultivation period was 19.5%. The values in the high and the low moisture plots were 23.4 % and 18.8 % respectively. As a result of measuring at 225days after planting, values of total leaf area and the rhizome growth rate in the high moisture plot were 1.8 times compared with the control plot. Curcumin content per rhizome dry weight was highest in the control plot, though the difference was statistically non-significant. On the other hand, the curcumin content per plant was significantly higher in the high moisture plot, 267.9 mg/plant.

INTRODUCTION

Turmeric (*Curcuma longa* L.) is a perennial herb from the ginger family (Zingiberaceae) cultivated throughout tropical and subtropical Asia. It is using in many countries as spice, dye and indigenous medicine. Turmeric is one of the key ingredients in many Asian dishes, imparting a mustard-like, earthy aroma and pungent, slightly bitter flavor to foods. Turmeric includes many functional components in the rhizome, and the most important ingredient is curcumin, a yellow pigment, which has anti-oxidant and anti-bacterial properties. In Japan, almost all raw materials of indigenous medicines are covered by import from overseas. Thus, studies on the domestic production of indigenous medicinal plants have flourished in recent years. Most of indigenous medicines use underground part of plants. Therefore, technique to increase the underground part with high content of functional ingredients is demanded.

Plant growth, rhizome yield and curcumin content change depending on the soil type, fertilization and light intensity, showed in previous studies (Hossain and Ishimine, 2005; Akamine et al., 2007; Hossain et al., 2009). Besides, irrigation method affects the growth and development of many tuber crops, e.g., potato (Deblonde and Ledent, 2001) and sweet potato (Gomes and Carr, 2001). In this present, we described the effects of relative soil moisture on the rhizome yield and curcumin content in turmeric.

MATERIALS AND METHODS

One seed-rhizome (ca. 30 g fresh weight (FW)) of turmeric was planted in a 30 L plastic pot on April 21, 2017. In this experiment, we set up three experimental treatment plots, such as the control plot (C plot): a mixture of akadama soil (small granule; ca. 3 mm in diameter) and humus (2:1, v/v)), the low soil moisture plot (L plot): a mixture of akadama soil (large granule; ca. 10 mm in diameter) and humus (2:1, v/v)), and the high soil moisture plot (H plot): a mixture of akadama soil (small granule), humus and vermiculite (1:1:1, v/v). For fertilization, Nitrogen, P_2O_5 and K_2O were given in the following substantial amounts at 3, 3, 2.5g per pot (Kobayashi et al., 2010), and after then no fertilizer was applied.

This experiment was carried out in the open field at Tokyo University of Agriculture, Kanagawa, Japan. Irrigation was given only twice after planting for the initial healthy growth of seed rhizomes and not after that. Soil moisture contents were measured it every month. The soil moisture sensor (EC-20, METER Group, Inc., USA) was inserted to the depth of 15 cm from topsoil and the soil moisture values (nominal volumetric water content percentage) displayed on a handheld reader (ECH$_2$O Check, METER Group, Inc., USA).

After planting at 120 and 225 days, measurements of plant height, total leaf area and rhizome FW were carried out. For curcumin analysis, rhizomes were sliced and dried, and then the slices were ground to a fine powder.

Curcumin was extracted from those powders in 80% methanol, all samples were filtered through 0.22 μm membrane filters before the injections. Curcumin content was determined by HPLC. Column (Sinergi Hydro-RP 150×4.6 mm, 4μm, Phenomenex, USA) was run at 40°C with a flow rate of 0.8 mL/min and monitoring at 425 nm. The eluent was used acetonitrile and ultrapure water at the ratio of 1:1. All the data are presented in triplicate. The means values were compared using ANOVA followed by Tukey's multiple range tests at the 5% level.

RESULTS AND DISCUSSION

The changes in the daily mean air temperature and the relative soil moisture content during the experimental period are shown in Figures 1 and 2, respectively. The average air temperature during the experimental period was 21.9 °C. From June to October in 2017, the temperature remained above 20 °C, and in July and August it was mostly over 25 °C (Fig. 1). The mean soil moisture content of the C plot during the cultivation period was 19.5%. The values in the H and L plots were 23.4 % and 18.8 % respectively. In the H plot, the value always showed higher compared with other treatment plots. Table 1 shows the results of the measurements conducted on the 120th day after the planting which is about half of the experimental period. Plant height, rhizome FW and rhizome growth rate were statistically non-significant. The total leaf area in the H plot was significantly larger compared with others.

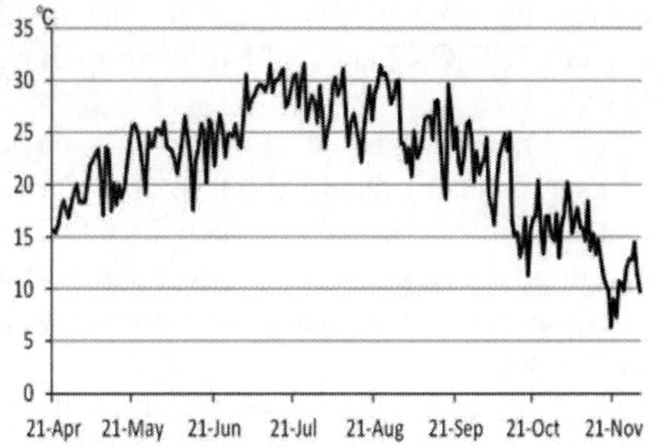

Figure 1. Daily mean of air temperature during the experimental period.

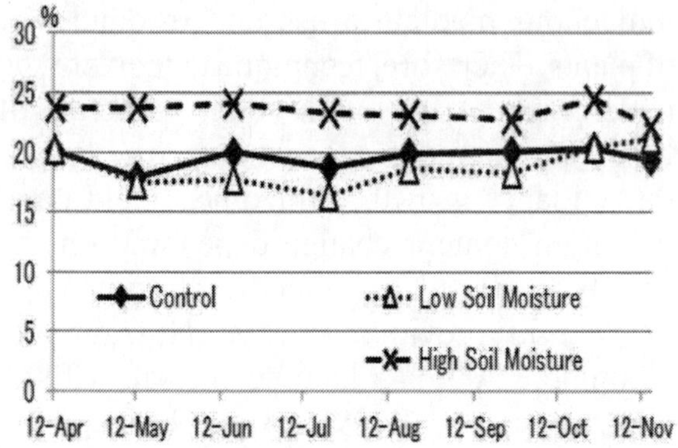

Figure 2. Soil moisture content (%) of each experiment plot during the experimental period.

Table 2 and 3 show the results of the final measurements conducted on the 225th day after the planting. Rhizome growth rate in the H plot was about 1.8 times, and the rhizome FW was doubled compared with the C plot (Table 2). Curcumin content per rhizome dry weigh was highest in the C plot, though the difference was statistically non-significant (Table 3). On the other hand, the content per plant was significantly higher in the H plot, 267.9 mg/plant.

Table 1. Effect of soil moisture content on the growth of turmeric (at 120 days after planting).

Treatment	Plant height (cm)	Total leaf area (m^2/plant)	Rhizome FW (g/plant)	Rhizome growth rate (%)*
Control	78.3	0.14 b	39.6	147.1
Low soil moisture	84.0	0.17 ab	40.6	128.3
High soil moisture	94.7	0.28 a	41.4	128.2

n=3,*The basis of the rhizome fresh weight at planting

Different letters indicate a significant difference at the 5% level by Tukey's test.

Table 2. Effect of soil moisture content on the growth of turmeric (at 225 days after planting).

Treatment	Plant height (cm)	Total leaf area (m^2/plant)	Rhizome FW (g/plant)	Rhizome growth rate (%)*
Control	98.0 b	0.30	305.5 b	1054.4 b
Low soil moisture	108.7 ab	0.33	401.9 b	1219.9 b
High soil moisture	114.3 a	0.53	601.2 a	1854.7 a

n=3 *The basis of the rhizome fresh weight at planting

Different letters indicate a significant difference at the 5% level by Tukey's test.

Root rot and enlargement suppression of underground part showed a tendency to occur in the field which is drainage failure or the high groundwater level. However, high soil moisture was the highest plant height and rhizome yield in this experiment (Table 2). We guess that it is most suitable for photosynthesis in the H plot, though the curcumin content (concentration or accumulation; mg /g DW) in rhizome was not affect by the soil moisture content (Table 3). Gill et al. (1999) showed that application of wheat straw mulch improved growth and yield of turmeric significantly but did not affect curcumin content. In addition, rhizome yield of turmeric increased shading and fertilizer application (Ferreira et al., 2016; Hossain et al., 2009; Akamine et al., 2007), but the curcumin content was almost the same. The differences in curcumin contents were caused by species and strains (Miyazaki et al., 2014). Anandaraj et al. (2014) also reported that the curcumin content was affected by the genetic background rather than environment factors. From these reports, it is thought that it is necessary to combine the soil moisture adjustment with high curcumin accumulation strains to increase the yield of turmeric rhizome and curcumin.

Table 3. Effect of soil moisture content on the content of curcumin in the rhizome (at 225 days after planting).

Treatment	Content(mg) (/g DW)	Content(mg) (/plant)
Control	3.01	158.2 ab
Low soil moisture	2.43	134.3 b
High soil moisture	2.52	267.9 a

n=3, Different letters indicate a significant difference at the 5% level by Tukey's test.

Literature Cited

Akamine, H., M.A. Hossain, Y. Ishimine, Y. Aniya, K. Yogi, K. Hokama and Y. Iraha. (2007). Effects of Application of N, P and K Alone or in Combination on Growth, Yield and Curcumin Content of Turmeric (*Curcuma longa* L.). Plant Prod. Sci. *10*:151-154. https://doi.org/10.1626/pps.10.151

Anandaraj, M., D. Prasath, K. Kandiannan, T.J. Zachariah, V. Srinivasan, A.K. Jha, B.K. Singh, A.K. Singh, V.P. Pandey, S.P. Singh, N. Shoba, J.C. Jana, K.R. Kumar and K.U. Maheswari. (2014). Genotype by environment interaction effects on yield and curcumin in turmeric (*Curcuma longa* L.). Ind. Crop Prod. 53:358-364. https://doi.org/0.1016/j.indcrop.2014.01.005

Deblonde, P.M.K. and J.F. Ledent. (2001). Effects of moderate drought conditions on green leaf number, stem height, leaf length and tuber yield of potato cultivars. J. Agron. *14*:31-41. https://doi.org/10.1016/S1161-301(00)00081-2

Ferreira, M.I., C.S. Marques, L.G.P. Pereira, U.M. Rodrigues, M. Massimiliano, V. Fabio and M.L. Chau. (2016). Exclusion of solar UV radiation increases the yield of curcuminoid in *Curcuma longa* L. Ind. Crops Prod. *89*:188-194. https://doi.org/10.1016/j.indcrop.2016.05.009

Gill, B.S., R.S. Randhawa, G.S. Randhawa and J. Singh. (1999). Response of turmeric (*Curcuma longa* L.) to nitrogen in relation to application of farmyard manure and straw mulch. J. Spices and Aromatic Crops. *8*:211-214. http://updatepublishing.com/journals/index.php/josac/article/view/296

Gomes, F. and M.K.V. Carr. (2001). Effects of water availability and vine harvesting frequency on the productivity of sweet potato in southern Mozambique. I. Storage root and vine yields. Exp. Agr. *37*:523-537. https://doi.org/10.1017/S0014479702001047

Hossain, M.A., H. Akamine, Y. Ishimine, K. Yamawaki, R. Teruya and Y. Aniya. (2009). Effects of Relative Light Intensity on the Growth, Yield and Curcumin Content of Turmeric (*Curcuma longa* L.) in Okinawa, Japan. Plant Prod. Sci. *12*:29-36. https://doi.org/10.1626/pps.12.29

Hossain, M.A. and Y. Ishimine. (2005). Growth, Yield and Quality of Turmeric (*Curcuma longa* L.) Cultivated on Dark-red Soil, Gray Soil and Red Soil in Okinawa, Japan. Plant Prod. Sci. 8:482-486. https://doi.org/10.1626/pps.8.482

Kobayashi, T., A. Miyazaki, A. matsuzawa, Y. Kuroki, T. Shimazaki, T. Yoshida and Y. Yamamoto. (2010). Change in Curcumin Content of Rhizome in Turmeric and Yellow Zedoary. Jpn. J. crop sci. *79*:10-15. https://doi.org/10.1626/jcs.79.10. (in Japanese)

Miyazaki, A., T. Kobayashi, Y. Aoki, S. Kurita, T. Kashiwagi, Y. Yamamoto and H. Hayakawa. (2014). Effects of Rhizome Yield and Curcumin Content on Curcumin Yield from Curcuma. Trop. Agric. Dev. *58*:163-168. https://doi.org/10.11248/jsta.58.163

Combined Proceedings IPPS 68:53-56. 2018.

Trying to Layer Cherry Blossom Tree 'Somei-yoshino'

Hiroaki Ohashi
Faculty of Agriculture, Ehime University
3-5-7, Tarumi, Matsuyama, Ehime, 790-8566, Japan

ohashi@agr.ehime-u.ac.jp

Keywords: Vegetative propagation, *Prunus*, *Cerasus*, *yedoensis*, cuttings, sakura, Oxyberon, IBA.

INTRODUCTION

'Somei-yoshino' cherry (*Prunus ×yedoensis* (Matsum.) Masam. et Suzuki) (syn. *Cerasus ×yedoensis*) is one of the most popular flowering trees in Japan, and numerous Japanese have a deep attachment to the flower and the trees.

During the autumn of 2018, 'Symbolic 3 'Somei-yoshino' trees at the Faculty of Science, Ehime University were scheduled to be cut down because of expansion work for a road. In March, I asked the following question. Would it be possible to raise clonal nursery stock from these trees?

Generally, the propagation of this cultivar is by grafting onto a rootstock of *C. speciosa* (Koidz.) H. Ohba. But the rootstocks were not prepared to this appropriate time of the grafting. Therefore, I tried propagation by cuttings and layering.

MATERIALS AND METHODS

Plant Materials

The tested trees were planted in the west side of the campus (Bunkyo-cho, Matsuyama, Ehime prefecture). Their presumed age was over fifty years. The 3 trees were planted in a line north-south, hereinafter, this is called north tree, center tree and south tree

Layering Methods

Layering treatment was applied to five branches on each tree (total 15 branches) on May 15, 16 and 21, 2018. The previous year's long branches with the long of 50-100 cm were selected for this management, but, only one current year's branch was managed in north tree.

The girdled section was packed with wetted sphagnum and wrapped with plastic film with a plastic straw in the upper part of the wrapping. This was finally covered by aluminum foil (Figures 1 and 2).

The layered branches had water poured through thc plastic straw, every 2-3 days to keep the sphagnum moist.

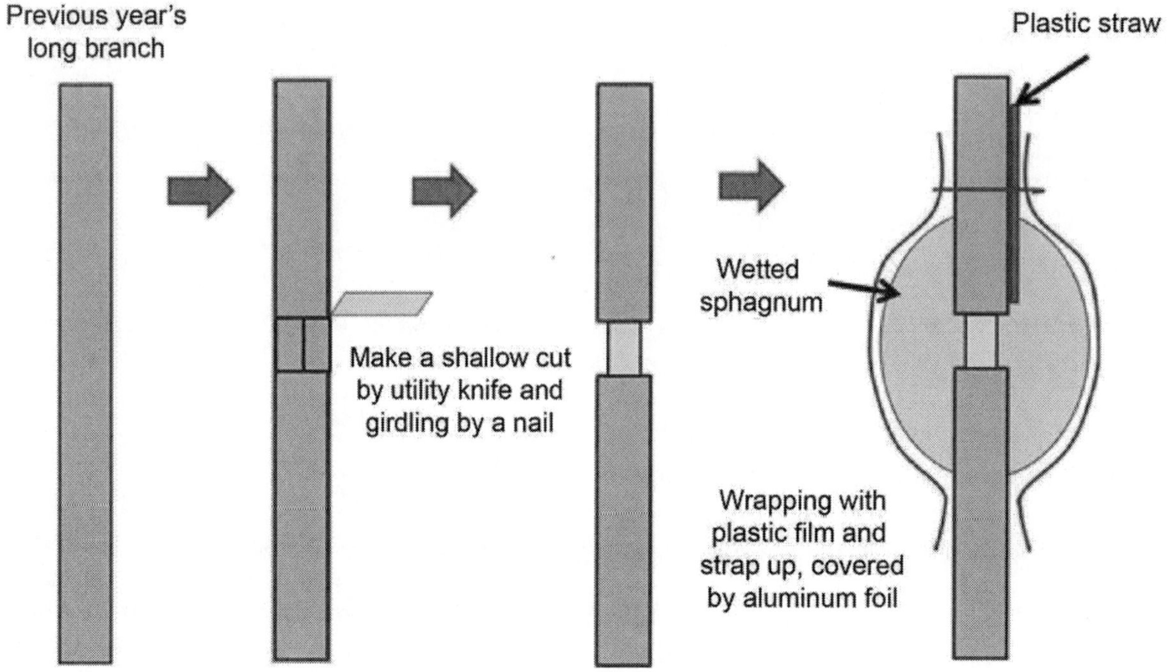

Figure 1. Outline of the layering managements to 'Somei-yoshino' cherry.

Figure 2. A layering managed previous year's long branch of 'Somei-yoshino' cherry.

For confirmation of rooting, one layering managed branch of the center tree had the aluminum foil cover removed on August 7, and the result showed vigorous rooting.

The rooted managed branch was removed from tree, planted in a plastic pot (a diameter of 15 cm, a height 17 cm, 1.9 L) by mixed soil, cultivated in 60-70% shaded plastic house, irrigated by 2 minutes mist at 90 minutes intervals in the daytime. The other 14 layered branches were removed on August 28, 2018, planted and growing in the same way.

Cutting Methods

For cutting, current year's long branches were removed from 3 trees on June 8, 2018. The branches were clipped to stem segment 3-7 cm each with more than 2 leaves. The basal leaf was removed, and the remaining leaves cut in half. The cuttings were sorted by with or without apical bud and the base soaked in tap water or the Oxyberon SL (indole-3-butyric acid; Bayer Cropscience Co., Ltd.) solution diluted 50 times with water for approximately 15 hours. After treatment, cuttings were put into propagation bed filled up Kanuma soil and handled as describe previously for rooted layered branches.

On August 15, 2018, these cuttings were observed, rooted cuttings were planted in plastic pots (a diameter of 7.5 cm, a height 6.5 cm) by mixed soil, thereafter, cultivated in the same plastic house.

RESULTS AND DISCUSSION

Layers on current year's branches withered and died. Layers applied to previous year's branches had 14/15 that survived and showed 64.3 % rooting (Table 1). The survival rate of layered branches with root systems was 93.3 %. Although I did not observe through the rooting process, there was considerable rooting by late July. Layered branches removed on August 7 showed vigorous rooting, and the roots of removed at August 28 rooted branches were already lignified (Figure 3).

In the future, it is necessary to examine that proper time of layering management and the period required for rooting. If the layering managements were had in early growth period e.g. mid to late April, the managed branches may be well rooted in mid- growth period e.g. June and become increasingly likely to a success rate of layering.

Table 1. Rooting of layered branches of 'Somei-yoshino' cherry 100 days after the initiating the layering process.

Mother tree	Survival rate (%)	Overall rooting (%)	Rooting in survival branches (%)	Notes
North tree	80.0	80.0	100	Current year died
Center tree	100	60.0	60.0	
South tree	100	40.0	40.0	
Total	93.3	60.0	64.3	

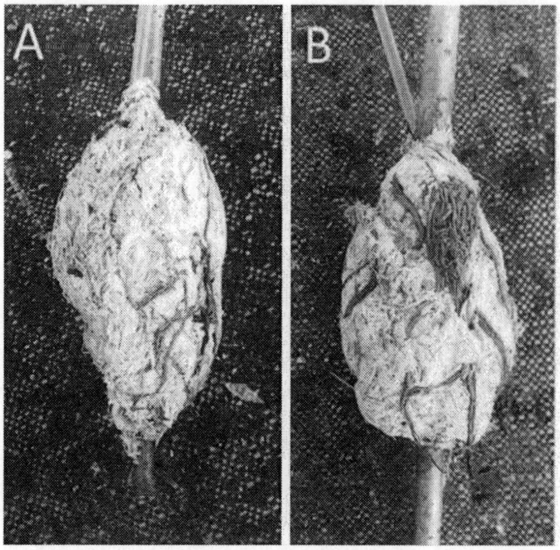

Figure 3. A look of rooting in layering managed branches A: One of the center tree's managed branches, August 7, 2018, 84 days after the layering managements, B: One of the north tree's managed branches, August 28 2018, about 100 days after the layering managements.

On cuttings, rooting was 70.2% in Oxyberon treatments and 66.4 % in tap water treatments (Table 2). The success rate using layering was similar to using cuttings. It appears that layering was practical and an easy propagation method for 'Somei-yoshino' cherry to obtain few large nursery trees in a short period of time.

'Somei-yoshino' layering is scarcely reported. The website 'Sakura-no-kai' (https://sakuramori.at.webry.info/200504/article_10.html), shows methods of layering of cherry blossom tree, the layering managements were had in late May – June, keep to June of the following year. In this study, managed branches were rooted nearly 3 months only, if it goes well, I appear that rooted branches were able to be planted the next early spring.

Table 2. Effects of Oxyberon soaking treatments and the types of cutting in the cuttage of 'Somei-yoshino' at 68 days after the cutting.

Soaking treatment	Type of cuttings	No. of cuttings	No. of rooted cuttings	Rooted rate (%)
Tap water	without apical bud	63	39	61.9
	with apical bud	59	42	71.2
	Total	122	81	66.4
Oxyberon SL solution diluted 50 times	without apical bud	63	42	66.7
	with apical bud	58	43	74.1
	Total	121	85	70.2

Combined Proceedings IPPS 68:57-61. 2018.

Effects of Light Quality on The Growth of Tissue Cultured Transplants

Syo Takeda[1], Arisa Noguchi[1], Wakanori Amaki[1], [2]Yoshiki Takahashi, [2]Takashi Norimoto, [3]Kiyohisa Kawakami and [3]Akemi Mizutani
[1]Department of Agriculture, Tokyo University of Agriculture, 1737 Funako, Atsugi, Kanagawa 246-0034, Japan
[2]Toshin Electric Inc., 4-5-10 Shibaura, Minato, Tokyo 108-0023, Japan
[3]Verde Co., Ltd., 244 Kitagaya, Wakamatsu, Toyohashi, Aichi 441-8123, Japan

amaki@nodai.ac.jp

Keywords: Micropropagation, LEDs, wasabi, Japanese horseradish, *Eutrema*, strawberry, *Fragaria*, sweet potato, *Ipomoea*.

Abstract

Tissue cultured transplants cultivated under two LEDs with different light quality and conventional fluorescent lamp as artificial light sources, were examined and compared the growth of plantlets. Culture conditions were 23±1°C, 16 hours illumination / 8 hours dark, PPFD at illumination was 100 µmol m^{-2} s^{-1} for all light sources. The light sources used for this examination were a cool white fluorescent lamp (FL) and two types of straight-tube type LEDs, Tecoled G2 (cool white: G2) and Tecoled G4 (warm white: G4). In strawberry, 'Akihime', there was no significant difference in total fresh weight (FW) and leaf number per plantlet under the three different light sources. However, for 'Tochiotome' and 'Toyonoka', the total FW and leaf number were larger under G4. Compared with FL, the petiole length tends to be shorter in both of LEDs. In sweet potato, total FW per plantlet was greater under G4 in 'Narutokintoki' and 'Beniharuka' and under G2 in 'Beniazuma'. The differences in response to the different light sources were relatively great depending on the cultivars in sweet potato. In Japanese horseradish, the value of all measured parameters was the maximum under G2, and the overall growth was also the most vigorous under G2.

INTRODUCTION

As an artificial light source for the tissue culture transplant production, most production facilities have used fluorescent lamps. However, due to the entry into force of the "Minamata Convention on Mercury" in Japan, legal measures were set for fluorescent lamps which are mercury-based products, for storage and disposal. Furthermore, from the basic energy plan of the Japanese government, conversion from fluorescent lamps to SSL (Solid State Lighting) such as light emitting diodes (LEDs) has been promoted, and the announcement of discontinuation of fluorescent lamp manufacturing is successive from manufacturers in Japan. In recent years, the use of LEDs has spread in various scenes of plant cultivation (Mitchell *et al*., 2015).

Many research reports have investigated the influence of monochromatic light on the growth and morphogenesis of in vitro cultured plants using LEDs (*e.g.* Shimizu *et al*., 2013; Alvarenga *et al*., 2015; Ramĭrez-Mosqueda *et al*., 2017; Silva *et al*., 2017). From the viewpoint of production workability, white light is desired as artificial light. However, most of the white LEDs are made to emit white light by covering the blue LED chip with some phosphors, and the light quality of the irradiation light is changed by the fluorescent substance (phosphor) to be used (Viršile *et al*., 2017). It has not been clarified how much such difference in light quality affects the quality of cultured transplants.

Then, tissue cultured transplants inoculated by Verde Co., Ltd. (Toyohashi city, Aichi prefecture) which is one of the representative tissue culture transplant production companies in Japan, cultivated under two LEDs with different light quality and conventional fluorescent lamp as artificial light sources, were examined and compared the growth of plantlets.

MATERIALS AND METHODS

For the experimental materials, we used 3 cultivars of strawberry (*Fragaria ×ananassa* (Duchesne ex Weston) Duchesne ex Rozier, 'Akihime', 'Toyonoka' and 'Tochiotome'), 3 cultivars of sweet potato (*Ipomoea batatas* (L.) Lam., 'Narutokintoki', 'Beniazuma' and 'Beniharuka') and 1 cultivar of Japanese horseradish (*Eutrema japonicum* (Miq.) Koidz. Syn. *Wasabia japonica* (Miq.) Matsum., 'Mazuma').

All of inoculated explants (9 explants per plant box) moved to the irradiation facility installed in the culture room of the Tokyo University of Agriculture (TUA) Atsugi campus (Atsugi city, Kanagawa prefecture) as soon as possible after transplanting to the plant box (made of polycarbonate, top side 7.5 cm, height 9.5 cm) containing the modified Murashige and Skoog (1962) media at Verde Co., Ltd., and started the light quality treatment. Culture conditions were 23±1°C, 16 hours illumination / 8 hours dark, PPFD at illumination was 100 μmol m $^{-2}$ s $^{-1}$ for all light sources.

The light sources used for this examination were a cool white fluorescent lamp (FL40SS-N / 37, Toshiba Lighting & Technology Co., Ltd. (Yokosuka city, Kanagawa prefecture); hereinafter referred as FL) which is usually used in Verde Co., Ltd., and two types of straight-tube type LEDs of Toshin Electric Co., Ltd. (Shibaura, Minato ward, Tokyo), Tecoled G2 (cool white; hereinafter G2) and Tecoled G4 (warm white; hereinafter G4). The respective emission spectra of three light sources measured by a spectrometer (USB2000 Fiber Optic Spectrometer, Ocean Optics Inc., Dunedin, Florida, USA) are shown in Figure 1.

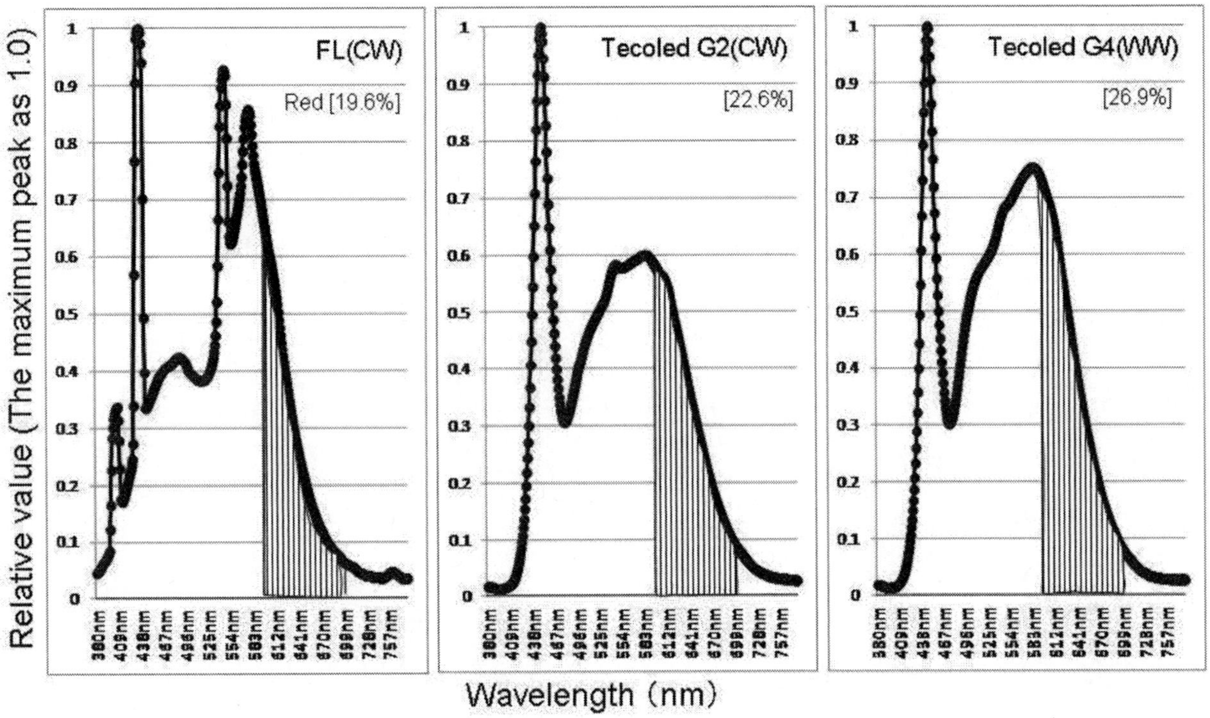

Figure 1. Respective emission spectra of three light sources. The value (%) shown in parentheses is the ratio of the red-light component (600-700 nm) to the total emitted light. CW: cool white, WW: warm white

After 3 to 4-week culture under each of light sources, the fresh weights of top and roots and the number of leaves per each plantlet were measured. Leaf lamina length and petiole length of the longest leaf, and the longest root length were further measured according to the plant species.

RESULTS AND DISCUSSION

Strawberry

In a cultivar 'Akihime', there was no significant difference in total fresh weight and leaf number per plantlet under the three different light sources. However, for 'Tochiotome' and 'Toyonoka', the total fresh weight and leaf number were larger under G4 (Table 1).

In the case of strawberry cultured transplants, when the petiole length is longer, the transplants shield the light from each other, which is undesirable. Compared with FL, the petiole length tends to be shorter in both of LEDs (Table 1), resulting in a compact appearance as a whole under LEDs.

Sweet potato

Total fresh weight per plantlet was greater under G4 in 'Narutokintoki' and 'Beniharuka' and under G2 in 'Beniazuma'. There was no significant difference in leaf number in 'Narutokintoki'. However, the leaf number of 'Beniharuka' and 'Beniazuma' tended to be larger in G4 (Table 1). In sweet potatoes, the differences in response to the different light sources were relatively great depending on the cultivars.

Table 1. Effects of light source on the growth of in vitro cultured transplants in strawberry and sweet potato.

Plant	Cultivar	Light source	Total FW (g/plantlet)	Leaf number per plantlet	Petiole length (cm)	Longest root length (cm)
Strawberry	Akihime	FL	0.43	4.9	3.9a	5.8a
		G2	0.45	4.4	2.7b	3.9b
		G4	0.44	4.8	2.4c	4.0b
	Tochiotome	FL	0.92a	5.8a	2.2	6.4a
		G2	0.73b	4.3b	1.7	5.1b
		G4	0.96a	5.4ab	1.9	5.7ab
	Toyonoka	FL	0.54b	5.6ab	2.3a	5.8b
		G2	0.47c	5.3b	1.7b	6.5a
		G4	0.88a	6.4a	1.9b	5.5b
Sweet potato	Narutokintoki	FL	0.81c	4.2	3.1b	12.3
		G2	0.96b	4.2	3.6a	13.0
		G4	1.15a	4.2	3.1b	13.0
	Beniazuma	FL	0.55b	4.1a	4.8a	15.8
		G2	0.71a	3.6b	3.7b	13.9
		G4	0.56b	4.2a	3.6b	15.3
	Beniharuka	FL	0.64b	6.9b	3.7b	20.1b
		G2	0.80a	7.1b	4.6a	24.3a
		G4	0.84a	7.8a	4.3a	19.2b

n = 10-30. Different letters within cultivar indicate a significant difference at the 5% level by Tukey's test.

Japanese horseradish

The value of all measured parameters was the maximum under G2, and the overall growth was also the most vigorous under G2 (Table 2).

From the above results, for strawberry, sweet potato and horseradish, it is considered that there is no serious problem to the production of tissue culture transplants even if it is replaced with the LED light source used this experiment from FL.

Table 2. Effects of light source on the growth of in vitro cultured transplants in Japanese horseradish.

Plant	Cultivar	Light source	Total top FW (g/plantlet)	Total root FW (g／plantlet)	Length of the largest leaf(cm)	
					Leaf lamina	Petiole
Japanese horseradish	Mazuma	FL	0.21b	30.0b	1.5b	2.4b
		G2	0.40a	60.0a	1.9a	2.9a
		G4	0.36a	41.7b	1.8a	2.8a

n = 10. Different letters within cultivar indicate a significant difference at the 5% level by Tukey's test.

However, when estimating the results in detail, it was found that strawberry and sweet potato with high growth under the G4 light which the red-light component is relatively large, and horseradish with vigorous growth under the G2 light which is relatively large in the blue light component. Thus, favorable light quality is different among plant species. In the near future, when switching the light source from fluorescent lamps to LEDs, it is necessary to select the light source by paying attention to this point.

Literature Cited

Alvarenga, I.C.A., Pacheco, F.V., Silva, S.T., Bertolucci, S.K.V. and Pinto, J.E.B.P. 2015. In vitro culture of *Achillea millefolium* L.: quality and intensity of light on growth and production of volatiles. Plant Cell Tiss. Organ Cult. *122*: 299-308. https://doi.org/10.1007/s11240-015-0766-7

Mitchell, C.A., Dzakovich, M.P., Gomez, C., Lopez, R., Burr, J.F., Hernandez, R., Kubota, C., Currey, C.J., Meng, Q., Runkle, E.S., Bourget, C.M., Morrow, R.C. and Both, A.J. 2015. Light-emitting diodes in horticulture. Horticultural Reviews *43*:1-87. https://doi.org/10.1002/9781119107781.ch01

Murashige,T. and Skoog, F. 1962. A revised medium for rapid growth and bioassays with tobacco tissue cultures. Physiol. Plant. *15*:473-497. https://doi.org/10.1111/j.1399-3054.1962.tb08052.x

Ramìrez-Mosqueda, M.A., Iglesias-Andreu, L.G. and Bautista-Aguilar, J.R. 2017. The effect of light quality on growth and development on in vitro plantlet of *Stevia rebaudiana* Bertoni. Sugar Tech. *19*: 331-336. https://doi.org/10.1007/s12355-016-0459-5

Silva, S.T., Bertolucci, S.K.V., da Cunha, S.H.B., Lazzarini, L.E.S., Tavares, M.C. and Pinto, J.E.B.P. 2017. Effect of light and natural ventilation systems on the growth parameters and chavacrol content in the in vitro cultures of *Plectranthus amboinics* (Lour.) Spreng. Plant Cell Tiss. Organ Cult. *129*: 501-510. https://doi.org/10.1007/s11240-017-1195-6

Simizu, M., Machino, Y., Akima, K. and Amaki, W. 2013. Effects of light quality on the plantlet production from PLB segment. Proc. 11th APOC, pp.245-249. (in Japanese with English abstract)

Viršile, A., Olle, M. and Duchovsks, P. 2017. Light emitting diodes for agriculture. In LED lighting in horticulture, S.D. Gupta, ed. (Springer-Verlag, Berlin), p.113-147.

PROCEEDING'S PAPERS

EASTERN REGION OF NORTH AMERICA

Dr. Charles W. Heuser, Jr., Regional editor

Sixty-eighth Annual Meeting - 2018

Newark, Delaware USA

Combined Proceedings IPPS 68:62-69. 2018.

Biocontrol for Propagation Greenhouses

Sinclair Adam
2010 Cornwall Rd, Ste 1, Lebanon, Pennsylvania 17042 USA

saa19@psu.edu

Keywords: Diseases, fungi, fungicides, bio-fungicides, insects, IPM, mites, thrips, fungus gnats.

INTRODUCTION

Biological control utilizes living organisms including insects, mites, fungi, or bacteria to control problem pests, and diseases. Utilizing beneficial control agents (BCAs) as insect and disease control procedures requires a different approach compared to pesticides and fungicides. To be successful, growers and managers need to deploy these BCAs early in the crop cycle, and not after an outbreak occurs. BCAs are used preventatively in most cases, and should be used when crops are young, pest numbers are low, and damage has not reached a critical level.

Bio control has been used extensively on greenhouse vegetable crops. Plants such as pepper and tomato when produced in greenhouses will often be longer term production cycles than many ornamentals grown as plugs or starter plants. Since only the fruit needs to be blemish free, these crops actually have a higher threshold for insect damage. Pesticides can be effective, but laws for pesticides are stricter than with ornamentals. Frequently pollinator insects will be part of the production system, and these beneficials are more compatible with BCAs than with many chemicals.

Some of the advantages in using BCAs over chemicals would include worker exposure to pesticides is reduced, reduced potential for spray injury, REI would be short to non-existent, minimum equipment needs for application, potential environmental (or green) marketing, and reduced selection pressure for resistance. Customers of the operation would also have lower exposure. Challenges would include rate of control, as BCAs will not work as quickly as a chemical, greater need for understanding the pest and the BCA life cycle, understanding the environmental requirements for success. Shelf life and handling can also be a challenge, as these beneficials need strict control of temperature in shipping and storage. One of the best management practices, would be to start on a limited basis, using one greenhouse. As experience is gained then it is possible to expand use to other greenhouses or facilities (Stack et al., 2016).

An increasing number of BCAs are on the market, and the sources for these organisms are expanding as well (Bale et al., 2008; van Lenteren et al., 2018).

Companies such as Creek Hill Nursery, North Creek Nurseries, Terra Nova Nurseries, and Longwood Gardens use biocontrol strategies for propagation and production. The Penn State (PSU) Flower Trials also utilizes biocontrol in both greenhouse production and in the PSU Flower Trials in the Field. We do use conventional pesticides but prefer to use bio control for the health of the pollinator insects and beneficials that are on the property.

BIO FUNGICIDES

Bio Fungicides are an effective control of pathogens when used preventatively prior to infection. RootShield® and RootShield® Plus have been used for a considerable time to control root rot diseases in greenhouse and nursery applications. This BioWorks® product controls *Pythium, Rhizoctonia, Fusarium, Thielviopsis, Cylindrocladium,* and *Phytopthora.* The active ingredient is *Trichoderma harzianum* (Figure 1), which are naturally occurring fungi. In RootShield® plus, there are two species (*T. harzianum* and *T. virens)* which work effectively against these problem diseases. Applied as a soil drench at potting, the *Trichoderma* fungus colonizes root systems and then prevents pathogens from attacking root systems (Dicklow, 2014).

A newer product with increased shelf life is now being sold by Marrone® Bio-Innovations, called Bio-Tam®. Bio Tam® contains *T. asperellum, and T. gamsii,* and will also control *Sclerotinia, Armillaria, and Rosellina* (Bogash, 2018). Both of these products are successfully used in the Penn State Flower Trials for production of the entries tested. In five years, there have been less than twelve containerized plants exhibiting root injury from root rot pathogens out of the thousands of containerized plants grown. RootShield® plus is added to the potting media when formulated, and the four and ½-in. pots which are planted in the final containers all are drenched with RootShield or Bio Tam® after plugs are planted in the 4.5-in. pots.

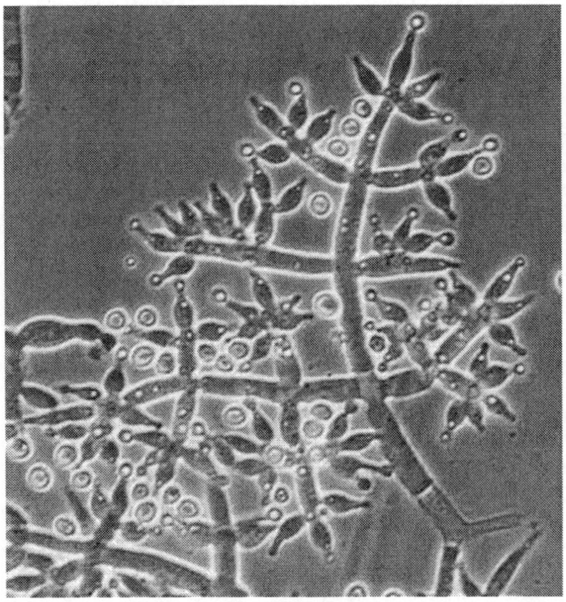

Figure 1. *Trichoderma*, Photo credit: Wikimedia.

Actinovate® SP has also been used successfully at the PSU Experiment station in Manheim, Pennsylvania. This bio fungicide has filamentous structure and works to control Powdery mildew, downy mildew, White mold, *Phytomatotricum,* and *Alternaria* as a foliar spray. Used as a root drench, it will control *Pythium, Phytopthora, Rhizoctonia*, and *Fusarium* (Stack et al., 2016).

Cease® (*Bacillus subtilis*). Cease®, and Stargus® are two products which have *Bacillus*, as the active ingredient. Both of these bio fungicides are used as protectants and should be applied before there is a pathogen present. The bacteria create a protection against infection and block the potential pathogens from entering the plant. Cease® has been successful in preventing infection by *Xanthomonas* with begonias at the PSU Flower Trials. Rotating this product with a copper fungicide has been a successful strategy in most growing seasons. Other products that utilize *Bacillus* work on the same principal and create barriers against bacteria and fungi (Dicklow, 2014).

Contans®, a product that contains *Coniothyrium minitans* works against *Sclerotinia* spp and attacks the apothecia and the sclerotia bodies that are part of white mold's biology. White mold has increased in Pennsylvania cut flower and high tunnel production recently, Contans® is labeled for greenhouse and field production.

Newly introduced in 2017, Botrystop™ (Bioworks®) is labeled for *Botrytis, Monilinia,* and *Sclerotinia* and aggressively outcompetes pathogens for nutrients in dead or dying tissue. As with the other bio fungicides listed, this one has a four-hour REI, and can be applied to a wide range of plants. Bio fungicides are successful control agents by parasitism, by rhizosphere competence, by antibiosis, induced metabolic change and plant growth promotion (Dicklow, 2014; Raudales and McGehee, 2017).

INSECT BENEFICIAL CONTROL AGENTS

Pest management in greenhouses has evolved into a complex process, using many of the tools in an IPM management toolkit. Using biocontrol agents (BCAs) has increased recently and BCAs would include predators, parasitoids, and parasites (Topliff et al., 2007). BCAs can be specialists or generalists depending on their diets. Aphids such as potato aphid, and green peach aphid are generalists, and attack a wide range of plants. *Aphis nerii* (milkweed aphid), *Chrysanthemum* aphid, and *Heliopsis* aphid (*Dactynotus*), are specialists working on a small group (*Asclepias* or *Heliopsis*) of plants (Figure 2). Similarly, BCAs can be specialists or generalists.

Figure 2. (A left) *Aphis nerii* on *Asclepias*, (B right) *Heliopsis* aphid.

Insects such as green lacewings, or lady bird beetles are generalists, and feed on a wide range of prey insects. *Aphidius ervi* is a specialists BCA, and only attacked the larger forms of aphids (Cloyd, 2015). *Hippodamia* is a generalist and will feed on a wide host range (Aristizabel and Arthurs, 2014). As aphids can build up rapidly due to asexual reproduction, prompt action in deploying BCAs is important (Chowder, 2007). An inundative release could be successful when numbers are building up rapidly, as with chrysanthemum aphid. May species of Lady bird beetles are present in the USA, and most of these are generalists. During the pupal stage of the life cycle, the lady bird beetle is vulnerable to attack from other species (such as dragonfly, assassin bugs, parasitic wasps, and ants) (Aristizabel and Arthurs, 2014).

Aphidoletes is a delicate midge which attacks aphids, and operates during long days, biting aphids on their knees and injecting a toxin (Stack et al., 2016). The wasp *Aphelinus* works well on potato aphid, and foxglove aphid, and can tolerates higher temperatures than *Aphidius* (Stack et al., 2016). Syrphid flies are excellent aphid predators in the larval stage, and while the adults feed on nectar and pollen, these populations can build up in the presence of diverse flowering plants. *Lobularia* plants provide good nectar and pollen sources for sustaining Syrphid fly populations (Shepherd et al., 2002). Green lacewings are also a generalist predator, and in the larval stage are very quick acting, but these insects usually require repeat applications to be successful (Stack et al., 2016). Aphid mummies indicate the presence of *Aphidius* wasps, which are tiny insects that are parasitoids of aphids. The hardened aphid exoskeleton is a hollow shell after the wasp has eaten its way out of the aphid (Stack et al., 2016). *Aphidius colemani* attacks the smaller aphids, like green peach or melon aphids. *Aphidius ervi* attacks larger aphids such as potato or foxglove and is about twice as large as *A. colemani* (Stack et al., 2016). *Aphidius* can be maintained in greenhouses using a banker system (Figure 3A). Plants of oats, barley, or other grassy species can be grown to rear Bird Cherry Oat Aphid which will feed exclusively on grass plants (Figure 3B). These plants are then deployed in the greenhouse to provide a food source for the wasps. When pest aphids drop below sustainable thresholds, the bird cherry oat aphid will sustain the *Aphidius* population (Wollaeger et al., 2015; Cloyd, 2015).

Figure 3. (A left) Banker system, (B right) grassy species can be grown to rear Bird Cherry Oat Aphid which will feed exclusively on grass plants.

A good reference guide is a big help in managing the BCAs in propagation greenhouses, and the New England Greenhouse Floricultural Guide is an excellent choice for in depth information on insect and disease challenges (Stack et al., 2016) (Figure 4). *Lepidoptera* larvae (caterpillars) have been successfully managed at the Penn State Flower Trials using Dipel® DF, or one of the other *B. thuringiensis* products. At the Penn State Flower Trials, we have been concerned with

Tobacco bud worm, and the Virginia tiger moth (or yellow wooly bear), due their predilection for petunia flowers. Naturally occurring predators such as wheel bugs, assassin bugs and birds also contribute to control in this outdoor setting.

Figure 4. New England Greenhouse Floriculture Guide 2017–2018.

Two spotted spider mites are typically found on the lower leaf surfaces of plants. These mites have shown resistance to a number of miticides, but good BCA solutions are available for control (Stack et al., 2016). One of these BCAs *Phyoseiulus persimilis* is an excellent control and moves rapidly through the plant canopy (Wollaeger et al., 2015; Cloyd, 2015). These mites work well at moderate temperatures, and consume eggs, nymphs, and adult Two Spotted Spider Mites (Cloyd, 2015). At temperatures over 80 °F using *Neoseiulus* or *Galendromus* mites could be a preferable control strategy (Stack et al., 2016).

Thrips have been showing resistance to a number of insecticides and are often difficult to control. One of the reasons for this is their behavior in plant canopies. Thrips tend to be found in flowers and buds making them hard to get good chemical coverage. This cryptic behavior of thrips makes it hard to get good control with insecticides (Greer and Diver, 2000). Fortunately, several good BCAs are available for control. Pirate bugs (*Orius* spp.) are excellent consumers of thrips especially the western flower thrips (*Frankliniella occidentalis*) (Stack et al., 2016). At the PSU Flower Trials, we have been using pirate bugs for four years, and the population continues to maintain itself without additional pirate bug releases. These BCAs do a fair job in controlling thrips, but in 2018 additional support was required. To accomplish this, *Steinernema feltiae* were deployed through the injector system to assist with control of the soil borne stages of thrips life cycle. *Amblyseius swirskii,* was also applied in August, at 1 sachet/plant. All three of these control methods worked very well in concert to drop the thrips levels by 75-80%.

While we used sachets of the *Amblyseius swirskii,* they can be applied to flats and plug trays by shaking the mites out of the shipping container or using an air blast delivery system (Dogramaci, et al., 2013). To assist with pirate bug support, both Black Pearl and Purple Flash ornamental peppers are placed in the trials program to provide a nectar and pollen source for the pirate bugs. These two bankers (or support plants) have been placed in each bed of the trials. Research has shown that both Black Pearl and Purple Flash do a good job providing flower nectar and pollen for the pirate bugs to feed upon (Wong and Frank, 2011; Waite, 2012).

Fungus gnats are frequently problematic in many propagation systems. The larval stage feeds on newly developing roots in mist or fog propagation systems (Stack et al., 2016). While a number of chemical control products are available, the use of *Steinernema feltiae* has become the preferred control procedure for many operations. Applications of *Steinernema* on a 14 to 21-day interval provides effective control (Stack et al., 2016). Dissolving the packet of *Steinernema* in water and applications through the injector system, watering can, or pressure sprayer provides good coverage. *Hypoaspis miles*, and rove beetles also provide control the fungus gnat populations in flats and plug trays. These BCAs work on the soil borne insect populations and can be compatible with some chemical (Cloyd et al., 2010).

CONCLUSIONS

BCAs are being utilized successfully at an increasing rate in propagation greenhouses and nurseries in the USA (Stack et al., 2016).

While challenges can be increased compared with conventional chemical use, the BCAs are a good alternative choice for propagators and producers of ornamental plants (van Lenteren et al., 2018). Using BCAs will take more time especially for scouting, but the benefits to consumers, employees, and the environment are significant (Bale, et al., 2008) (van Lenteren et al., 2018). Using BCAs in both propagation greenhouses, and at the Penn State Flower Trials, have shown to be successful approaches to insect and disease control.

Literature cited

Aristizabal, L.F., and Arthurs, S.P. (2014). Convergent lady beetle. UF IFAS EENY 592. http://entnemdept.ufl.edu/creatures/BENEFICIAL/convergent_lady_beetle.html

Bale, J.S., van Lenteren, J.C., and Bigler, F. (2008). Biological control and sustainable food production. Philosophical Transactions R. Soc. Lond. B. Sci. Feb 27:*363* (1492) 761-776. https://www.ncbi.nlm.nih.gov/pmc/articles/PMC2610108/

Bogash, S.M. (2018). Personal communications.

Chowder, D.W. (2007). Impact of release rates on the effectiveness of augmentative biological control agents. J. Insect Sci. 7:15. https://www.ncbi.nlm.nih.gov/pubmed/20307240

Cloyd, R. (2015). Biological control agents guide. Greenhouse Management. http://www.greenhousemag.com/article/biological-control-agents-guide/

Cloyd, R., Timmons, N.R., Goebel, J.M., and Kemp, K.E. (2010). Pesticides and rove beetles are they compatible? Greenhouse Product News Jan 2010. https://gpnmag.com/article/pesticides-and-rove-beetles-are-they-compatible/

Dicklow, M.B. (2014). Bio fungicides. UMass Amherst Extension GH Crops and Floriculture https://ag.umass.edu/greenhouse-floriculture/fact-sheets/biofungicides

Dogramaci, M., Kakkar, G., Kumar, V., Chen, J., and Arthurs, S. (2013). Swirskii mites – featured creatures UF IFAS EENY 565. https://edis.ifas.ufl.edu/pdffiles/IN/IN100100.pdf

Greer, L. and Diver, S. (2000). Greenhouse IPM sustainable thrips control. ATTRA. https://attra.ncat.org/attra-pub/summaries/summary.php?pub=50

Raudales, R., and McGehee, C. (2017). Bio fungicides for control of root diseases on greenhouse grown vegetables. E Gro Edible Alert. https://egrouni.com/pdf/E207.pdf

Shepherd, M., Black, S.H, and Kerns, C. (2002). Flower flies. USDA Forest Service Pub.

https://www.fs.fed.us/wildflowers/pollinators/pollinator-of-the-month/flower_flies.shtml

Stack, L.B., Dill, J., Pundt, L., Raudales, R., Smith, C., and Smith, T. (2016). New England greenhouse floricultural guide. New England Floriculture, Inc.

Topliff, L.A., Pinkston, K.N., von Broembsen, S.L., Schnelle, M.A., and Smolen, M.D. (2007). Using biocontrol agents in the commercial greenhouse. Oklahoma Coop. Ext. http://www.biofac.com/Urban_Greenhouse/Biological_Agents/body_biological_agents.html

Van Lenteren, J.C., Bolckmans, K., Kohl, J., Ravensberg, W.J., and Urbaneja, A. (2018). Biological control using invertebrates and microorganisms: plenty of new opportunities. BioControl *63*:39-59. DOI 10.1007/s10526-017-9801-4

Waite, M.O. (2012). New strategies to improve the efficiency of the biological control agent *Orius insidiosus* in greenhouse ornamental crops. M.S. Thesis Univ of Guelph. https://atrium.lib.uoguelph.ca/xmlui/bitstream/handle/10214/5080/Waite_Meghann_201212_Msc.pdf;sequence=5

Wollaeger, H., Smitley, D., and Cloyd, R. (015). Commercially available biological control agents for common greenhouse insect pests. MSU Bulletin 3299. http://msue.anr.msu.edu/uploads/resources/pdfs/FINALtoBOOKSTORE_BCAFactSheets.pdf

Wong, S., and Frank, S. (2011). Black pearl pepper banker plant for biological control of thrips in commercial greenhouses. SARE Final Report 2018. https://projects.sare.org/sare_project/gs10-089/

Some useful websites:

- Association of Natural Biological Producers: http://www.anbp.org/
- Biobest Biological Systems: https://www.biobestgroup.com
- Side Effects Manual: https://www.biobestgroup.com/en/side-effect-manual
- Biological Control: A Guide to Natural Enemies in North America: http://www.nysaes.cornell.edu/ent/biocontrol/)
- Buglady Consulting – Biological Control Services: http://www.bugladyconsulting.com
- Koppert Biological: www.koppert.com
- New England Greenhouse Update: www.negreenhouseupdate.info
- University of Massachusetts Extension, Greenhouse Crops and Floriculture Program: http://www.umass.edu/umext/floriculture/

- University of Vermont, Entomology Research Laboratory: http://www.uvm.edu/~entlab/Greenhouse%20IPM/greenhouseipm.html

- IPM Laboratories New York: https://www.ipmlabs.com/plant-pest-management

- Penn State University Nematodes: https://ento.psu.edu/extension/factsheets/parasitic-nematodes

- Penn State University Broad Mites: https://extension.psu.edu/broad-mites-an-example-of-using-biocontrols-for-management

- Vendors of Beneficial Organisms in North America: Jen White and Doug Johnson, Univ of Kentucky 2010. http://www2.ca.uky.edu/entomology/entfacts/ef125.asp

Combined Proceedings IPPS 68:70-75. 2018.

Coming Up Short: Groundcover Reblooming Daylilies

Darrel Apps
Garden Adventures, 29 Dewey Ave., Wild Rose, Wisconsin 54984 USA

darrel_apps@hotmail.com

Keywords: *Hemerocallis*, breeding, hybrids, hybridization.

INTRODUCTION

I made my first daylily crosses in 1969. My long-term goal was to produce daylilies that rebloomed and extended the flowering time. In the early 1970s, I was inspired by the breeding work of Walter Jablonski of Merryville, Indiana. In 1975 he introduced 'Stella de Oro' which eventually became the most popular daylily sold in the USA, Canada, and many European countries. It remains a popular daylily because of the following reasons:

1) Rebloom: often more than 10 weeks of bloom.
2) Early bloom: it starts before most other cultivars and is a harbinger of late spring
3) Winter hardy; to USDA Zone 4.
4) Rapid increase: 10:1 in container production when grown in full sun with adequate water and high fertility regimes
5) Good height for different container sizes—sold in quart, gallon, and 2-gal containers.
6) Bright mass gold color—self colors usually have the best curb appeal.
7) Extended bloomer: flowers open for nearly 24 hr.
8) Self-cleaning small (2.75 in.) blossoms.
9) Rust resistance.

Rebloom is often misunderstood in the daylily registration process and cultivars are registered with the "Re" symbol even though they are not reliable rebloomers. Also, rebloom differs depending on the zone in which the plant is grown. Nursery marketing (knowingly and unknowingly) uses the mis-representation to their advantage.

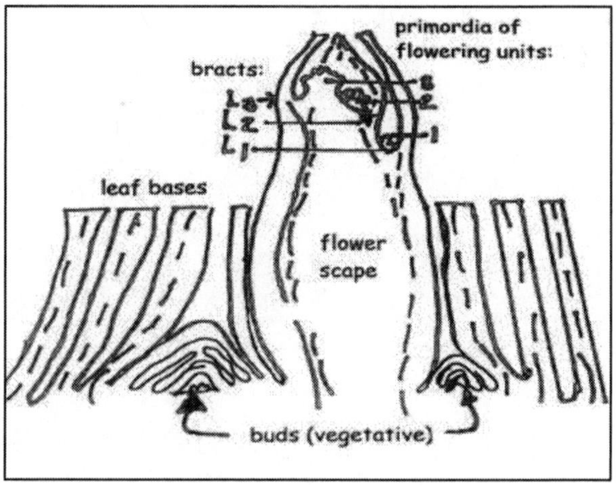

Figure 1. Anatomy of a dormant daylily

Figure 1 shows a typical dormant daylily with flower scape and vegetative buds. Plants such as 'Stella de Oro' have additional flower buds and have sequence blooming for several weeks. However, after flower primordia develop and flower, vernalization is required before more flower primordia develop.

MY DAYLILY BREEDING PROGRAM

My daylily breeding work has explored several different routes in order to find hardy dormant repeat bloomers. In this paper they'll be discussed as steps.

Step 1: Identifying rebloom in species and early cultivars

In 1963 daylily breeder Warner introduced a miniature rebloom daylily 'Bitsy'. Data on this daylily from Longwood Gardens, Kennett Square, Pennsylvania showed that it had 3 periods of bloom starting in early June and ending in September. Figure 2 shows much of the genealogy of this hybrid. The species involved are: *Hemerocallis aurantiaca, H. fulva* 'Europe', and *H. minor*. The rebloom appears to come from the reblooming evergreen species *H. aurantiaca*.

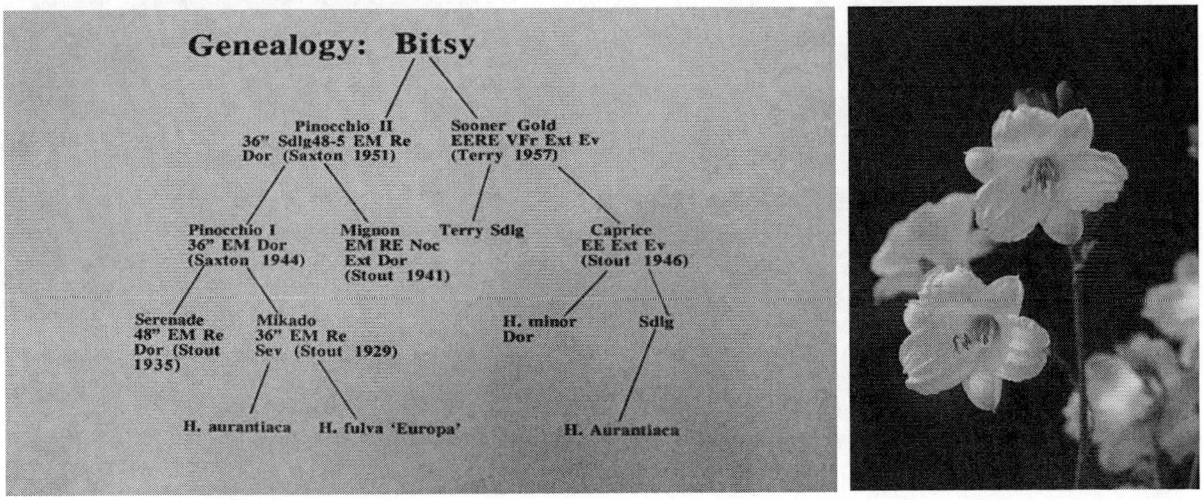

Figure 2. A (left) the genealogy of *Hemerocallis* 'Bitsy', B (right) 'Bitsy' flower.

After breeding diploid daylilies for several years, it became apparent that the evergreen gene was dominant to the deciduous gene and followed simple mendelian inheritance. Thus, the progeny of an evergreen daylily and a dormant daylily are all evergreen and heterozygous plants. Crossing these F_1 hybrids would produce about ¼ deciduous and ¾ evergreen plants. The matrixes are shown in Figure 3.

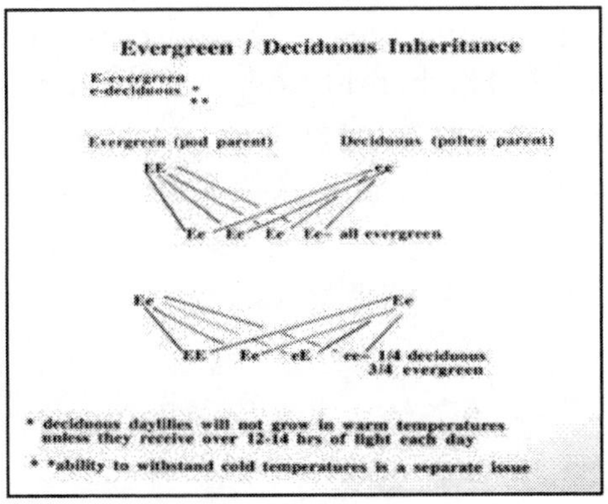

Figure 3. Evergreen and deciduous inheritance in daylily.

I visited Walter Jablonski's hybridizing during several different years in the early 1970s. He mentioned several different plants in the 'Stella de Oro' pedigree. Among them *H. lilio-asphodelus* which may be responsible for the extended bloom time. *Hemerocallis minor* was another background plant which may account for the small stature of 'Stella de Oro'. He also said he had used the cultivar 'Bitsy'.

Step 2: Experimental crosses to search for rebloom

After obtaining 'Stella de Oro' (Figure 4), I crossed it to most of the cultivars I had in my collection (about 50 different hybrids). The one rebloomer that came out of these crosses was 'Happy Returns' (Figure 4). It first blossomed in 1983 and was registered in 1986 and introduced by Wayside Gardens in 1987 (IPPS member Roy Klehm did the production).

Figure 4. Reblooming daylilies (left) 'Stella de Oro', (right) 'Happy Returns'.

Another hybrid came from an entirely different line of breeding. I had noticed some rebloom in two cultivars, 'Golden Chimes' and 'Perennial Pleasure', and crossed them together. The hybrid from that cross was red (it did not rebloom) and it was crossed with 'Little Grapette' which produced the reblooming 'Pardon Me'.

Using these two plants with some other rebloomers I was able to produce the reblooming 'Rosy Returns' which was an early color break from other rebloomers that were mainly yellow and gold. The pedigree of 'Rosy Returns' is shown in Figure 5.

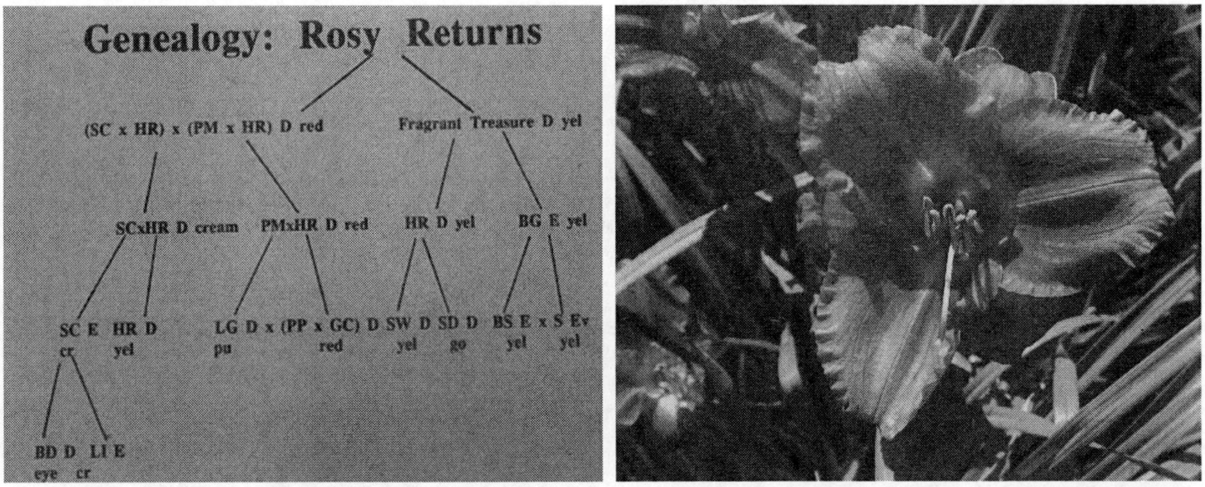

Figure 5. (A left) The genealogy of 'Rosy Returns'. The letters represent the cultivars in the genealogy of 'Rosy Returns'. Additionally, the letters "D" and "E' are used to designate dormant and evergreen. Cultivars letter designations: BD-'Buffy's Doll', LI-'Little Infant', SC' 'Sugar Cookie', HR-Happy Returns', PM 'Pardon Me', LG 'Little Grapette' , PP 'Perennial Pleasure', GC 'Golden Chimes' SW. 'Susie Wong', SD 'Stella de Oro' BG 'Brocaded Gown', BS 'Buttermilk Sky', and S 'Sabie'. (B right) 'Rosy Returns'.

Step 3: Breeding and selecting new rebloom cultivars motivated by a branded program

In the early 2000s I helped develop a couple of daylily brands with Denny Blew of Centerton Nursery, New Jersey, and now continue to work with Bob Blew of the same firm. One of the brands was for reblooming daylilies and identified as Happy Ever Appster® Daylilies.

It is also important to recognize other hybridizers that have contributed to my breeding program:

Rebloom genes from other hybridizers:

- Warner: 'Bitsy' 1963
- Jablonski: 'Stella de Oro' 1975
- Simpson: 'Lemon Lollypop' 1985
- Millikan: 'Sunny Honey' 1989
- Huben: 'Early and Often' 2001
- Carpenter: 'Connie Can't Have It' 2004
- Herrington: 'Little Gold Nugget' 2005
- Derrow: 'Adena Blizzard' 2014

Below are examples of cultivars I introduced and are included in the brand Happy Ever Appster® (see Figure 6 for selected images):

- 'Apricot Sparkles'
- 'Big Time Happy'
- 'Dynamite Returns' Passionate Returns™ daylily
- 'Endless Heart' NJ Earlybird Cardinal™ daylily
- 'Happy Days are Here Again'
- 'Happy Enchantment'
- 'Happy Returns'
- 'Red Hot Returns'
- 'Romantic Returns'
- 'Rosy Returns'
- 'Scentual Sundance'
- 'Stephanie Returns'
- 'Sunset Returns'
- 'When My Sweetheart Returns'

Figure 6. Examples of cultivars I introduced and are included in the brand Happy Ever Appster® daylilies. Top row left to right: 'When My Sweetheart Returns', NJ Earlybird Cardinal™ 'Endless Heart', 'Happy Days are Here Again'; Bottom row: 'Red Hot Returns', 'Scentual Sundance', 'Stephanie Returns'.

Step 4: Breeding self-colors for mass display and curb appeal

Daylilies in bright colors and usually of one color have the best curb appeal. Generally bright reds (Figure 7), bright pinks, whites, yellows, and oranges provide the best curb appeal. Patterned daylilies and those with eyes often are not attractive from a 100 ft or more and do not have curb appeal; however, they are useful for decorating smaller gardens. Several seedlings are shown that represent some of my recent breeding work toward curb appealing daylilies.

Evaluating, producing, and marketing a new cultivar (brand) involves more than a plant breeder's input. Large nurseries often use focus groups representing gardeners, retailers, l andscapers, and growers to help them in the evaluation process. Growers need to be consulted to determine special production needs, growing methods, and timing schedules. Marketing with patents and trademarks needs to be carefully planned for the best economic outcome.

Some criteria I use in the evaluation process of groundcover daylilies:

- Clean self-colors with curb appeal (pinks, reds, white, yellow, and orange)
- Rebloom (8 weeks plus)
- Rust resistance
- Early bloom (June 1 in Zone 6)
- Hardy to Zone 4
- Rapid increase greater than1:5
- Short plant less than 18 in.
- Extended opening greater than 16 hr
- Self-cleaning, 3-4 in. flowers

Figure 7. Bright red selections with curb appeal.

Step 5: A trend toward smaller homes and gardens (less time for maintaining big gardens)

The main focus of daylily breeders has been with large plants and large flowers. These plants usually require deadheading each day. However, there appears to be a trend for more compact plants for smaller spaces and for container gardening.

In order to compete with long blooming annuals there needs to be greater effort in producing more rebloom in daylilies. As in all ornamental plants, there needs to be continued improvement in new colors and color combinations.

I'll end this by saying: My dream is to see millennials and generation Z customers walking out of garden stores with "6-packs" of unique quart sized daylilies all in bloom on June.

Combined Proceedings IPPS 68:76-79. 2018.

Alcohol-Based Rooting Hormones: Do They Burn Cuttings?

Eugene K. Blythe[1] and James T. Ray[2]
[1]Mississippi State University, Coastal Research and Extension Center, South Mississippi Branch Experiment Station, Poplarville, Mississippi 39470; [2]Mississippi State University Extension Service, Gulfport, Mississippi 39502

[a]blythe@pss.msstate.edu, [b]tim.ray@msstate.edu

Keywords: IBA, *Lantana*, *Artemisia*, *Ficus*, *Impatiens*, *Pelargonium* ×*hortorum*, *Trachelospermum jasminoides*, *Chrysanthemum*.

INTRODUCTION

Auxins have been used for cutting propagation since the mid-1930s, with indole-3-butyric acid (IBA) and 1-naphthaleneacetic acid (NAA) being the most common (Hartmann et al., 2002; Thimann and Koepfli, 1935; Zimmerman and Wilcoxon, 1935). Commercial root-promoting products ("rooting hormones") are available in various formulations: liquid concentrates, water-soluble salts and tablets, powders (talc), and gels (Boyer et al., 2013). Being growth regulators, commercial root-promoting products sold or used in the United States must be registered by the Environmental Protection Agency, falling under the Federal Insecticide, Fungicide, and Rodenticide Act (FIFRA). Technical grade IBA and the potassium salt of IBA (K-IBA) are not registered for sale and use as plant growth regulators. Auxin solutions may be applied to stem cuttings by several methods: basal quick-dip, foliar spray, total immersion, extended basal soak extended soak and other less common methods (Blythe et al., 2007).

In the United States of America, four products are currently registered for use in preparing auxin solutions for cutting propagation (Boyer et al., 2013). Dip 'N Grow (Dip 'N Grow Inc., Clackamas, Oregon) and Wood's Rooting Compound (Earth Science Products Corp., Wilsonville, Oregon), which contain IBA and NAA, are alcohol-based, dilutable concentrates. Hortus IBA Water Soluble Salts (Phytotronics Inc., Earth City, Missouri) and Rhizopon AA Water Soluble Tablets (Phytotronics Inc.) are dissolvable, contain IBA, and form K-IBA when dissolved in water. Alcohol is not required for preparing solutions of water-soluble salts but may be used if desired. Along with acting as a co-solvent, alcohol can also act as a surfactant, disinfectant, and/or preservative, and will evaporate if left in an open container.

There have been anecdotal reports that the use of alcohol in auxin solutions can cause "stem burn" on stem cuttings. However, no formal research has been conducted to adequately establish occurrence of tissue damage on stem cuttings using alcohol-based auxin solutions. Therefore, the objective of our research project was to assess potential phytotoxic effects of alcohol on stem cuttings of several herbaceous and woody plant taxa using methods of applications representative of those used in the nursery industry.

MATERIALS AND METHODS

Cuttings of six herbaceous and woody taxa were obtained from actively growing production and landscape plants at the South Mississippi Branch Expt. Station in Poplarville, Mississippi: *Artemisia afra*, *Ficus benjamina*, *Impatiens* (interspecific) 'Coral', *Pelargonium ×hortorum* 'Mary Helen', and *Trachelospermum jasminoides*. Cuttings of *Lantana* 'New Gold', were obtained from plants that were not actively growing due to summer heat but were included as an example of cuttings that were of less than optimal condition for propagation. Cuttings of *Chrysanthemum* Mammoth™ 'Yellow Quill' were supplied by Ball Horticultural Company (Figure 1). Cuttings were freshly prepared to a uniform size appropriate for the species, with basal leaves removed from cuttings of some taxa.

In Expt. 1, cuttings received a 1-sec basal dip to a uniform depth of 0.5 in. in a solution containing Hortus IBA Water Soluble Salts at 0, 1000, or 2000 ppm IBA. In Expt. 2, cuttings received a 5-second immersion in a solution containing Hortus IBA Water Soluble Salts at 0, 100, or 200 ppm IBA and allowed to dry. In both experiments, solutions containing each of the three rates of IBA were prepared with 0%, 25%, or 50% isopropyl alcohol for a total of 9 treatment combinations per experiment.

Treated cuttings were stuck into 50-cell rooting trays containing Fafard 3B mix using a completely randomized design with 33 cuttings per treatment. Cuttings were placed under intermittent mist (10 sec/10 min during daylight hours). Evaluation concluded at 30 days (*Artemisia* and *Chrysanthemum*) or 50–55 days (all other taxa) after sticking. Data collected included mortality, visual evidence of basal stem burn, and visual evidence of leaf burn. The binomial response data were analyzed using generalized linear models.

RESULTS

In Expt. 1, solutions containing up to 50% alcohol were safe for cuttings of *A. afra*, *C.* Mammoth™ 'Yellow Quill', *F. benjamina*, *Impatiens* 'Coral', *P. ×hortorum* 'Mary Helen', and *T. jasminoides* using a basal quick-dip application. Limited mortality and stem burn were present on cuttings of *L.* 'New Gold' with treatments containing 25% and 50% alcohol; therefore, when using a solution of IBA for cuttings of *L.* 'New Gold', no alcohol or only a low rate of alcohol should be used in the solution. Cuttings from actively growing and healthy lantana cuttings are preferred for propagation.

In Expt. 2, solutions containing up to 50% alcohol were safe for cuttings of *C.* Mammoth™ 'Yellow Quill', *F. benjamina*, and *I.* 'Coral' using the total immersion method. Stem and leaf burn occurred on cuttings of *A. afra*, *L.* 'New Gold', and *P. ×hortorum* 'Mary Helen'; therefore, treatment of cuttings of these three taxa should be limited to solutions with no alcohol. Although alcohol burn did not occur on surviving cuttings of *T. jasminoides*, solutions containing 50% alcohol did increase cutting mortality; therefore, alcohol content should not exceed 25% in solutions used to apply IBA to cuttings of this crop.

Figure 1. Experimental plants: Top row: *Artemisia afra*, *Ficus benjamina*, *Impatiens* (interspecific) 'Coral'; Middle row: *Pelargonium* ×*hortorum* 'Mary Helen', *Trachelospermum jasminoides*, *Lantana* 'New Gold'; Bottom row: *Chrysanthemum* Mammoth™ 'Yellow Quill'.

CONCLUSIONS

When using a basal quick-dip, auxin solutions containing up to 50% alcohol are safe for stem cuttings that are in good condition. Some crops may be sensitive to alcohol-based rooting solutions applied to the entire cutting.

In these cases, test crops for sensitivity to your alcohol solution and be aware that cuttings in suboptimal condition may be more sensitive.

Literature Cited

Blythe, E.K., Sibley, J.L., Tilt, K.M., and Ruter, J.M. (2007). Methods of auxin application in cutting propagation: A review of 70 years of scientific discovery and commercial practice. J. Environ. Hort. *25*:166–185.

Boyer, C.R., Griffin, J.J., Morales, B.M., and Blythe, E.K. (2013). Use of root-promoting products for vegetative propagation of nursery crops. Kansas State University Agricultural Experiment Station and Cooperative Extension Service, Publication MF3105, December 2013. http://www.ksre.ksu.edu/bookstore/pubs/MF3105.pdf

Hartmann, H.T., Kester, D.E., Davies, F.T., Jr., and Geneve, R.L. (2002). Hartmann and Kester's Plant Propagation: Principles and Practices. 7th ed. Prentice Hall, Upper Saddle River, New Jersey.

Thimann, K.V., and Koepfli, J.B. (1935). Identity of the growth-promoting and root-forming substances of plants. Nature *135*: 101–102.

Zimmerman, P.W., and Wilcoxon, F. (1935). Several chemical growth substances which cause initiation of roots and other responses in plants. Contrib. Boyce Thomp. Inst. *7*: 209–229.

Combined Proceedings IPPS 68:80-84. 2018.

The United States National Arboretum and Its Azaleas: The Last Ten Years

Barbara L. Bullock

Curator of Rhododendrons and Azaleas, U.S. National Arboretum, 3501 New York Avenue, NE. Washington, DC 20002 USA

Barbara.Bullock@ars.usda.gov

Keywords: History, USDA, B.Y. Morrison, Glenn Dale, *Rhododendron*.

INTRODUCTION

The United States National Arboretum (USNA) azalea collection is a reference collection of documented new and old varieties, species, and cultivars. Within the collection is the group of Glenn Dale azaleas that were planted between 1946 and 1947 on the south face of Mt. Hamilton by the Arboretum's first director, Benjamin Y. Morrison, and breeder of the Glenn Dale azaleas. In late May 2011, a very generous donor provided the Friends of the National Arboretum with $1 million to support the staffing needed to address routine maintenance of these azaleas, thus preserving these plants for the public and for the historical significance of their role in Benjamin Y. Morrison's breeding work. In 2013, the Agricultural Research Service provided funds to rejuvenate the 67-year-old planting and to remove excess saplings and invasive weeds. The process of rejuvenating azaleas will be discussed further in this paper.

One of the main visitor attractions at the USNA in the spring, is the Glenn Dale hybrid azaleas, introduced by the Arboretum's first director, Benjamin Y. Morrison (1891-1966) (Figure 1). The USNA may have the most complete collection found in the United States. *USDA Agriculture Monograph No. 20, The Glenn Dale Azaleas,* issued in October 1953 and authored by Morrison, provides some of the background story behind the Glenn Dale azalea breeding program. The actual story is longer and more complex.

Figure 1.
Benjamin Yeo Morrison (1891-1966) speaking at the Morrison Azalea Garden dedication, May 3, 1954.

While weeding and restoring the Glenn Dale azaleas, over 190 labels with Glenn Dale (Bell Station) accession numbers, also known as Bell numbers, scratched on them were discovered. These reveal more information about the Glenn Dale azaleas that were planted there between 1946 and 1947. The author, her volunteers and staff found, mapped, and researched these labels, revealing new facts about the azaleas and the breeding and selection program that produced them. For more information, a detailed account of the author's findings was published in *The Azalean, the Journal of the Azalea Society of America*, in two parts: *Glenn Dale Azaleas on Mt. Hamilton—The Long and Winding Road to Today, Parts I & II*, in Spring and Summer 2017 issues; Vol. 39, Nos. 1 and 2 (Figure 2).

Figure 2. Two-part article in The *Azalean*, Journal of the Azalea Society of America, Vol.39.

Growing within the azalea collection today is one of the most diverse and comprehensive documented collections of azalea hybrids, species, and cultivars to be found anywhere. Our collection of Glenn Dale azaleas is centered around the Morrison Azalea Garden. Located to the northeast of the Morrison Garden on Mt. Hamilton is the area we refer to as the "Loop", a network of trails and beds in a woodland setting that features the Henry Mitchell Cultivar Walk. Within this area are planted extensive collections of Kurume, Robin Hill, Back Acre, Satsuki, Pericat, Harris, Aromi, and North Tisbury cultivars. These are mainly arranged as sub-collections in group-themed beds within the azalea collection.

Other azalea hybrids that are not as well-represented in the collection are often planted within beds of similarly colored cultivars for comparison purposes. These include the Encore® azaleas, Beltsville, Linwood, Matlack, Marshy Point, Mauritsen, Hirado, Holly Springs, Knap Hill, Belgian-Glenn Dale, Pride, Pryor, Vuyk, Sander, Southern Indian, Weston, and Henry T. Skinner hybrids. Many of these were widely distributed in the 20th century, but are not so common now, as newer cultivars make their way into the market. If a cultivar is lost in the trade, it may still be growing at the USNA. Propagators may request cuttings for reintroduction to the trade, for breeding, or for research if they cannot be found elsewhere.

After the dedication of the Morrison Garden in 1952, and Morrison's subsequent retirement in 1954, the azalea collection began to take on its present-day arrangement, and by the mid-1970s the position of curator was created. Beginning in 1958, the USNA actively sought to collect all azaleas that would grow in the mid-Atlantic region. (Incidentally, this was also the year *Azaleas* by Frederic P. Lee was published by The American Horticultural Society.) And so, beginning in 1958, azalea cultivars began being planted into the collection by their color. The color groups currently found within the azalea collection are the reds, pinks, salmons (orange-pinks, yellowish-pinks), purples, whites and bicolors.

For example, within our salmon (yellowish-pink) bed, we have planted several salmon-colored Glenn Dales such as Coralie, Ambrosia, Colleen, Jubilee, Lullaby,

Opera with over one hundred other cultivars of similar color of various groupings and flanked them with low-growing azaleas raised from wild collected seed we brought back from Japan, *Rhododendron indicum*. Arranging the plantings this way enables the staff to educate the public on the range of forms, breadth of colors, flower shapes and sizes, time of bloom and heights found within a single-color palette.

In April 1971, the Frederic P. Lee Garden was dedicated to author and devoted friend of the National Arboretum. Following a suggestion of Lee's, The Lee Azalea Garden features mostly late-blooming Satsuki hybrids that grow in our mid-Atlantic region in hardiness Zone 7B. Over 130 cultivars of Satsuki can be found growing in the garden. Other late blooming hybrid groups planted adjacent to the Lee Garden are the North Tisbury, Back Acres, and the Encore® azaleas.

#005 (May be 'Viking')

#016 (May be 'Pirate')

#024 (May be 'Dimity')

#034 (May Be 'Zephyr')

#038 (May be 'Mayflower')

#054 (Probably 'Ambrosia')

#044 (May be 'Dayspring')

#050 (May be 'Caress')

#073 (May be 'Limerick')

Figure 3. A sample of Glenn Dale azalea hybrids.

There are only a few Hirado group azaleas due to their lack of hardiness. We have seven cultivars and among these 'Shirokujaku' stands out with its white, four-inch diameter, single flowers and dark evergreen foliage. We have interplanted these with the late blooming North Tisbury hybrids, and scattered specimens of the Kyushu azalea, *R. kiusianum*.

All of these sub-collections are tied together and nicely interpreted through signage made possible by the friends of the late Henry Mitchell, garden writer and another friend of the National Arboretum. Currently within the azalea collection of over 1,990 living accessions, over 1,560 accessions are *Rhododendron*. We have 530 accessions of Glenn Dales with 388 different cultivars represented (Figure 3 shows a sample of Glenn Dale hybrids).

Our next largest grouping after the Glenn Dale and Satsuki groupings are the Kurumes with 121 accessions representing 106 different cultivars.

Rejuvenating azaleas begins with removing the largest oldest canes, about one third, from each plant (Figure 4).

The most optimal time to do this is in March (or early spring) before the new growth begins to emerge. After completely removing entire branches to the ground, remove all leaf litter that has been building up so that new growth is not suppressed. This should be done again after the following autumn season until new growth has become well-established. In less than three years, the azaleas will be full and vigorous.

Since the renovation of the Glenn Dale Hillside, we have begun replanting the bare areas with Glenn Dale cultivars which we raise from cuttings taken from the Morrison Azalea Garden. Work is accomplished by group projects with help from staff, volunteers, volunteer work groups, and interns funded by the Friends of the National Arboretum.

Figure 4. One year after rejuvenation (left image); three years after rejuvenation (right image).

A pair of Bald Eagles has set up residence in a tulip poplar on Mt. Hamilton high above the Glenn Dale azaleas since 2015 which means a portion of the Glenn Dale Azalea Hillside is closed when the eagles are raising their young (late December to mid-July). To view the eagles, live go to www.dceaglecam.org.

Visit www.usna.usda.gov/abe (Arboretum Botanical Explorer) a plant search/finder for information on mapped inventoried USNA plants. The USNA mobile app is now working for both Apple iOs and Android phones. Search "National Arboretum" in iTunes for our Mobile App to find plants while you're visiting the USNA.

ACKNOWLEDGEMENTS

Barbara L. Bullock is a Horticulturist at the U. S. National Arboretum in Washington D. C. and has been Curator of the extensive Azalea and Rhododendron collections there since January 1990. Much appreciation goes to Botanist, Stefan Lura for his careful review of this paper.

The author would also like to express sincere gratitude to all the people past and present for volunteering their talents and time in the garden. And a special thanks to the Friends of the National Arboretum for their ongoing and consistent support of legacy collections at the National Arboretum.

Combined Proceedings IPPS 68:85-95. 2018.

Propagation of Herbaceous Native Perennials

Neil Diboll
Prairie Nursery, Inc., P.O. Box 306, Westfield, Wisconsin 53964 USA

Ndiboll@prairienursery.com

Keywords: Seed dormancy, cuttings, division, wildflowers, sedges, grasses, rushes, germination.

INTRODUCTION

Herbaceous native perennials include wildflowers, grasses, sedges, and rushes. Most can be readily propagated from seed. Some exhibit complex seed dormancies and are more easily propagated vegetatively by root division or stem cuttings. This article will focus on propagation of wildflowers and grasses using seed, as this is a commonly used, but often misunderstood method of producing native herbaceous perennials. The methods described herein are based upon our forty-five years of experience at Prairie Nursery in propagating a wide range of native plants from seed.

SEED TREATMENTS WE USE TO OVERCOME SEED DORMANCY

Most native perennials require that their seed be pretreated to break dormancy prior to seeding. There are four basic types of seed treatments or planting methods that we use to overcome seed dormancy:

1. Dry stratification: Seed is exposed to freezing temperatures for 30 or more days.
2. Moist stratification: Seed is mixed with a damp inert substrate and stored in a refrigerated environment at 34–36°F (1–2°C). The seed should not be frozen, as this may damage the cell walls and destroy the seed.
3. Scarification: Seed with hard seed coats are scratched with sandpaper to allow moisture to penetrate the seed and initiate the germination process.
4. Hot water: Seeds that are stimulated to germinate by wildfires are treated with near-boiling water.

Dry stratification

Many native seeds require exposure to cold temperatures as a protective mechanism, so they do not germinate in fall and have their tiny seedlings killed over winter. The term "seed stratification" originated many years ago when wildflowers seeds were originally pretreated by planting them in layers of damp, clean sand, and refrigerating them to mimic the effects of winter. Many native seeds require exposure only to cold

IPPS Vol. 68 – 2018

temperatures without the addition of moisture to break dormancy. The process of treating seeds with cold temperatures to break dormancy is referred to as "dry stratification."

Most of the prairie grasses and many native prairie wildflowers require simple dry stratification. Seed can be dry stratified by placing it in a refrigerator or freezer for 30 to 90 days prior to seeding.

Large quantities of seed can be stored in an unheated building over winter in rodent-proof metal containers.

Commonly grown native prairie wildflowers and grasses that require only dry stratification to break seed dormancy are found in Table 1.

Table 1. Commonly grown native prairie wildflowers and grasses that require only dry stratification to break seed dormancy.

Latin name	Common name	Time (days)
Wildflowers:		
Asclepias tuberosa	butterflyweed	30
Coreopsis lanceolata	lanceleaf coreopsis	30
Desmodium spp.	prairie tick trefoils	30
Echinacea purpurea	purple coneflower	30
Heliopsis spp.	ox eye sunflowers	30
Monarda spp.	bergamot, beebalm, etc.	30
Ratibida spp.	coneflowers	30
Rudbeckia spp.	black eyed susans	30
Solidago rigida	stiff goldenrod	30
Grasses:		
Andropogon spp.	bluestems	30
Bouteloua spp.	grama grasses	30
Elymus spp.	wild ryes	30
Koeleria macrantha	junegrass	30
Panicum virgatum	switchgrass	30
Sorghastrum nutans	indiangrass	30
Sporobolus heterolepis[1]	prairie dropseed	30

[1]**Note:** Prairie dropseed only germinates under cool temperatures in the field. It is best sown directly in early spring as soon as the soil can be worked, or in fall as a dormant seeding.

Moist stratification

Species that benefit from moist stratification

Many prairie flowers and most woodland wildflowers require moist stratification to break dormancy and yield high rates of germination. For example, shootingstar (*Dodecatheon meadia*) has a zero rate of germination when dry stratified, but after 30 days of moist stratification it will germinate at close to a 100% rate.

Different species require varying lengths of moist stratification to break dormancy. Lupine (*Lupinus perennis*) requires only 7 days. After 1 week of treatment, it often begins to germinate while still in the refrigerator. Members of the genus *Iris* require 90 days or more of moist stratification to yield good germination. Dormancy in most prairie wildflowers can be broken with 30 days with this treatment. Some prairie species that benefit from moist stratification of their seeds are shown in Table 2 (Figure 1 shows a few images of these plants).

Figure 1. Examples of plants requiring moist stratification for germination, top row: *Callirhoe involucrata*, *Camassia scilloides;* bottom row: *Eryngium yuccifolium, Ruellia humilis.*

Table 2. Species that benefit moist stratification[1,2].

Latin name	common name	Time (days)
Wildflowers		
Allium spp.	wild onions	30
Amorpha canescens (low shrub)	leadplant	30
Asclepias spp.	most milkweeds	30
Aster (*Symphyotrichum*) spp.	asters	30
Baptisia spp.	false indigos	90
Callirhoe spp.	poppy mallows	30
Camassia scilloides	wild hyacinth	30
Cassia hebecarpa	wild senna	30
Dodecatheon meadia	shootingstar	30
Echinacea pallida	pale purple coneflower	30
Eryngium yuccifolium	rattlesnake master	30
Eupatorium spp.	joe pye weeds, boneset	30
Helenium spp.	sneezeweeds	30
Helianthus spp.	sunflowers	30
Iris species	wild iris, blue flag	90 to 1 year
Lespedeza spp.	bushclovers	30
Liatris spp.	blazingstars	30
Lobelia spp.	cardinal flower, lobelias	30
Lupinus perennis	wild lupine	7
Parthenium integrifolium	wild quinine	30
Penstemon spp.	penstemons, beardtongues	30
Ruellia humilis	wild petunia	30
Silphium spp.	compassplant, prairie dock	30
Solidago spp.	most goldenrods	30
Tradescantia spp.	spiderworts	30
Verbena spp.	vervains	30
Vernonia spp.	ironweeds	30
Veronicastrum virginicum	culver's root	30
Zizia spp.	golden alexander species	30
Grasses, sedges, and rushes:		
Carex spp.	sedges	30
Calamagrostis canadensis	Canada bluejoint grass	30
Spartina pectinata	prairie cordgrass	90–150
Scirpus spp.	rushes, bulrushes	30

[1]**Note:** Many of the species listed above requiring moist stratification can be direct sown into flats and will germinate well in a greenhouse with regular mist or watering to maintain consistent soil moisture. Although nontreated seed will typically germinate at significantly lower rates, the seeding rate can be increased to ten seeds per plug to compensate for the lack of pretreatment. This increases seed costs, but saves significant labor and infrastructure required to carry out the moist stratification process.

[2]**Beware:** Some species and genera, such as shootingstar, spiderworts, false indigos, *Eryngiums*, and irises will not germinate without the requisite period of moist stratification pretreatment.

Seed can be moist stratified by mixing it with an equal or greater volume of slightly damp inert material. Oak and pine sawdust work admirably for this purpose. It is easy to work with, absorbs moisture and transfers it to the seed well. The relatively high acidity of the sawdust limits the growth of bacteria during the stratification process. Vermiculite, perlite, and peat most can also be used as the inert material.

The inert matter should be only lightly dampened prior to mixing with the seed. If water can be wrung out of the sawdust or peat moss by squeezing it, then it is too wet. Vermiculite and perlite should be moistened in a bowl or colander, so that excess water will drain off. Mix the seed and inert matter together thoroughly, place in a zip top plastic bag labeled with the species and date and place it in the refrigerator for the specified amount of time for the species being treated.

Another method of moist stratifying seed is to plant the seed directly into flats, cover them with plastic wrap to retain moisture in the soil, and store them in a refrigerator or walk-in cooler. If such facilities are not available, the flats can be seeded in fall and stored over winter in an unheated building or greenhouse. Make sure that the flats are protected from damage by mice and other animals during winter storage.

Timing of moist stratification pretreatment

The initiation of moist stratification should be timed so that the seed will be removed from the refrigerator at the appropriate time of year for optimal germination. Cool season plants can be started in a cool greenhouse in late winter or early spring. Warm season plants can be started once the air temperature reaches the high 70s or low 80s °F.

Scarification

Seeds with hard seed coats often require scarification, or scratching of the outer seed surface, to allow penetration of water into the seed itself to initiate the germination process. This can be accomplished by placing a single layer of seed in the bottom of a wooden box and rubbing it with sandpaper wrapped around a wooden block or sandpaper holder. Rub the seed with the sandpaper just hard enough to scratch the outer surface, being careful not to grind the seed into flour! Light pressure is usually sufficient to scarify all but the most resistant seeds.

Some genera, such as *Baptisia, Ceanothus,* and *Iris*, require scarification followed by moist stratification or a hot water treatment (Figure 2). Following scarification, the seed should be moist stratified as described in the directions above or treated with hot water (below).

Figure 2. *Baptisia australis* (left) and *Ceanothus americanus* (right) that require scarification.

Hot Water

A few species are known to benefit from treatment with hot water, which mimics the effect of a wildfire. Some seeds have dormancy mechanisms that require exposure to high temperatures, signaling that a fire has recently occurred and there will be open soil available for germination and growth of new seedlings. The prairie shrub, New Jersey tea (*Ceanothus americanus*) exhibits higher germination following exposure to hot water, followed by 30 days of moist stratification.

Place the scarified seed to be treated in a bowl. Heat water in a teakettle to boiling, then turn off the heat and allow the water to cool for a minute or two. Pour the hot water over the seed and allow it to cool to room temperature. Pour off the water, and the seed can be seeded directly, or in the case of New Jersey Tea, mixed with a damp inert material and moist stratified for 30 days prior to seeding.

Other growers have reportedly had good results using the hot water treatment with the genus *Baptisia*, followed by placing the seed in the freezer for a short period, until ice crystals begin to form on the wet seed (about 1 hr or less). One grower uses this treatment three times in succession (hot water followed by near freezing) to obtain higher rates of germination on this notoriously recalcitrant genus.

OTHER CONSIDERATIONS IN NATIVE SEED PROPAGATION

Fleshy fruited seeds

Many woodland wildflowers have fleshy pulp on the outside of their seeds. The pulp often possesses compounds that can prevent seed germination, and therefore must be removed prior to sowing. Collected when ripe, the flesh surrounding the seed is usually soft and can be easily removed. Wash the seed with water while rubbing the seed carefully across a screen with openings smaller than the seed (a ¼ in. screen works for most species). The flesh will go through the screen and the seeds will remain on top where they can be collected. If the flesh is hard, allow it to soften for a week or longer in a bucket, storing the seed in a cool, damp place until it softens. Woodland wildflowers with fleshy fruits are found in Table 3.

Table 3. Woodland wildflowers with fleshy fruit.

Latin name	Common name
Actaea spp.	red baneberry, white doll's eyes
Aralia spp.	spikenard, wild sarsaparilla
Arisaema spp.	jack in the pulpit, green dragon
Caulophyllum thalictroides	blue cohosh
Cornus canadensis	bunchberry
Hydrastis canadensis	goldenseal
Panax quinquefolium	ginseng
Polygonatum spp.	Solomon's seals
Smilacina spp.	Solomon's plume, starry Solomon's seal, Canada mayflower

Double-dormant seeds

Some species, especially members of the rose and lily families, exhibit a phenomenon known as "double dormancy." These seeds require exposure to two consecutive winters in the soil before they will germinate. Some species will "germinate" in the first year, but all of their development occurs underground, and no visible leaves are produced. The seedlings emerge in the spring after the second winter, almost 2 years after seeding.

The seeds of these species are typically sown fresh, directly in beds in the ground, or allowed to overwinter in flats, either in a cooler or in a greenhouse at ambient temperature. During the growing season, the flats are kept in a cool greenhouse or shade house. They are then allowed to experience a second winter in the flat, stored in a cooler over the winter or in an unheated greenhouse. The seed will then germinate the following spring.

Because this is such a lengthy process, many growers accelerate the process by seeding plug trays or flats and placing them in a cooler for 2–3 months, starting in early fall. The trays are then removed from the cooler in spring and placed in a warm location for 2–3 months, mimicking summer. Following this period, the flats are again stored in the cooler for 2–3 months, and then set out in the greenhouse to germinate. This process saves a full year.

Some of our best-known wildflowers produce seeds that are double-dormant (Table 4) (Figure 3).

Table 4. Species with double-dormant seeds.

Latin name	Common name
Allium tricoccum	wild leek
Caulophyllum thalictroides	blue cohosh
Polygonatum spp.	Solomon's seal
Rosa spp.	rose species
Smilacina species	Solomon's plume, starry Solomon's seal
Trillium grandiflorum	large flowered trillium
Uvularia grandiflora	bellwort

Figure 3. Double-dormant seed plants: *Polygonatum biflorum* (left) and *Trillium grandiflorum* (right).

Timing of seed sowing and pretreatment

Different species germinate in nature at different times of the year. Most summer-blooming prairie flowers and grasses are "warm season" plants and germinate best at temperatures near or above 80 °F (27 °C). The warm season prairie grasses are best seeded in mid to late spring or early summer, and not in fall. The exceptions include the cool season native grasses, which do best when seed in fall or early spring. Spring-blooming prairie and woodland flowers are "cool season" plants, and typically germinate in early spring at cool temperatures in the 60 and 70s °F (15 to 21 °C).

The following prairie grass genera germinate well at warm temperatures (Table 5).

Table 5. Prairie grasses that germinate well at warm temperatures.

Latin name	Common name
Andropogon spp.	bluestems
Bouteloua spp.	grama grasses
Elymus spp.	wild ryes
Panicum spp.	switchgrass, panic grasses
Schizachyrium scoparium	little bluestem
Sorghastrum nutans	indiangrass
Spartina spp.	cordgrass, saltgrass (after extended moist stratification)

Cool season grasses and sedges typically germinate best when sown in early to mid- spring when temperatures are cool. They can also be seeded in fall as a "dormant" seeding and will germinate the following spring when conditions are optimal. Prairie cordgrass, although a warm season grass, germinates best when sown in fall because it requires an extended period of moist stratification to break dormancy. The following prairie grasses germinate best at cool temperatures (Table 6).

Table 6. Prairie grasses and grass-like plants germinate best at cool temperatures.

Latin name	Common name
Calamagrostis canadensis	bluejoint grass
Carex spp.	sedge species (following 30 moist stratification)
Koeleria macrantha	junegrass
Hierochloe odorata	vanilla sweet grass
Sporobolus heterolepis	prairie dropseed

Planting freshly collected seed

Certain wildflowers of both prairies and woodlands germinate well when their seeds are sown fresh, immediately after collecting in summer. This is particularly true of woodland wildflowers that possess *elaiosomes*, a fleshy, strap-like appendage that is attached to the exterior of the seed. Rich in lipids and proteins, elaiosomes attract ants, which harvest the seeds and take them back to their nests. After the ants have consumed the elaiosomes, they take the seed to their waste disposal sites and "plant" them in this nutrient rich environment. This symbiotic relationship benefits both parties and has been observed in several different species of ants and plants.

In some cases, the seeds may enter a state of "deep dormancy" if allowed to dry out, in which it becomes resistant to germination. Once a seed has entered deep dormancy, it typically requires exposure to cool, moist conditions for an extended period to induce germination. Planting the seed fresh, immediately after harvest, is recommended for the following species and genera of woodland wildflowers (Table 7).

Table 7. Planting the seed fresh, immediately after harvest, is recommended for the following species and genera of woodland wildflowers.

Latin name	Common name
Asarum canadense	wild ginger
Caulophyllum thalictroides	blue cohosh
Hydrastis canadensis	goldenseal
Jeffersonia diphylla	twinleaf
Sanguinaria canadensis	bloodroot
Trillium spp.	trilliums

Some other species that do not possess elaiosomes but generally benefit from seeding immediately after collection include the following (Table 8).

Table 8. Other species that generally benefit from seeding immediately after collection.

Latin name	Common name
Actaea spp.	red baneberry, white doll's eyes
Claytonia virginiana	spring beauty
Hepatica spp.	hepaticas
Mertensia virginiana	Virginia bluebells
Tiarella cordifolia	foamflower
Uvularia grandiflora	bellwort (double dormant)

When planted fresh in summer when they ripen, these seeds generally will not germinate until the following spring, or the second spring if they are double dormant. Planting the seed immediately after collection prevents it from drying out and allows the process of internal "after-ripening" to proceed under conditions similar to those in nature. The seeds can be sown directly into the ground or into flats. Keep the seeded flats in a cool shade house and avoid exposure to high temperatures and dry conditions during the summer. In fall, move the flats to a secure, unheated building or cooler that is protected from rodents that might damage the flats. Bring them out into the greenhouse in late winter or early spring to initiate the germination process.

General rule of thumb for seeding woodland wildflowers: When in doubt, plant the seed fresh and allow it to experience the natural seasonal cycles.

Spring blooming prairie flowers

Certain spring-blooming prairie flowers will often germinate in late summer or early fall when their seed is planted immediately after being collected in summer. The seedlings will develop in fall, in preparation for their most active growth period early the following spring. Species and genera whose seed will often germinate shortly after sowing in summer include the following (Table 9) (Figure 4 shows some examples).

Table 9. Spring-blooming prairie flowers that germinate in late summer or early fall when their seed is planted immediately.

Latin name	Common name
Anemone patens	pasque flower
Delphinium spp.	larkspurs
Geum triflorum	prairie smoke
Lupinus perennis	lupine
Ranunculus spp.	buttercups
Tradescantia spp.	spiderworts
Viola spp.	birdsfoot violet, prairie violet, etc.

Figure 4. Spring-blooming prairie flowers that germinate in late summer or early fall, from left to right: *Geum triflorum, Tradescantia ohiensis, Viola pedata.*

PROPAGATION BY CUTTINGS AND ROOT DIVISIONS

Some native species are extremely slow growing from seed, or seed is rarely available. For instance, producing plants of the genus *Dodecatheon* (shootingstars) requires up to 5 years, as members of this genus are only active for a few months in spring. Their growth rates are painfully slow, making propagation by seed both expensive and time-consuming. However, members of this genus are easily propagated by root divisions, which can be dug and divided either in early fall (they go dormant in mid-summer) or in very early spring before these spring ephemerals initiate new growth. Once they begin to emerge, they should not be disturbed, since they are very sensitive to root damage when actively growing.

Members of the genus *Phlox* and *Viola* have exploding seed pods that make seed collection challenging, to say the least. To make matters worse, most members of these two genera are indeterminate, and bloom over a period of time so only a few seeds come ripe at any given time. Seed collection is laborious and costly.

Seeds of some native species are not reliably available in the marketplace, making propagation by cuttings the only method. Others, especially members of the Lily Family with double dormant seeds, can be readily propagated by dividing their rhizomes or corms, rather than performing multiple cold and warm seed pre-treatments. Most of our native *Liliaceae* are also slow growing from seed, making divisions a faster and more economical method of production. The following species are commonly propagated by taking cuttings (Table 10).

Table 10. Cutting propagated species.

Latin name	Common name	Notes
Coreopsis rosea	rose coreopsis	
Fragaria virginiana	wild strawberry	Runners root readily
Phlox spp.	phloxes	
Salvia azurea	blue sage	
Sedum ternatum	wild stonecrop	
Solidago caesia	blue stemmed goldenrod	
Solidago odora	anise scented goldenrod	

Some species are easily and most economically propagated by division of roots, rhizomes, bulbs, or scales (Table 11) (Figure 5.).

Table 11. Species economically propagated by division of roots, rhizomes, bulbs, or scales.

Latin name	**Common name**
Arctostaphylos uva-ursi	bearberry
Asarum canadense	wild ginger
Carex pensylvanica	Pennsylvania sedge
Dicentra cucullaria	dutchman's breeches
Dodecatheon spp.	shootingstars
Erythronium americanum	trout lily
Ferns (although some growers propagate ferns using spores)	
Hepatica spp.	hepaticas
Iris spp. dormancy results in poor germination	irises: difficult from seed
Polygonatum spp.	Solomon's seal
Podophyllum peltatum	mayapple
Smilacina spp.	Solomon's plumes, Canada mayflower
Trilliums spp.	trilliums (some divide better than others)
Uvularia grandiflora	bellwort
Viola spp.	violets

Figure 5. Examples of species propagated by division from left to right: *Asarum canadense, Podophyllum peltatum, Smilacina racemosa.*

SUMMARY

By following these procedures and using quality seed from a reliable supplier, the mysteries of propagation of native species from seed can be unraveled. With a little experience, reliable results can be achieved in growing our beautiful native wildflowers and grasses for fun and profit!

Combined Proceedings IPPS 68:96-98. 2018.

Hog Compost as a Substrate Amendment: Preliminary Report

Winston Dunwell[1], Dwight Wolfe[1], Edwin Ritchey[2], Virginia Travis[1], and Christine Bradley[3]
University of Kentucky, [1]Department of Horticulture, [2]Department of Plant and Soil Science, [3]Department of Entomology, Lexington, KY 40546 USA

winston.dunwell@uky.edu

Keywords: Nursery production, nitrate, nitrogen, substrate analysis, alternative substrates.

INTRODUCTION

Composts made of manure, bedding and animals are available in abundance in Kentucky. Mechanical composting directly captured from a hog production facility floor mixed with woodchips and automatically turned in windrows under cover (Figure 1) will create a low moisture, low readily degradable organic matter. Suggesting the finished compost would have lower transportation costs and should provide value as a soil conditioner (Cook, et al., 2015). A west Kentucky hog producer has tried to market compost in retail consumer packaging. The hog compost has been certified for organic growing. He has found that a quality compost cannot compete with prices for less consistent composts in the marketplace. He has determined bulk use for soil conditioning and as a substrate amendment have potential as market outlets (O'Bryan, 2018). Hog compost will be tested for use as a substrate amendment in container production of ornamental plants.

MATERIAL AND METHODS

Substrates of 100% pine bark and 85% pine bark and 15% hog compost by volume were mixed for 15 minutes in a Cube Cart-Away mobile cement mixer (Figure 2) on 12 July 2018. Samples of each substrate were sent to the University of Kentucky soils laboratory for analysis.

Figure 1. Automated composting windrows.

'Smaragd' arbovitae (*Thuja occidentalis*) were planted in 7-gal containers at 30 containers for each substrate with 15 randomly selected for pour-through sampling.

Pour-through (Dunwell, 2013) sampling was done September 5, 2018 following eight weeks of cyclic timed irrigation of 10 minutes each from Agridor 4463-20 spray emitters in each container at 1:00 pm. and 4:00 pm.

Figure 2. Cube Cart-Away mixer

RESULTS AND DISCUSSION

The arborvitae plants were allowed to grow without additional fertilizer for 8 weeks. Dramatic color differences were observed. The plants in the 100% pine bark were chlorotic while the plants in the 85% pine bark:15% compost were green (Figure 3).

Figure 3. Substrate foliage color comparison.

Pour-through results for electrical conductivity (µS/cm) (Figure 4) and pH (Figure 5) were significantly higher for the compost amended substrate as would have been expected from the substrate test from samples at mixing (Table 1). Pine bark samples with just 1 ppm Nitrate-N versus 139 ppm nitrate-N for the pine bark/compost indicates fertilization of straight pine bark substrates at planting is necessary.

Table 1. Substrate test at mixing.

		Conductivity nitrogen-N		Nutrient concentration (ppm)									
Media	pH	µS/cm	ppm	P	K	Ca	Mg	B	Na	Cu	Fe	Mn	Zn
PB[1]	4.8	440	1	8.9	55.4	29.8	8.1	0.1	10.4	0.2	14.2	5	2.9
PB/ Compost	5.8	4,240	139	287.1	714.3	165.8	93.4	0.4	227.5	3.4	44.5	20.2	49.6

[1]PB=pine bark

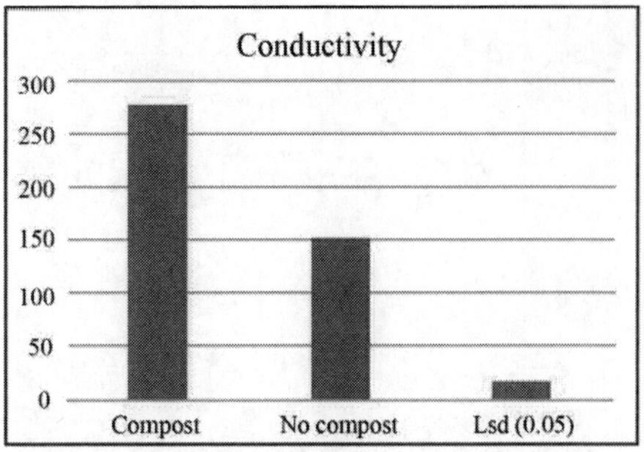

Figure 4. Substrate conductivity comparison.

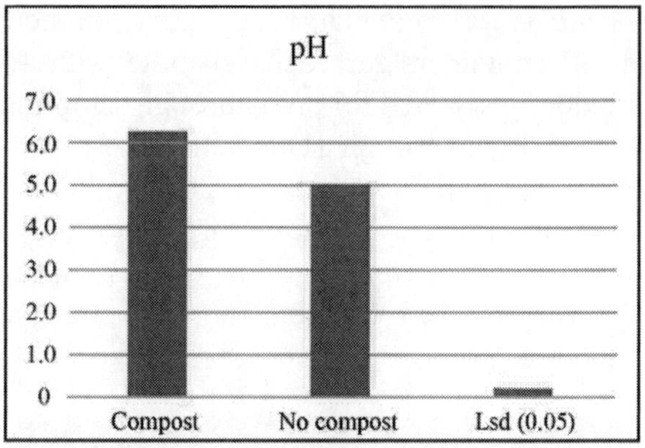

Figure 5. Substrate pH comparison.

ACKNOWLEDGMENT

This work is partially funded by the University of Kentucky Nursery/Landscape Fund, Kentucky Horticulture Council and Kentucky Agricultural Development Board Fund

Literature Cited

Dunwell, W.C., Wolfe, D., and Grable, C. (2013). Influence of time on measuring container fertility by the pour-through extraction. HortScience *48*(9) Supplement—2013 ASHS Annual Conference; pp S299-S300.

Cook, K. L., Ritchey, E.L., Loughrina, J.H., Haleya1, M., Sistania, K.R., and Bolstera C.H. (2015). Effect of turning frequency and season on composting materials from swine high-rise facilities. Waste Manag. *39*:86-95

O'Bryan, Jerry. 2018. Personal Communication.

Combined Proceedings IPPS 68:99-107. 2018.

Cold Hardiness: Successes and Failures at the University of Delaware Botanic Garden

John Frett
Plant and Soil Sciences, University of Delaware, 152 Townsend Hall, Newark, Delaware 19716 USA

jfrett@udel.edu

Keywords: plant collections, germplasm, evaluation, winter hardiness, witchhazel, palms, *Camellia, Magnolia, Itea, Osmanthus, Quercus, Viburnum, Agave*.

INTRODUCTION

Botanic gardens and arboreta have always played an important role in the maintenance and testing of novel germplasm. By their very nature, the diverse collections of plant material provide germplasm to the trade, an opportunity to assess cultural requirements and information regarding the adaptability of diverse plants to varying environments. Structured programs such as the Plant Collections Network sponsored by American Public Gardens Association and the USDA-Agricultural Research Service seek to coordinate a network of gardens to build a national collection of plants and facilitate access to these plants and information generated by the program. Whether collections in public gardens are part of a larger regional, national or international effort their value is increasing rapidly as the nursery industry evolves to maintain economic viability. Nurseries that maintain diverse inventories have significantly reduced their diversity to increase efficiency and maximize profits. Many nurseries that historically produced amazingly diverse plant lists have either greatly reduced the selection or even gone out of business. The result of this economic reality results in an ever-increasing role for public gardens to maintain diverse collections and provide basic information on the adaptability of the collections.

MATERIALS AND METHODS

The University of Delaware Botanic Gardens (UDBG) is located in Newark, Delaware at the transition of the coastal plain and piedmont region of the Eastern United States in USDA Zone 7a. The soil is a silt-loam that may have minor drainage issues, particularly in areas lacking topography. All information regarding plant survivability is based on plants planted in the ground, not containerized specimens. Several microclimates exist within the UDBG. There are "courtyard" areas adjacent to Townsend Hall that are surrounded on three sides by the building, only open to the east. Most potentially marginally hardy plant material is first evaluated in these areas. If the material proves to be reasonably cold hardy in these areas, plants are established in more exposed areas of the UDBG for final evaluation.

This paper focuses on broadleaf evergreen plants that may be questionably hardy in Zone 7a. It is not a complete list of accession in the UDBG collection in a particular genus, rather those that are marginally hardy based on the literature. Most all succulents are planted in raised beds with the courtyards. The soils in these beds was removed to a depth of 1m. The soil was mixed with approximately 75% sharp sand and the beds were filled with the sand-soil mix.

Weather data was collected by an automated weather station, part of the Delaware Environmental Observing System; the station name is Newark, DE-Ag Farm, immediately adjacent to the UDBG. The weather data is available at: http://www.deos.udel.edu/.

Plants are evaluated for survivability as either alive or dead, dead plants are deaccessioned from the plant records. Occasionally, more detailed notes are made regarding winter damage. There are no attempts to provide additional winter protection, even in the first year of establishment.

While winter hardiness is a combination of many factors, not limited to drainage, physiological conditions leading into the winter, and rate of temperature change, the USDA hardiness map uses the average annual extreme minimum temperature as a relative measure of winter hardiness. As a point of reference, the annual extreme minimum temperature will be used in this work.

RESULTS

As to be expected, there is great variation of the lowest annual temperature. During the period of this study, the lowest reported temperature was -5°F and the warmest winter recorded 19°F as the lowest annual temperature (Figure 1). Temperatures approximated 0°F in the winters of 1996, 2005, 2014 2015, and 2018 during the period of this report. These years, along with 1994 were the greatest challenge to plants in the collection. The average annual extreme minimum temperature for this period was 7.4°F.

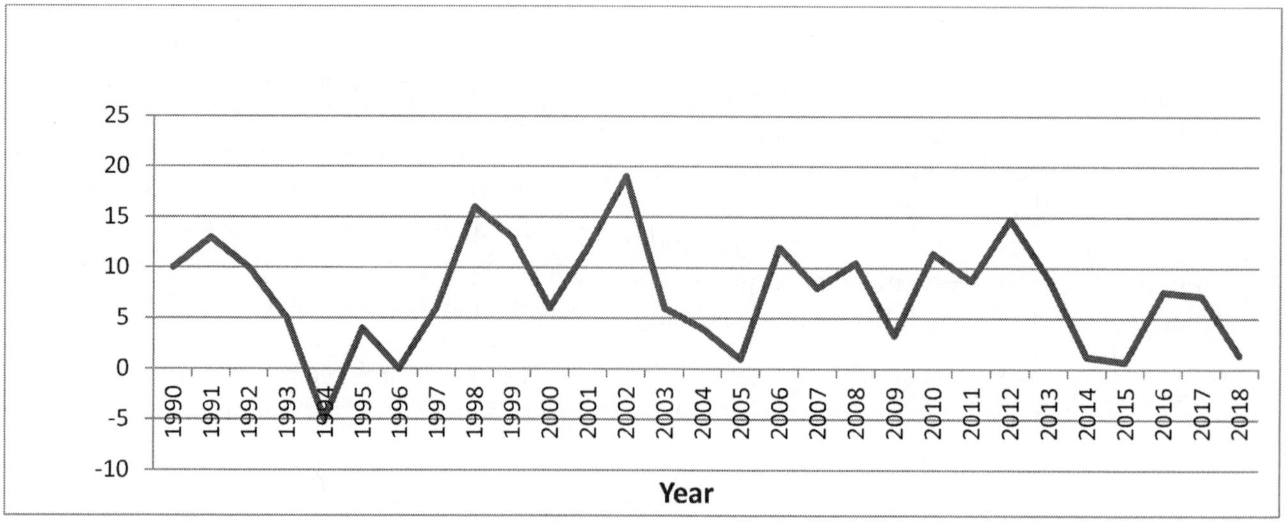

Figure 1. Lowest annual temperature at UDBG by year.

One major genus in the UDBG's collection is *Camellia*. The major emphasis of this collection is at the species level, not the cultivar level. Results of survivability of these plants are presented in Table 1. The Ackerman hybrids have been consistently hardy. Flowers begin in October-November and continue until a killing frost. Unopened flower buds never open the following spring, rather are frozen by the cold during winter. The *C. japonica* specimen was a seedling that originated from a Morris Arboretum collection trip to Korea. The plant survived 1994 (-5°F) unscathed with all flower buds opening in the spring. It has never had any winter foliar or flower bud damage. *Camellia sasanqua*, represented by the species and a single cultivar, has proven reliably hardy. Undoubtedly, some of the many cultivars would suffer winter damage or die if grown in the collection. The various species that have survived typically survive the winters with little to no damage. *Camellia rosaeflora* has died in three separate plantings, established in different years.

Table 1. Survivability of Various Species of *Camellia* in the UDBG.

Species	Year accessioned	Year deaccessioned
C. 'Winter's Charm'	1994	Alive
C. 'Winter's Interlude'	2007	Alive
C. 'Winter's Joy'	2007	Alive
C. 'Winter's Waterlily'	2007	Alive
C. chekiangoleosa	2011	2015[1]
C. chrysanthoidies	2011	2012
C. crapnelliana	1999	2000
C. cuspidata	2011	Alive
C. edithae	1999	2000
C. euryoides	2015	2016
C. fraterna	1999	2000
C. furfuracea	2011	2012
C. × *hiemalis*	2011	Alive
C. japonica	1991	Alive
C. japonica 'Spring Promise'	2009	Alive
C. longicarpa	1999	2000
C. lutchuensis	2011	2014[1]
C. octopetala	2011	Alive
C. oleifera	1995	Alive
C. rosaeflora	2011	2014[1]
C. sasanqua	1995	Alive
C. sasanqua 'Long Island Pink'	2005	Alive
C. saluenensis	1999	2000
C. sinensis	2006	Alive
C. transnokoensis	2011	2013
C. truncata	2012	Alive
C. yuhsienensis	2012	Alive

[1]Lowest temperature near 0°F for this year.

Another priority collection are plants in the witchhazel family (Hamamelidaceae). The winter performance of these plants is presented in Table 2. In general, the *Distyllium* species and cultivars have performed well. Several of the cultivars have been established outside of the protected courtyards. Even in these more open situations plants survive the winter with no to minor winter burn and/or dieback in severe winters. *Loropetalum chinense* 'Roseum' was received as a cutting, grown on, and planted in 1994 and survived -5°F its first year with significant dieback. It typically will partially defoliate in most winters and suffered dieback in severe winters but the original plant has survived 24 years. *Parrotiopsis* (deciduous), *Sinowilsonia* (deciduous), *Sycopsis* (evergreen), and *Sycoparrotia* (primarily deciduous) have all thrived in the collect with no winter damage.

Table 2. Survivability of various species of *Hamamelidaceae* in the UDBG.

Species	Year accessioned	Year deaccessioned
Distylium 'Athen's Tower'	2018	Alive
D. 'PIIDIST-II' Blue Cascade™	2014	Alive
D. 'PIIDIST_V' Cinnamon Girl™	2015	Alive
D. 'sPg-3-007' Spring Frost™	2015	Alive
D. 'Vintage Jade'	2014	Alive
D. buxifolium	2018	Alive
D. myricoides	1999	Alive
D. racemosum	1999	Alive
Loropetalum chinense 'Roseum'	1994	Alive
L. chinense 'Chang Nian Hong' Ever Red® fringe flower	2015	2018
L. chinense 'Shangi-hi' Purple Diamond® fringe flower	2015	Alive
Parrotiopsis jacquemontiana	1997	Alive
Sinowilsonia henryi	2013	Alive
Sycopsis sinensis	1989	Alive
×*Sycoparrotia semidecidua*	1989	Alive

The evergreen species of *Itea* were obtained from several nurseries. The winter performance is listed in Table 3. Both *I. oldhamii* and *I. yunnanensis* were received as *I. chinensis*. *I. oldhamii* is less hardy than the other species, dying in the winter of 2015 after surviving two other winters of near 0°F. *I. ilicifolia*, has twice died the first winter after planting.

Table 3. Survivability of various species of *Itea* in the UDBG.

Species	Year accessioned	Year deaccessioned
I. chinensis	1992	Alive
I. ilicifolia	2015	2016
I. oldhamii	1999	2016
I. yunnanensis	1997	Alive

The survivability of several *Osmanthus* species and cultivars is listed in Table 4. Only *O. fragrans* and *O. fragrans* var. *aurantiacus* have suffered partial defoliation to significant dieback after severe winters with once specimen dying after 0°F even though plants are grown with winter protection in the courtyard. Most other specimens are grown in more exposed landscape situation. *O.* ×*fortunei*, planted in an open field, suffered complete defoliation and significant dieback after -5°F.

Table 4. Survivability of various species of *Osmanthus* in the UDBG.

Species	Year accessioned	Year deaccessioned
O. americanus	1997	Alive
O. armatus	2012	Alive
O. decorus 'Beki Kasapligil'	1999	Alive
O. delavayi	2012	Alive
O. fragrans	1997	2018
O. fragrans var. *aurantiacus*	2012	Alive
O. × *fortunei*	1966	Alive
O. × *fortunei* 'San Jose'	1998	Alive
O. hererophyllus 'Goshiki'	1997	Alive
O. hererophyllus 'Gulftide'	1998	Alive
O. hererophyllus 'Kembu'	1997	Alive
O. hererophyllus 'Purpureus'	1998	Alive
O. hererophyllus 'Sasaba'	2012	Alive
O. megacarpus	2012	2013[1]

[1]Death of plant due to mechanical damage.

The survivability of various palm species is presented in Table 5. Several of the accessions have not yet experienced a winter. To date, *Rhapidophyllum* has thrived, never showing winter damage. *Sabal minor* has survived for 12 years but produces little new growth. *Sabal lousiana* has survived 5 years and has increased in size nicely.

Table 5. Survivability of various species of palms in the UDBG.

Species	Year accessioned	Year deaccessioned
Butia capitata	2015	2016
Rhapidophyllum hystrix	1998	Alive
Sabal minor	2006	Alive
Sabal minor 'Chipola Dwarf'	2018	Alive
Sabal minor 'McCurtian'	2018	Alive
Sabal lousiana	2013	Alive
Sabal uresana	2018	Alive
Sabal palmetto	2010	2011
Trachycarpos fortunei	2011	2012
Trachycarpus fortunei 'Wagnerianus'	2018	Alive

The data for the genus *Magnolia* represents, primarily, the evergreen species with a few select deciduous species/cultivars. This is only a portion of the *Magnolia* species/cultivars grown in the collection. Several of the evergreen species were formerly listed as *Michelia* but are listed here as *Magnolia*. The survivability data is presented in Table 6. *M.* 'MicJUR01' and *M.* 'Free Spirit' (both formerly *Michelia*) did suffer some dieback during the near 0°F winters of 20014, 2015 and 2018 but plants recovered well and flowered in subsequent years. *M. figo* has grown well and survives cold winters with minor foliar burn but flowers well in the spring. All *M. grandiflora* cultivars have grown well with minimal to no winter damage. Even *M. grandiflora* 'Little Gem' has not suffered winter damage since the early establishment years. *M. insignus* dieback every year until it ultimately died. *M. yuyuanensis* never shows winter damage, is fully evergreen and flowered for the first time in spring of 2018.

Table 6. Survivability of various species of *Magnolia* in the UDBG.

Species	Year accessioned	Year deaccessioned
M. 'Caerhays Belle'	1997	Alive
M. 'MicJUR01' Fairy Magnolia® Blush	2012	Alive
M. campbellii var. mollicomata	1994	1996[1]
M. figo	1992	Alive
M. x*foggii*	1996	1996[1]
M. grandiflora 'Brakens Brown Beauty'	1994	Alive
M. grandiflora 'D. D. Blanchard'	1993	Alive
M. grandiflora 'Glen St. Mary'	1994	Alive
M. grandiflora 'Little Gem'	1993	Alive
M. grandiflora 'MGTIG' Monrovia's Greenback™ magnolia	1998	Alive
M. grandiflora 'Samuel Sommer'	1993	Alive
M. grandiflora 'Victoria'	1993	Alive
M. grandiflora 'Copper Top'	2018	Alive
M. grandiflora 'Edith Bogue'	1989	Alive
M. grandiflora 'North Star'	1993	Alive
M. grandiflora 'Russett'	1994	Alive
M. grandiflora 'Southern Charm' Teddy Bear® magnolia	2015	Alive
M. insignus	2012	2015[1]
M. maudiae	2013	Alive
M. yunnanensis 'Free Spirit'	2013	Alive
M. yuyuanensis	2012	Alive

[1]Lowest temperature near 0°F for this year.

Many of the *Mahonia* in the UDBG collection have grown well and survived for many years (Table 7). Most of these are grown with the protection of the courtyard which has mediocre drainage. Several of the species died within two to three years of planting during winters that were rather mild, suggesting the drainage may be responsible for the demise of the plants. *M.* ×*media* cultivars grow with mixed results. Those that have survived have grown well, producing flowers that begin in November and continue into December and January.

M. eurybracteata 'Soft Caress' died the 1st year after planting when temperature approached 0°F. New specimens were planted in 2018, along with *M.* 'sPg-15-1' to determine the suitability of these popular, fine textured cultivars.

Table 7. Survivability of Various Species of *Mahonia* in the UDBG.

Species	Year accessioned	Year deaccessioned
M. 'Arthur Menzies'	2000	Alive
M. 'sPg-15-1' Beijing Beauty™	2018	Alive
M. aquifolium	1998	Alive
M. bealei	1978	Alive
M. duclouxiana	1999	Alive
M. eurybracteata 'Soft Caress'	2015	2016
M. fortunei	1999	Alive
M. gracilipes	1999	Alive
M. japonica	2012	Alive
M. × *lindsyae* 'Cantab'	2000	2003
M. lomariifolia	1999	Alive
M. mairei	1999	Alive
M. × *media* 'Charity'	2000	Alive
M. × *media* 'Hope'	2000	2003
M. × *media* 'Lionel Fortescue'	2005	Alive
M. × *media* 'Underway'	1999	2002
M. × *media* 'Winter Sun'	2000	Alive
M. napaulensis 'Maharajah'	1999	2001
M. nervosa	1999	2003
M. pinnata 'Ken Howard'	1999	2001
M. piperiana	1999	2001

The oaks which are represented in Table 8 are either evergreen or species of questionable hardiness, other species in the collection are not represented. Most of the deciduous species have grown well without any winter damage. *Quercus myrsinifolia*, obtained from Morris Arboretum, has grown well with only minor defoliation in particularly severe winters. *Quercus virginiana* was wild collected in the vicinity of Wilmington, NC and has grown well but has some defoliation to near complete defoliation in severe winters but has never demonstrated dieback.

Table 8. Survivability of Various Species of *Quercus* in the UDBG.

Species	Year accessioned	Year deaccessioned
Q. aliena	1993	Alive
Q. dentata	1993	Alive
Q. geminata	2016	2017
Q. incana	2007	2010
Q. laurifolia	1991	Alive
Q. lyrata	1991	Alive
Q. myrsinifolia	1994	Alive
Q. nigra	1991	Alive
Q. nutallii	1991	Alive
Q. laurifolia	2016	2017
Q. virginiana	2008	Alive

Survivability for several genera/species of "succulent" plants are presented in Table 9. All of these plants have been grown in an amended soil, approximately 75% sand by volume. To date, none of the *Agave* have survived more than two years. *Dasylirion acrotrichum* and *Nolina microcarpa* are the only species in their respective genera to survive multiple years to date. Most *Opuntia* species grown have survived. The *Yucca* species are more tender representative grown in the UDBG but have survived well. Many of these plants have yet to grow through a winter.

Table 9. Survivability of various species of succulents in the UDBG.

Species	Year accessioned	Year deaccessioned
Agave 'Blue Glow'	2018	Alive
Agave 'Mr. Ripple'	2016	2018
Agave americana	2008	2010
Agave havardiana	2018	Alive
Agave ×loferox	2018	Alive
Agave polyacantha	2016	2017
Agave univittata	2016	2018
Agave victoriae-reginae	2006	2008
Agave virginica	2016	2018
Dasylirion acrotrichum	2008	Alive
Dasylirion berlandieri	2018	Alive
Dasylirion leiophyllum	2018	Alive
Dasylirion longissimi 'Toothless Spoon'	2006	2008
Dasylirion texanum	2015	2018
Dasylirion wheeleri	2008	2010
Hesperaloe parviflora	2008	Alive
Nolina microcarpa	2006	Alive
Nolina nelsonii	2006	2008
Opuntia cacanapa 'Ellisiana'	2017	2018
Opunti fasilaris 'Baby Rita'	2016	Alive
Opuntia humifusa	2010	Alive
Opuntia polycantha	2016	Alive
Yucca constricta	2008	Alive
Yucca harrimmaniae	2015	Alive
Yucca treculeana	2016	Alive

The UDBG has a significant collection of viburnums. Only the broad-leaved evergreen and a few marginally hardy species are represented in Table 10. Most of these are growing in the protected courtyard while the rest of the collection is distributed throughout the garden. *Viburnum davidii* struggled for the years that it survived. It appeared as the summer heat and humidity stressed the plants as much as the cold of winter. There was not a significantly cold winter during the period the plant was in the garden, yet it died, further suggesting that summers were as stressful as winters. Several other species survived for only a few years and died following relatively mild winters suggest that the plants never established well. These are good potentials for reevaluation.

Table 10. Survivability of various species of *Viburnum* in the UDBG.

Species	Year accessioned	Year deaccessioned
V. atrocyaneum	1999	2000
V. awabuki 'Chindo'	1999	Alive
V. cinnamomifolium	2013	Alive
V. cylindricum	2005	Alive
V. davidii	1998	2002[1]
V. foetidum	2003	2004
V. harryanum	1999	2001
V. ×*hillieri* 'Winton'	2013	Alive
V. japonicum	1997	Alive
V. obovatum 'Mrs. Schiller's Delight'	2013	2015
V. obovatum 'Reifler's Dwarf'	2014	Alive
V. propinquum	1988	Alive
V. tinus	1995	Alive
V. utile	1997	Alive

CONCLUSIONS

Many factors other than the lowest temperature determine if a plant will survive the winter. The rapidity of transition from hot to cold and back, temperature fluctuations during the winter, soil moisture, stress going into the winter and more will impact winter performance. Reported here is only the lowest temperature as an attempt to gauge winter hardiness.

Plants that are well established, growing vigorously, also tolerate a challenging winter better than those recently planted. Many of the plants lost during these trials where lost in the first year. Some of these years were particularly cold, others were not. Replanting specimens that died after the first year would give a better measure of the reliability of these plants, particularly when plants are challenged the first winter after planting. In some cases, this has been done, in others it still needs to be repeated.

Many of these evaluations occur in a protected microclimate, as illustrated by the survival of pinapple guave, *Feijoa sellowiana* (USDA Zone 8-10) for eight years in the same conditions. Future efforts will focus on establishing additional plants into more typical, exposed conditions throughout the garden. Those plants that were evaluated outside of the microclimate of the courtyard have proven their adaptability to "typical" landscape conditions in Newark, DE.

The UDBG continues to plant new plants of questionable hardiness to test their survivability under field conditions in the mid-Atlantic region. A major goal of the garden is to develop a broadly diverse collection of plants that serve as an illustration of what is possible to grow, and the diversity of plants available beyond the typical nursery trade. It is a resource to the trade, the public and the university community.

Combined Proceedings IPPS 68:108-115. 2018.

Nomenclature: The Game of the Name

John Friel
Emerald Coast Growers, 7400 Klondike Rd, Pensacola, Florida 32526 USA

frielster@frontier.com

Keywords: Plant names, binomial nomenclature, Naamlijst, trademarks, patents.

INTRODUCTION

First things first: I'm an English major dropout with no formal horticultural training. I began my horticulture career in 1979 as a truck driver for the erstwhile Green Leaf Enterprises. I wasn't a stranger to green things; I had a garden, and houseplants on every available horizontal surface, as everyone did in the 1970s. That helped. And my English/writing background helped as my responsibilities evolved. By the mid-1980s I was representing the company at trade shows, writing catalogs and newsletters, and speaking at various events.

Most of what I know about nomenclature I've learned by trying to get the names right for catalogs. I like to think I've succeeded; I've had this marketing gig with two large propagators, naming and describing hundreds of different taxa. I've chaired the Perennial Plant Association's Nomenclature Committee and served on their Board, including two terms as President.

At both companies, customers told us they used our catalog as a reference work. Which is flattering, but also very wrong: A catalog is a sales tool. Decisions on arranging and naming are based on your customers' needs first, taxonomical precision second.

You probably know someone who hates Latin names—like my late neighbor John, a crusty old guy who asked what he should plant on a bank near the road. I replied, well, there's *Hemerocallis*, or *Ceratostigma plumbaginoides*—and he got this scornful, pained look and made a hand-waving gesture. If you had seen us but weren't close enough to hear our conversation, you'd have thought I'd farted, and he was fanning away the stink.

I said, "John, you know what a *Begonia* is, and a *Geranium*, right? Well, those are Latin names!" It didn't help. So instead of *Hemerocallis* and *Ceratostigma* we talked about daylilies and leadwort, and he was happy.

John died a couple of years ago of liver cancer. I can just picture him when the doctor said, "John, I have bad news: You have *Hepatocellular carcinoma*." I wasn't there, of course, but I'll bet John scrunched up his face and fanned the Latin away.

If you know someone like John, please, be kind. Latin names can be difficult. Even professionals make mistakes: I've seen a sign at a very good garden center identifying their English ivy as "*Hendra helix*". Obviously, it should say *Hedera*. I should know, I've sold hundreds of thousands of *H. helix*, God forgive me.

Some seemingly simple names, even abbreviated, are just difficult for some. In New Jersey I passed a farm market with a wagonload of "muns" for sale.

So please, be gentle with civilians (i.e., non-plants folk), because some of them seem to need all the help they can get, dealing with our products. My favorite picture from the 2018 Philadelphia Flower Show is of a big, viciously spiny gold barrel cactus labeled, "Please Do Not Sit" (Figure 1).

Figure 1. Big, viciously-spiny gold barrel cactus.

ROOTS OF MODERN NOMENCLATURE

Like my neighbor John, unlike everyone reading this, many people find Latin names confusing. What they don't realize is that the two-part Latin names we know represent a dramatic simplification compared to how botanists used to name plants. The Swedish genius Carolus Linnaeus (1701-1778), born Carl von Linne, gave us two revolutionary concepts: He perfected binomials, and he organized plant groups based on the number and arrangements of reproductive organs. This was pretty radical, even scandalous in the 18th century.

The binomial system, which we've all used for 300 years, was brilliant and, compared to what came before it, very simple. Example: the common carnation. Linnaeus called it *Dianthus caryophyllus*, and the name still stands. It's still not an easy mouthful for a non-plantsperson, but before Linnaeus, that lovely flower was known as *Dianthus floribus solitariis, corollis lacero-partitis, squamis calycinis ovatis acutis*. Names weren't just names, they were descriptions. That string means, roughly, "The *Dianthus* with one flower, fringed petals, and round, pointed sepals." In Latin, it even works as Gregorian chant.

Plant names and inebriated marsupials

The cascade of life as Linnaeus drew it up goes:

Kingdom
  Phyllum
    Class
      Order
        Family
          Genus
            Species

Around 1990, "Domain" was added above Kingdom. This is one of those scientific lists where you make up a mnemonic device to help you remember it, like "HOMES" for the Great Lakes, or "My Very Elegant Mother Just Served Us Nine Potatoes" to help remember our solar system's planets in order, from the sun out to poor defrocked Pluto.

There are several phrases that'll help you recall this list, but my favorite is "Drunken Kangaroos Punch Children On Family Game Shows." That I can remember.

Schizachyrium phylogeny

Let's apply that list to a real live plant introduced by North Creek Nurseries - *Schizachyrium scoparium* 'Standing Ovation'

Domain: *Eukaryota*

Kingdom: *Plantae*

Phylum: *Tracheophyta*

Class: *Liliopsida* (syn. *Monocotyledon*)

Order: *Cyperales*

Family: *Poaceae* (formerly *Gramineae*)

Genus: *Schizachyrium* (formerly *Andropogon*)

Species: *scoparium* (formerly *A. scoparius*)

Cultivar: 'Standing Ovation'

None of this is chiseled in stone: Some taxonomists replace the word "Domain" with "Superkingdom." Others divide Kingdom into Infrakingdoms and Sub-kingdoms. Some use "Branch" or "ramus" between Subkingdom and Infrakingdom. And of course, there are Clades; this is also known as cladistic nomenclature. Rhymes with sadistic.

There are also tribes, subtribes and supertribes. Take grasses: The USDA says there are 337 genera of grasses in the world. Modern taxonomy recognizes 771, distributed among 12 subfamilies, 6 super-tribes, 51 tribes, and 80 subtribes.

So much for the belief that "Latin names never change!" I heard a speaker tell that old lie back in spring—not at a plant talk, but at a birding conference. He was explaining how "common" names like Cooper's hawk and Lincoln's sparrow can be confusing because they're not common everywhere. But I digress.

And as long as I'm digressing, here's something that bugs me: "Genus" and "genera." I especially hear "genera" misused a lot, sometimes from a podium. They're not interchangeable, they're singular and plural. "Genuses" is not a word, and there's no such thing as "a genera." Genus / genera = mouse / mice. It really is that simple. Thanks, I feel better now.

Domains: Bacteria, Archaea, and Eukaryota

These are the three domains that encompass every known living thing in the Tree of Life, as defined by American microbiologist Carl Woese. The first two domains are all microorganisms; Eukaryota covers everything else, from lichen to blue whales and giant redwoods and *Homo sapiens*. We've segued from Linnaeus to Darwin in our plant organizing schemes. The new taxonomy is organized along theoretical evolutionary lines, and the science itself is still evolving.

In the 1950s German entomologist Willi Hennig developed a mathematical approach, Phylogenetics, to determining the most likely family tree of a group of organisms based on characteristics. It was a perfect fit when the revolution in DNA studies and genome mapping arrived, because DNA gives scientists a whole new world of characteristics to study—far beyond what the naked eye or even the microscope can detect.

Tree of Life: Plantae

I think Linnaeus could find his way around a modern chart of the Kingdom *Plantae*, but I doubt strongly that he'd recognize one important word: "bootstrap." I learned, while preparing for this talk, that I'm not as smart as I thought, at least in some ways. If you've had kids, they've probably had one of those toys that teach shapes. It's either a ball or a bench, with differently-shaped holes. There are little blocks – squares, triangles, ovals, stars – and each block will fit through just one opening. When I try to wrap my mind around modern taxonomy, I feel like one of those toys. I'm trying to stuff information into my brain, and the holes in my head don't match the shape of the pieces of data I'm trying to insert.

I get it that a "clade" is a single branch on the tree of life, i.e., a monophyletic group of closely related organisms. But other terms have me utterly stumped. Synapomorphies? Grex? What the heck is a grex? I'm stumped by these and numerous other terms. Give me drunken kangaroos any day.

I also learned that I'd misinterpreted the term "bootstrapping," which I thought meant "guessing," basically, as in the old saying, "Pick yourself up by your bootstraps," i.e., starting from scratch.

A high bootstrapping percentage indicates a high level of confidence that a plant is correctly placed in a phylogeny. It means they've run the numbers repeatedly, or tested a hypothesis from different directions, and arrived at the same result. Mea culpa.

Ch-Ch-Changes we've seen

Remember when *Tritoma* became *Kniphofia*? No, you don't. Taxonomists settled on that switch in 1938. But many of us didn't change our catalogs until the 1980s.

Remember when *Chrysanthemum* was changed to *Dendranthema*, and then back to *Chrysanthemum*? The first change happened in 1961, based on a study published by a Russian taxonomist, but the crap didn't hit the fan until 1989 when the RHS adopted the new name. It went back to *Chrysanthemum* in 1995 by popular demand, sort of, but not everybody liked that change either. Dutch taxonomists especially preferred *Dendranthema*.

Eupatorium* and *Gaura

RHS, MoBoT, and the Naamlijst don't show *Gaura* as having changed to *Oenothera*; nor does GRIN. PhytoKeys, a new one to me, says *Gaura* is "deeply nested within one of two major clades of *Oenothera*."

Some sources treat *Eupatorium* and *Eutrochium* as synonyms; others say it's now *Eutrochium* (formerly *Eupatorium*.) Not every species has left *Eupatorium*, but the one most in the trade, *E. purpureum*, Joe Pye weed, is now considered (by some) as a *Eutrochium*.

***Banksia* and *Dryandra*: more name changes**

We in the industry can be slow to accept change. In rare cases, new discoveries may lead to a sort of simplification, but still cause confusion. Example: Two Australian genera, *Banksia* and *Dryandra*, are now all *Banksia*. This change was proposed in 2011 by two taxonomists, one Australian and one American. It appears to be mostly, though not universally, accepted already. That's pretty fast—but there's been pushback.

Alex George is another Australian botanist who has researched and published widely in those two genera, and he argues strongly against the change. He complained, "They have changed the names of 100 species (and subspecies and varieties) and…confused, upset and inconvenienced many people including the public, scientists and the nursery trade, and all their…day-to-day activities that involve using the names of these plants."

He continued, "Biologists around the world are increasingly critical of taxonomic results that are dominated by molecular data. DNA is but one component of a biological organism's physical form…"

Linnaeus wrote, "In natural science, the basics of truths must be confirmed by observations." Sorry, Carl, we don't live that way anymore. Today's "truths" are arrived at via algorithms running on supercomputers analyzing DNA sequences—things that cannot be "observed" in the literal sense. After the IPPS meeting in Delaware, I had lunch with program chair Ron Strasko and fellow speakers Dr. Darrel Apps, Sinclair Adam, and Dale Hendricks. Dr. Apps told us, "I get so mad at taxonomists! Linne said, 'We must make things simple so that people can communicate.' And here they are doing the exact opposite!"

Eponymous grasses

I recently did a talk on grasses, with a section on "eponymous grasses" as a sort of category—i.e., genera named for people. *Muhlenbergia*, for example, honors Gotthilf Heinrich Ernst Muhlenberg, sort of a home-boy for me; he was a minister and naturalist in Lancaster, Pennsylvania. *Deschampsia* is named for a French surgeon, Unlucky Louis Deschamp. *Banksia hookeri* honors two legendary scientists: Sir Joseph Banks, the first European to collect this plant in Australia; and Sir William Hooker, an English botanist, director of Kew and also a plant explorer. But that's another talk for another time.

The taxonomist who did the *Banksia/Dryandra* work talks about "classifying organisms in a way that reflects their evolutionary relationships." I hope that doesn't mean choosing names that have no meaning in the real world, because the human element is a wonderful way to name a plant and to honor a human being simultaneously.

Stachys hummelo

This is the Perennial Plant Association's Perennial Plant of the Year for 2019 (Figure 2). The consensus seems to be that the proper name for this one is *Stachys officinalis*. In my catalog, it's listed (for now) as *Stachys monieri* because that's what the Naamlijst calls it and the Naamlijst is still the Perennial Plant Association's bible.

Figure 2. Is it *Stachys monieri*, *Stachys officinalis*, or *Betonica officinalis*? Depends on the authority.

Ibuprofen

Got a headache yet? We apparently need a new Linnaeus. Meanwhile, I propose a new drug-related nomenclature model, at least for commerce, which allows different people to discuss the same thing using different terms to suit different needs. Let's take a common pain reliever. Microbiologists can draw schematics of its molecular structure, which to me looks like something built by a plumber on LSD, or they can discuss $C_{13}H_{18}O_2$. Chemists can call it Isobutylpheol propanoic acid. Pharmacists rely on its generic name, Ibuprofen, and a guy with a headache goes looking for it by its trade name, Advil—or, in England, Arthrofen, and in Austria, Brufen, not unlike the colloquial names we assign plants.

***Ptilotus exaltatus* Joey® pink mulla mulla**

A green parallel, in reverse order: To a marketer—and therefore to a gardener, this Australian annual is "Joey." That was Benary Seed's sales pitch: "Just call it Joey." To Aussies, it's pink mulla-mulla. To a grower, it's *Ptilotus exaltatus*. To a taxonomist, it's found in the Angiosperm and Eudicot clades, in the Order Caryophyllales and the Family Amaranthaceae—useless information to most of us.

An interesting sideline about this plant: Two friends, both in the industry, told me—completely independently—that they were in the market for a new home. Both said that one of their criteria for choosing a house is—apologies to the squeamish, this is a tad indelicate—they had to have enough privacy to pee outdoors, right off the deck. It's a guy thing.

One friend unfortunately ended up with close neighbors, really close. He planted screening, so no one could actually see him urinating outdoors, but he had a bat-eared nosy old neighbor who could hear what he was up to, and she'd yell at him.

The solution? He planted Joey. When you tinkle on your pink mulla-mulla, it makes no sound. Because with *Ptilotus*, the P is silent.

ICBN vs. USPTO

Trademarks present a nomenclatural conundrum. It annoyed me at first, but I'm at peace with it now. It actually keeps things cleaner than some other games people can play.

What Conard-Pyle did with the Knock Out Rose "family" was ingenious. They patented the individual plants under cultivar names like 'Radrazz' and 'Radral', in single quotes as God and Linnaeus intended. "Rad" is short for William Radler, the breeder. Such code names are all but useless in commerce. Simultaneously, they trademarked a group of names under the Knock Out umbrella—attractive, memorable, marketable names.

For years, the International Code for Botanical Nomenclature and the International Code of Nomenclature for Cultivated Plants (ICBN and ICNCP) forbade giving codes and other "nonsense names" to plants, but the rule was ignored so often that the registrars deleted it.

A patent protects a plant's "inventor" for 20 years. A trademark lasts 10 years but can be renewed indefinitely. So, when the patent on the original red Knock Out® Rose expires in January 2019, anybody can propagate it and sell it as *Rosa* 'Radrazz' (Figure 3), without paying a royalty. That's a cultivar name, free for anyone to use as per ICNCP. But if you want to sell it as Red Knock Out®, the name people ask for at the garden center, that'll cost you a marketing fee. Conard-Pyle created a good name, trademarked it, and have rigorously policed their trademark to keep it from becoming a generic name.

Figure 3. *Rosa* 'Radrazz'.

The ICBN and the ICNCP are sort of gentlemen's agreements, with no force of law behind them. Patent and Trademark law are exactly that: enforceable law. And the United States Patent & Trademark Office (USPTO) doesn't care about the ICNCP. Ergo, alphanumeric codes, punctuation marks—that's all OK now, which is kind of a shame but there it is.

This is true of nearly every plant category. Mums, roses, perennials, shade trees, fruit trees, raspberries, and more are often patented with cultivar names that would have been deemed improper in the past.

There are lots of patented plants, over 29,000 as of late 2018, but there really aren't a lot of perennials so good that a trademark (TM) will be a big deal after the patent runs out. The Knock Out rose is one such plant; *Geranium* 'Rozanne' and *Pennisetum* 'Fireworks' are two others. Those of you who deal in woody plants can no doubt name a very different list. When I look at a lot of other TMs out there in perennials, I can't help but think that something better will come along.

CONCLUSION (RELUCTANT)

Nomenclature assigns a plant a place in the world. When our understanding of a plant increases, sometimes it's clear that it was misplaced, and has to move. Molecular studies and supercomputers aren't going away. Genies are notoriously reluctant to go back into their bottles.

Kevin Thiele, the Australian who decided *Dryandra* is really *Banksia*, said, "No science should reject new understanding simply because we're comfortable with the old." I can't argue with that. The term "settled science" is practically an oxymoron.

Much has changed in this field, and there are many more shoes to drop, but my advice remains the same: Choose an authority, and stick with it—until it fails, which it will. When that happens, consult the others and use your best judgement.

SOURCES

Online:

MoBOT: Missouri Botanical Garden
www.missouribotanicalgarden.org

RHS Plant Finder:
www.rhs.org.uk/plants/search-Form

GRIN: Germplasm Resources Information Network. www.ars-grin.gov

Naamlijst:
www.internationalplantnames.com

PhytoKeys: www.phytokeys.pensoft.net

Print (Yes, you can still look things up in books! And sometimes it's faster!):

AHS Encyclopedia of Perennials, Graham Rice Ed. A distinctly English bias in plants described.

Naamlijst van Vaste Planten and Naamlijst van Houtige Gewassen (List of names of perennials and list of names of woody plants)

Color Encyclopedia of Ornamental Grasses (Darke)

Combined Proceedings IPPS 68:116-117. 2018.

Seed Germination in Swamp Privet (*Forestiera acuminata*)

Robert Geneve and Sharon Kester
University of Kentucky, Department of Horticulture, Lexington, Kentucky 40546 USA

rgeneve@uky.edu

Keywords: Seed dormancy, native plants, Oleaceae.

INTRODUCTION

Swamp privet (*Forestiera acuminata*) is an underutilized member of the Oleaceae. Swam privet is native to wet areas of the lower Midwest and southeast (USDA Zones 5 to 8). It forms a large shrub or small tree. Plants are dioecious and female plants produce an ovoid, blue-black drupe. Its ornamental potential has not been explored, but it could serve as a non-invasive replacement for privet (*Ligustrum*) as a deciduous hedge. Its foliage is a good, clean dark to dull green and there is usually good yellow fall color. Swamp privet could make a good addition to the plants available for rain garden and bioretention areas.

Oleaceae is an interesting family related to seed germination and dormancy. The family is comprised of 15 genera. Seed dormancy types within temperate genera of the Oleaceae range from having no dormancy (*Ligustrum*), physiological dormancy (*Fraxinus*) to morphophysiological dormancy (*Chionanthus*) (USDA, 2018). There is no current information on propagation in swamp privet, therefore the objectives of this study were to elucidate dormancy and germination conditions for swamp privet.

MATERIALS AND METHODS

Fruits were collected at the blue-black stage and the outer fleshy fruit mesocarp tissue removed by washing. Cleaned fruits containing endocarp and botanical seed were placed in petri dishes containing moistened vermiculite. Dishes were placed directly into 22 or 25°C incubators or cold stratified at 5°C for 3 or 6 weeks before germination at 25°C in 16 hr light. Seeds that failed to germinate after 6 weeks of warm stratification were cold stratified for 3 weeks.

RESULTS AND DISCUSSION

Seeds were endospermic and contained a fully developed embryo that filled the length of the seed. The germination pattern was epigeal with the radicle emerging first followed by elongation of the hypocotyl to raise the cotyledons above the growing substrate (Figure 1).

The highest germination was in seeds placed directly at 22°C (63%). Seeds cold stratified for 3 or 6 weeks did not germinate to the same extent as seeds only exposed to warm stratification. Germination was higher at 22 compared to 25°C.

Also, non-germinated seeds from this group that were cold stratified for 3 weeks subsequently germinated at near 100%. This suggests that fresh swamp privet seeds have conditional dormancy.

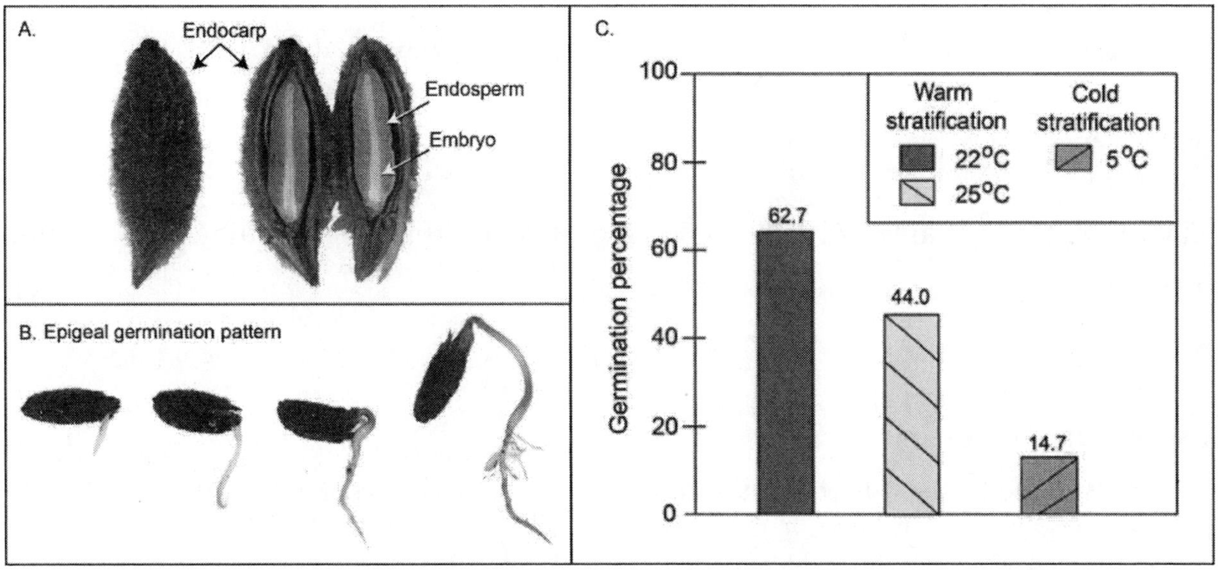

Figure 1. (A) fruit and seed morphology and (B) germination pattern in swamp privet. (C) Seed germination after 6 weeks following warm or cold stratification in swamp privet.

Literature cited

USDA Forest Service Agriculture Handbook 727. The Woody Plant Seed Manual. 2018. Government Printing Office. Wash. DC. https://archive.org/details/TheWoodyPlantSeedManual.

Combined Proceedings IPPS 68:118-125. 2018.

Propagation Down Under - Back to Basics New Zealand Style, Nothing Too Fancy - It's Just the Way We Do It

Lindsey Hatch
78 Jericho Road, R D 2 Pukehohe, Pukehohe East 2677, New Zealand

hatch.clan@xtra.co.nz

Keywords: seeds, cuttings, division, ferns, spores, leaf cuttings, root cuttings, propagation methods.

INTRODUCTION

Over the years I have seen many methods of propagation from the high tech to the simplest low tech you could have and interestingly enough some of the least thought-out have been the best methods that have the best results; maybe not the most productive but certainly some of the best return for money spent (Figure 1). The following are some of the methods used by our nursery and what we achieve.

Figure 1. Tools of the trade.

Before I get too much into depth about propagation methods, I should give a rundown of New Zealand's climate and zones. New Zealand has two main islands with many small islands surrounding its coastlines. The South Island is a temperate climate zone and holds the records for largest land mass, wettest and driest parts of the country, and also coldest and hottest with an average of 4,000–10,000 ml of rain on the west coast and as little as 0–250 ml in the south east. One town, Alexandra in central Otago, holds most of the records. In an average year it can have a low of -10 °C (14 °F) and as high as 35 °C (95 °F) with 0–250 ml of rain. It also has a mountain range that goes pretty much the length of the Island.

The North Island has a subtropical climate zone with less extremes, more people and productive horticultural land. The North Island average temperatures being 11 °C (52 °F) in winter and 20 °C (68 °F) in summer. Rainfall can be less predictable depending on what coast you are on or how close you may be to hills, rangers and the couple of mountains as these all create their own little microclimates; rainfall average can vary from 600–1600 ml in any given year. Our own little slice of paradise is in the upper North Island in one of New Zealand's most productive soils. Pukekohe grows a high percentage of the country's onions, potatoes and brassica for local and export consumption. Our average rain fall is 1100 ml over approximately 125 days with 2000 sunshine hr. Typical summers are warm and humid with a couple of dryer months and winters cooler and wet with a few frosts in colder years.

Now about us. **Joy Plants** founded in 1960s by Terry and Pam Hatch is a family business currently employing only family. Over the years we have had as many as three workers but at present it is Terry, Pam, Lindsey, and Sarah a granddaughter. We have a 24-acre property (9 hectares), of which about 3 acres is nursery (Figure 2), the rest is bush, gardens, and pasture—yes you can make a living of this amount of nursery. We grow mainly New Zealand natives, perennials, and anything else we think we would like to try. Most product is grown on speck and sold to landscapers and general public. Ninety percent of plants grown are propagated on site and sold in the North Island.

Figure 2. Joy Plants general nursery shot.

PROPAGATION METHODS

At our nursery, leaf cuttings, rhizome cuttings, root cuttings, stem cuttings, division, tissue culture, fern pups, spores, and seeds are used. Why's, what's, and when's of the propagation methods are outlined below.

Leaf cuttings

Why? These are normally done when we propagate more plants from small amounts of material. This is not necessarily the fastest method to produce a plant, but it may be a quicker way to bulk up your initial parent stock which may be used to produce other types of propagation material (Figure 3).

When? Leaf cuttings are taken when leaves are at their best and will depend on plants being produced. Many species may be able to have material taken all year while others in the case of bulb species may only have a 3- to 4-week window.

What plants? *Begonia, Peperomia, Eucomis,* and *Haemanthus*.

Figure 3. Leaf cuttings of *Peperomia*.

Rhizome cuttings

Why? These are normally made when we have excess rhizome left over from doing division work and pieces of rhizome that are large enough to cut into 2- to 3-cm slices. They are placed on potting medium and covered with pumice sand (Figure 4).

When? These can be taken any time, but most commonly done in spring when most division work is done, and the excess material can be put aside and worked on.

The rhizomes are then put in trays and placed in a cool shady part of the nursery and left till small plantlets start growing. These can be removed once roots form. In some cases, the rhizome can be replaced and in many cases more plantlets are produced until the rhizome rots away.

What plants? *Agapanthus, Iris, Bergenia, Cordyline,* and *Farfugium.*

Figure 4. Rhizome cuttings of *Farfugium japonicum* 'Aureomaculatum'.

Root cuttings

Why? These are made when there are good roots from many perennials and some tree and shrub species that are easy to produce plants from. This method can produce good plants that may be ready in the following growing season. Because many species are able to be produced in this way it is a good method to bulk up numbers (Figure 5).

When? These are taken when trees and shrubs are lifted in winter and perennials in winter and spring. For best results roots should be the best and normally the largest of the plants roots that can be taken without reducing roots to much which will affect stock plant growth for the growing season.

What plants? *Ajuga, Acanthus, Pulmonaria,* and *Stokesia.*

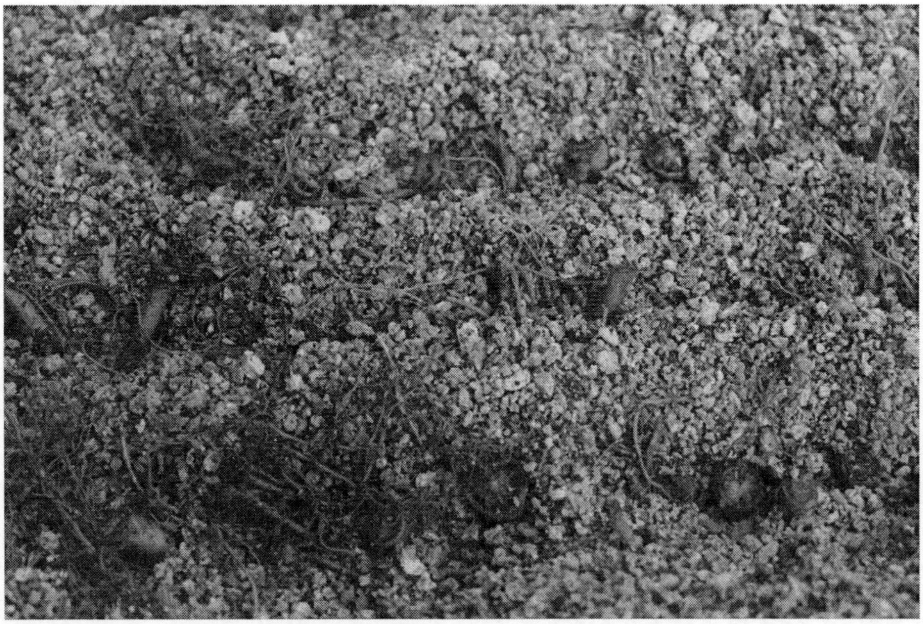

Figure 5. Root cuttings of *Stokesia laevis*.

Stem cuttings

Why? This method for many plants is the easiest method for many to produce cultivars of trees shrubs and perennials. The downfall of this method is that a reasonable number of stock plants are required to get large amounts of cutting material to produce larger numbers of cuttings

When? These will be taken all year round and will depend on species and even cultivar and type, e.g., softwoods normally in summer months, semi hardwood in late summer or early autumn, and hardwood in winter and early spring.

What plants? *Coprosma, Corokia, Dianthus, Salvia, Hebe, Melicytus*, and *Vinca* just to name a few.

Tissue culture

Why? This method is the most hi-tech method of propagation and changed propagation more than any other of the modern-day methods in my view. The issue is, in many cases, this is the most expensive method to bulk up plants, however the good points are that it has cleaned up many plants that had been very difficult to produce by other methods due to viruses or when very little plant material was available. It also has made shifting plant material to and from counties where issues of importation can restrict what can be brought in, e.g., soil material and plant pests (Figure 6).

When? These can be taken out at any time cultures are ready, but we tend to prefer our cultures to be done in early summer and autumn when temperatures are warming up and cooling down.

What plants? *Bergenia, Farfugium*, orchids, *Lobelia*, and almost anything you can think of has or is being propagated by this method including tree, food, and flower crops.

Figure 6. Tissue culture plantlets of *Lobelia aberdarica*.

Division

Why? This method is most common for named perennials and grass cultivars. An easy method of propagating large clumps of plant material into multiple plants with the use of knives, forks, and spades (Figure 7).

When? This is done best as plants just start moving mainly early spring to early summer, but it is possible with some groundcover species to be divided all year round. This is particularly true when it comes to New Zealand natives like *Leptinella* and *Lobelia* species.

What plants? *Carex, Iris*, ferns, daisies, *Ajuga, Hosta, Clivia* and so much more in the perennial lines.

Figure 7. Division of *Lomandra*.

Spores and pups

Why? This is in some cases the only way to propagate many fern species other than division; pups (Figure 8) are found on some perennials and is an additional method used by us to produce these particular plants.

When? These can only be taken when ready and something that may need constant observation to make best use of your materials. In most cases spores will need to be harvested and dried and then sown within days of spore drop. Pups on the other hand will need placing in propagation medium. Once harvested these are best done in spring to get optimal results and the longest possible growing time with the best temperatures.

What plants - spores? Most fern species we mainly grow including *Adiantum, Asplenium*, and *Blechnum.*

What plants - pups? *Asplenium bulbiferum, Neomarica*, and *Ajuga.*

Figure 8. Pups on a leaf of *Asplenium bulbiferum.*

Seed

Why? This method is great for numbers; but for us, this is important when growing natives for revegetation work because we have more diversity which has become more important in recent years with several new pest and diseases having arrived on our shores. The larger the diversity the higher chance of having more resistance occurring than with clonal material. Extinction of plants is always a threat when new pests are found on a species and this can have major consequences not just for one species or genus but a whole chain of plants, fungus, and insect's—whole eco systems can be put at risk. Seed has the advantage of potential new introductions with ornamental or medical possibilities this is why seed production would be my favorite propagation methods.

When? These would be harvested preferably when ripe although some seed can be harvested a little premature, this is not ideal but may be necessary if this is the only opportunity to collect. Other reasons to collect when premature may include if left to fully ripen seed may be eaten or the fall season makes harvesting impossible. Seed sowing will vary, but in many cases, we will sow within days of harvest unless it can be stored and not all seed harvested are need for production at that time. Some of our New Zealand native plant seeds are what we call "green seed" which if not sown very soon after harvesting it loses viability while others will hold better in propagation trays for 2 or more years before germination or before the need to tube. Many species we have found germinate better if left to go through cycles of hot, wet, cold, or dry which if sown and placed outside will break dormancy better than any storage methods in many cases.

This is due to inhibitors in the seed that may require several different temperatures for best results.

What plants? Most New Zealand native species, exotic trees, and many perennial species and bulbs (Figure 9).

Figure 9. Seed cleaning of the native *Pennantia corymbose*.

Hybridization

Why? Because we can and we love to see what results may turn up. How else are we to get new cultivars and someone has to do it? But the main reason is to improve on what we have with better forms, new colours, and better cultivars for a wider range of climates.

When? They are in flower or the opportunity arises.

What plants? Whatever takes your fancy, plants that Joy Plants have worked on are *Alstromeria, Agapanthus, Clivia,* belladonna, *Bergenia, Erysium, Eucomis, Helleborus, Iris, Libertia, Magnolia, Phlomis, Primula, Watsonia*, and a number of additional bulb species (Figure 10).

Hybridization leads into plant selection which I guess is one of the reasons we do it but also the how, when and what fore's. How does one select a new cultivar? It depends on what you are after. When should you make a selection? I think when you have reached the end goal, certainly not until you have improved on what was already available.

SUMMARY

The best time to propagate something is when the opportunity arises or when someone offers you the material. Take the opportunity or get out of the propagation shed; it's your loss not mine.

Figure 10. Examples of our hybridization program: Top row left to right: *Helleborus, Clivia*; Bottom row left to right: *Magnolia, Bergenia*.

Combined Proceedings IPPS 68:126-130. 2018.

The Big Picture: Plants, Soil, Climate + Work That Matters

C. Dale Hendricks
Green Light Plants, LLC., 1834 Flint Hill Road, Landenberg, Pennsylvania 19350 USA

dale@greenlightplants.com

Keywords: Carbon cycle, carbon dioxide, environment.

INTRODUCTION

In this paper I seek to enhance understanding of the roles that plants play in water, soil, weather, and climate and to link this to new science and understanding with emerging opportunities for horticulturalists.

CARBON CYCLE

First, we'll look at the carbon cycle and plants many roles in it. A look at global atmospheric CO_2 levels as illustrated here reminds all reality-based people to note that it is a greenhouse gas and has an insulating effect—trapping heat from the sun. Looking closer at this "Keeling Curve" as it's come to be known one can see a steady rhythm of fluctuation within each year (Figure 1). It turns out that the great majority of the Earth's land surface is in the northern hemisphere so when it's springtime here and plants grow, and deciduous trees sprout leaves the levels fall—literally breathing in gaseous carbon. In the autumn here, the plants growth slows greatly in the temperate zones and the "inbreathing" of carbon from the Southern hemisphere is less, making these steady gyrations. It's worth noting that every year up to 1/6 of atmospheric CO_2 is cycled through plants.

Some have claimed that we could mitigate fossil fuel impacts on climate merely by doubling photosynthesis.

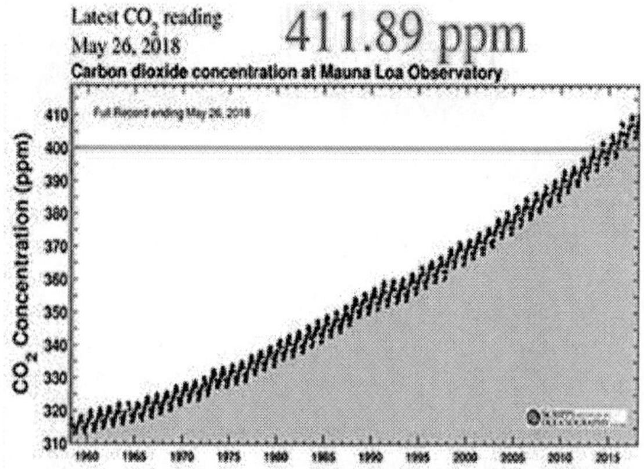

Figure 1. A look at global atmospheric CO_2 levels.

On land the great majority of the carbon is in the soil with 63% in soils, 15% in land plants, and 22% in the atmosphere (Figure 2). It also turns out that many agricultural and pasture soils have lost a great amount of their carbon—often ½ or more—since they were converted to these uses.

Let's look at how current practices are, often inadvertently, causing loss of carbon in terrestrial soils.

Water carries a huge amount of soil down streams and rivers. When it rains on bare ground this can be a huge problem. When soil is covered with dead plants/mulch the problems is somewhat diminished. When soils are covered in deep rooted, diverse and perennial vegetation very little is lost, and soil/plants act as a living water filter (Figure 3). All the brown muddy waters flowing down our streams are exporting carbon and while this happens much is oxidized and returned to the air and some remain as silt. The point is that a water friendly landscape is a carbon friendly landscape and whatever we can do to move toward deep rooted perennial and diverse living groundcover the better.

Another invisible pathway for carbon to leave soil is tillage and bare ground. Soils are exposed and greatly oxygenated, this stimulates the growth of bacteria which in turn digest a lot of carbon, as these bacteria die their minerals become plant available, good for a shot of fertility but during this process—called mineralization—a lot of CO_2 is breathed back into the atmosphere, effectively burning up our soils. This process can be quite rapid.

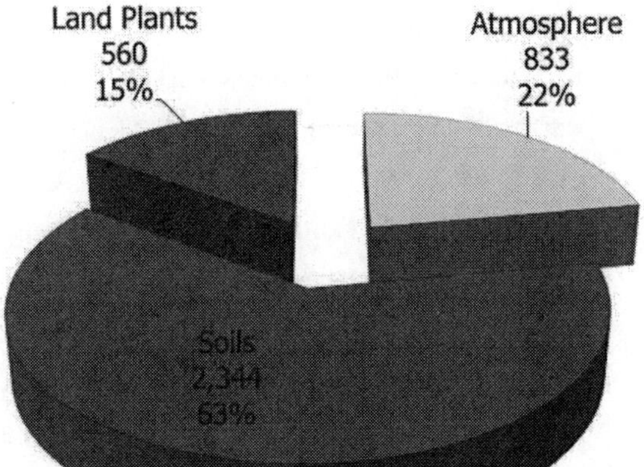

Figure 2. Carbon in the environment.

Many pesticides and fungicides have antibiotic activity and these materials can add NOx (nitrous oxide) and CH_4 (methane), potent greenhouses, to the atmosphere.

Figure 3. Water carries a huge amount of soil down streams and rivers and perennial vegetation acts as a living water filter.

Let's turn to how both native peoples and natural systems have built rich, high carbon soils and how we can use these processes to guide our actions.

The US Midwest soils are rich, dark, fertile and deep or at least they were at the time of early European contact. Let's look at how these have been created and, and how, on many regeneratively managed farms and ranches, still are being quite rapidly rebuilt.

1) These systems were deep rooted diverse perennial prairie and savannah systems. Perennial plants lose their vegetation above ground every season and this mulch/duff builds up soil.

2) Perennial plants lose ⅓–½ or so of their roots every year, the dead roots also aerate and add to the humus layers. If unplowed or not left bare a good portion remains for several years as "labile" that plants use as fuel. These first two forms of carbon generally cycle quickly back to the atmosphere in a process known as soil respiration.

3) About 85%–90% of plants have mycorrhizal relationships. These "root fungi" and plant root relationships represent a great cooperatively based evolutionary leap which has been very beneficial for both parties as well as the rest of the world. The hyphae of the fungi provide orders of magnitude greater surface area than roots alone can, making lots more water and nutrients available to plants. Mycorrhizal fungi can access nutrients plants need (and indeed request—as they are in constant communication!) by secreting enzymes that can for example dissolve rocks to make phosphorus available (Macfarlane, 2016; Simard, 2016). Plants trade these benefits by exuding their photosynthesis derived sugars—also mostly carbon—as well as an array of communication substances known as secondary metabolites; phenols, terpenes, etc. This mutual tango is also called the "liquid carbon pathway", dissolved organic carbon trading or bi-directional flow (Jones, 2018). As these liquid carbons pass through the bodies of fungus then often bacteria, they form complex and longer lasting carbon chains known as humus. This process can fix significant amounts of carbon from the atmosphere into productive, nutrient rich water sponge like soils. These carbon molecules often are quite long lasting and also contain glomalin—a glue like rich carbon substance that is tremendous at water storage.

4) In most of the most fertile soils, grasses, perennials and animals worked together. Grazing, like anything else can be done poorly—ok or well, it depends on management. Predators kept bison and other ruminants bunched up and on the move. This often resulted in herds moving onto a new area, grazing the best parts and moving on. When ranchers/farmers seek to mimic this process they purposefully let areas rest then move animals on, graze only to ½ or so of total vegetation is removed then move them again. This allows the plants to recover quickly as well as sloughing off newly unneeded roots. The addition of moderate disturbances from hoof action—again if done well (think the Goldilocks principle here…) can provide spaces for water to pool as well as addition of solid and liquid nutrients from the animals can be most beneficial. It is crucial to note here that the great majority of farming/grazing is not done well and often depletes soils and future prospects (Machmuller et al., 2015). More soil organic matter makes more rain: New satellite data shows just how important is plant-soil evapotranspiration and how it lasts longer than once believed (Newport, 2017).

5) Prairie fires set both by human hunters and naturally by lightning also contribute to soil building processes. If fires are not excessively hot or grazing has left moderate amounts of fuel, fires will leave a good dusting of charcoal. This dusting of long-lasting recalcitrant carbon, now known as biochar, often lasts hundreds to thousands of years. Biochar acts as glue, holding water and nutrients in plant available forms (weak bonds, surface area) while providing habitat for beneficial fungi and bacteria (Mao et al., 2012).

We've seen the dynamic, mutualistic and exciting way that science is helping us understand plant mediated, soil-based living carbon banking, now it's time to turn our attention to the plant/soil and weather/climate interactions.

We've long known about the heat island effect—first described in urban areas. As they are paved and built over, they grow noticeably hotter. Mostly because the sun's heat directly warms up exposed surfaces—and this can include bare ground, desert, fallow fields, etc. Any land surface that is not fully covered in living and photosynthesizing vegetation will see the sun's energy turned into "sensible heat" or heat we can feel. However, sun shining on photosynthesizing plants does not add this heat; it is converted into chemical energy and later released as the cooling effect of transpiration. So, the heat island can be pictured as a rising column of hot, often low humidity air that has been observed to be repellent of rains, clouds and thunderstorms. What is new is there is growing evidence for the idea that the effects we can feel at smaller personal scales—say on our skin—also appear to be major drivers of weather and even climate at larger scales (Pearce, 2018; Wright, 2017).

Scientists are just beginning to understand the significant driver that both this effect and its opposite—what many are calling the Biotic Pump Theory have on weather and climate. As water evaporates from plants and trees via transpiration the cooling effect writ large comes into force. It's worth noting here that on a planetary scale about 95% of the planetary heat load is mediated/moved via the most important greenhouse gas of all—water vapor—and roughly 4% is moved or affected via carbon dioxide—the latter is still a major force but is by no means all of the story. Where and when clouds and rains occur have huge amounts to do with how water is held—or not—on the land, the vegetation regime and the cooling effect of plants are huge. Mist then clouds rise and the resultant cooling causes a pressure gradient, this low pressure then draws and attracts more rain. In this way inner continental areas like the Amazon Rain Forest more or less both make their own rain and attract and recycle rains that may have originated over oceans thousands of miles away. Trees and other plants also have much more surface area than say a flat ocean and hence more opportunity for cooling transpiration and cloud formation. This can partially explain way we now see drier areas getting hotter and drier and wetter areas like say here, getting wetter… To quote Bill Mollison: "everything gardens." In other words, humans, plants, and ecosystems create conditions for their long-term thriving—or at least they are capable of it… thinking now of us humans…

In this talk we've seen that plants—when assembled into ecosystems—either naturally or assisted by conscious design, can thrive, bank carbon, clean water and provide habitat for humans and many other creatures. Plants looked at or managed individually or in monocultures have requirements or needs, and when assembled into holistic systems they can all contribute, benefit from mutualistic relationships and contribute to the whole. To wrap up, ecosystem restoration, diverse perennial crops, edible landscapes, agroforestry, food forests, permaculture, green roofs and more are not only fashionable and increasingly popular but may help point the way forward to a time when motivated and hardworking people seeking meaningful work will join us. Sounds like fun.

Literature cited

Jones, C. (2018). Light farming: restoring carbon, organic nitrogen and biodiversity to agricultural soils. http://amazingcarbon.com/JONES-LightFarmingFINAL(2018).pdf

Macfarlane, R. (2016). The secrets of the wood wide web. https://www.newyorker.com/tech/annals-of-technology/the-secrets-of-the-wood-wide-web

Machmuller, M.B., Kramer, M.G., Cyle, T.K., Hill, N., Hancock, D., and Thompson, A. (2015). Emerging land use practices rapidly increase soil organic matter. Nature Comms. https://www.nature.com/articles/ncomms7995

Mao, J.-D., Johnson, L., Lehmann, J., Olk, D.C., Neves, E.G., Thompson, M.L., and Schmidt-Rohr, K. (2012). Abundant and stable char residues in soils: implications for soil fertility and carbon sequestration. Environ. Sci. Technol. *46*:9571–9576.

Newport, A. (2017). More soil organic matter makes more rain. Beef Producer https://www.beefproducer.com/management/more-soil-organic-matter-makes-more-rain?NL=FP-002&Issue=FP-002_20170119_FP-

Pearce, F. (2018). Rivers in the sky: how deforestation is affecting global water cycles. YaleEnvironment360. https://e360.yale.edu/features/how-deforestation-affecting-global-water-cycles-climate-change

Simard, S. (2016). How trees talk to each other TEDSummit. https://www.ted.com/talks/suzanne_simard_how_trees_talk_to_each_other/transcript?language=en

Wright, J.S., Fu, R., Worden, J.R., Chakraborty, S., Clinton, N.E., Risi, C., Sun, Y., and Yin, Lei. (2017). Rainforest-initiated wet season onset over the southern Amazon. PNAS *114* (32) 8481-8486.

Combined Proceedings IPPS 68:131-136. 2018.

Evolution of Edibles in Plant Retailing

Clive Larkman

Larkman Nurseries Pty Ltd, P.O. Box 567, Lilydale VIC 3140, Australia

clive@larkmannurseries.com.au

Keywords: garden centers, food crops, marketing, edible plants, edimentals, herbs.

INTRODUCTION

We are all here due to an involvement in an industry that goes back many thousands of years. Over that time there have been a multitude of changes in why, what, and when people buy plants. I intend to look briefly at the long-term history and how it has shifted around the demand and use of edible plants. I will then review the more recent trends over the past 50 years and how they relate to human culture, the economy and social fluctuations. Retailing is a complex subject at best and is impossibly unpredictable in edible horticulture. To set the record for this paper the class of edible plants encompasses all the plants used in the home: edible, medicinal, and fragrant.

PLANT RETAILING

Plant retailing, like all retailing requires a group of people wanting to buy a product from a range of people wanting to sell that product. Like other key retailing sectors, plant selling is a wide ranging and stable part of the world economy. Theoretically it goes back as far as the history of human settlement. Originally, man was a nomadic hunter gatherer, travelling the land hunting and collecting food. Over the millennia there have been a range of factors that have pushed human communities in one direction or another. The first, and greatest influence, on settlements was the cultivation of plants. Initially it was simply a matter of cultivating desirable plants and removing unwanted species. Next step was to "fence" off the area prior to actually "moving" and "planting" specific plants. The ability to cultivate plants enabled man to settle down in one spot and cease the nomadic life. At the same time, it demanded that man settled down. The nature of plant cultivation meant that humans no longer had to chase food to feed the family, but it required they stay in one place to feed the plants.

Initially plant growing was limited to farmers and men producing crops to sustain the village. They grew what was easy, what was needed and generally what was indigenous to the region. It was a simple form of horticulture that lasted for thousands of years. Over time the ability to grow a wider range of plants enabled farmers to sell to a more diverse customer base. Individual farmers also selected better yielding forms, which created a demand for artificially propagated plants. However, the production of the plants was limited to the grower's ability to propagate from their own stock. They collected their own seed and over time developed the ability to asexually propagate their best forms. These selections demanded the highest prices and stimulated mass production. Now the ability to sell was restricted by their ability to transport the plant.

It was at this stage that the basics of plant retailing started to develop. Once a grower could select good selections and then produce a plant that replicated their characteristics, they could sell the plants as well as the produce. From this the plant retailing industry was born. At first it was the basic concept of one grower supplying another with no attempt to influence purchases through presentation (i.e., marketing). Plants were basically grown and sold for use as producers of food and ancillary products.

This was the system for hundreds of years with the only real change being when communities started to recognize layers of wealth. The evolution of classes (in all cultures) created some people with ample assets and others who struggled to survive. When combined with the new culture of land ownership the levels of wealth evolved with the ability to demonstrate this wealth through the presentation of "better ornamental" gardens. The working class had gardens (if they could afford land) which were primarily used to produce food, medicines, and home beautification (flowers, etc.) products.

The wealthy landowners had more than enough money to purchase the basic necessities of food and health plants, so they started to buy plants and to pay staff to create attractive gardens. They needed plants that were more ornamental than edible.

They also started to search for plants that were more exotic and needed growing skills. It was around 2000 years BC that the aristocracy in western Asia and eastern Europe started to develop complex and ornate gardens. As the trends moved west to Europe gardening also developed in eastern Asia where structured gardens were the mainstay. This moved through China and into Japan where highly manicured gardens representing the natural landscape were the aim.

It is hard to compare communities from Asia, Europe, Africa, the Americas, and Australasia as the records of what was grown is minimal. Also, the cultures are quite different with some attaining a high level of development and other remaining quite tribal. A wealthy lord from central Europe had quite different tastes, needs, and resources to a tribal leader from central Australia. The styles reflected much of the nature of the culture in a region and tended to be similar across the area. Over the past two centuries landscapes have devolved into a multitude of styles each with its own set of average to excellent examples. The better gardens generally had a greater range of unusual plants. As the retail sector grew so did the diversity and availability. The concept of retailing varied greatly across cultures but in most cases has evolved to the same position.

The desire to have what others do not quickly created the concept of supply- and demand-based pricing. Originally this focused on jewelry, precious stones etc. but quickly covered all aspects of human production. The wealthy would produce beautiful and exotic gardens and the working class would soon try to emulate them. New and exotic plants were the realm of the rich and the desire of the poor. In some parts of the world they even became a form of currency.

CREATING AN INDUSTRY

Originally the producer was also the retailer, and this remained the case until the 20th century when retailing and growing became advanced and segregated. Retailing focused on presentation and range with a different set of skills to the grower. Growing required propagation skills and an understanding of what particular plants needed to maximize growth rate and improve quality. Since then the division has formalized as commercial and retail. The commercial tends to supply landscapers, farmers and retailers and the retail supplies the home gardener. When reviewing the history of gardening most research is based on the large ornamental gardens of the wealthy and the government. These are really the realm of the commercial grower. For the purpose of understanding retail, it is best to look at the small businesses supplying the home gardener. They have different needs and shop in a different way giving rise to two separate selling groups. From these developments the nursery industry arose with three sectors, growing, retailing and allied trades.

For most of the previous 5000 years it was the need to grow food/medicinal plants that drove the nursery sector. In the 1960s large supermarkets took the place of small local stores as the main supplier of food products. Until then many families relied on the "father" to grow a range of herbs, veggies, and fruit to supplement what was sold at the food retailers. Suddenly, the "mother" could buy what she needed as and when she wanted so "father's" products weren't needed any more.

"Father" turned his talents and spare land into ornamental gardens to improve the home. For the first time in history decorating the exterior of the home became a major industry for the masses. These new gardeners wanted rare and unusual plants as well as the common attractive varieties. They needed to find retailers selling all sorts of plants from a wide range of growers. They wanted to have bigger, better, prettier or whatever made their friends and neighbors envious. This desire drove retail nurseries to search for more "new" and "rare" products - ornamental rather than edible/usable.

As I said above the changes in ornamental gardening has had the biggest direct effect on plant retailing. Edible plants are what bring many buyers into the nursery, but it is changes in what is "fashionable in the garden" that has kept the consumer talking about plants. We must accept that ornamental gardening is a fashion-based industry. As such it is governed by all the principles of fashion retailing. It is both predictable and unpredictable. From the outside it seems that plant selections and garden styles change at will. Strappy, perennial, bright colors structured and informal are just some of the trends over the past few decades. However, if you look at the changes and compare them with what is happening in other fashion sectors you can predict where plant styles will go.

Garden styles like other fashion items, take inspiration from all parts of life. With television and the internet so prevalent any new formats can quickly move around the world. Prior to the internet, garden styles moved around the world as designers and plant hunters travelled and brought the designs home. Now the consumer sees what they like in another country or region and then request that the designers to use it. This has resulted in rapid shifts in design and also a situation where there can be several very different styles all leading the fashion trend.

The world of edible plants is also based on fashion—the fashion of food. This is a far more recent trend and is much harder to define. In many countries food fashion has varied little in a thousand years. It is changed only with improvements in the raw food supply. In other countries, like Australia, where the population has grown through mass immigration from other cultures, the food styles are extremely diverse and changeable.

Food fashion is still a new process and it is a little like music where a new style hits the consumer and is the "in thing" until the next style hits. Like music, as more styles develop so do the fusion styles and the tastes of the consumer. Also, as the individual matures so does their taste for different foods. This is not really surprising as both music and taste are part of our basic senses and have a strong effect on emotional stability. The way we deal with them is a key to our enjoyment of plants and food.

As an industry we are again at the mercy of what appear to be random shifts in demand. This causes excess and shortages in supply and rewards those willing to keep looking for what is new. The word "new" being very subjective as many "new" plants are very old to some cultures yet new to others (Figure 1). The winners in the new world of edible plants are those that have an understanding of edible plants as a source of food, as a part of an ornamental home garden and as status symbol in a shifting culture.

Figure 1. New is again the key but new is not new—it is just exotic. Top left to right: turmeric—*Curcuma longa*, wasabi—*Eutrema japonicum*; Bottom left to right: mushroom plant—*Rungi klossii*, yam bean—*Pachyrhizus erosus*.

In Australia and other western nations, food has become a national past time. Cooking and eating are now a social event. Gardeners have a desire to grow their own food as it gives them greater range and "cleaner" food. They can control what is used on it and are comfortable that no nasty chemicals have been used. All of a sudden, we are growing our food in our garden, we are cooking it in the garden, and we are eating it in the garden. We freely move from the house to the garden. This has brought a new generation of people into the garden and they are now looking to learn about ornamental and edible plants. They want their edible plants to look good and are looking for more of what I call "edimentals"—plants that are primarily ornamental but are also edible. We have returned to wanting edible/medicinal plants as well as ornamental ones.

Since man first started growing and retailing plants the range being produced has continued to expand. However, the greatest expansion has been in the last 150 years with the majority of the increase in the last 25 years. The development of better pots, pot media and fertilizers has helped plant growers gain the skills needed to take plants from their natural environment to the home garden. For most of the history of our industry edible plants changed slowly and the biggest shifts in plant types have been in ornamental styles.

Ornamental plants have evolved as we are able to move plants around the world and as we have improved our breeding techniques. Over the last 20 years a "new" plant or cultivar sold just because it was "new". This trend has eased, and gardeners are looking for a bit more in their plants. With the internet becoming a ubiquitous part of society, plant growers now want the story behind the plant (Figure 2). They also want a bit more than a season of pretty flowers. Hence the drift back to edibles.

Up until the mid-twentieth century gardening and plant growing were the domain of the male. At that time the shift in why people garden occurred (from food to ornamental) and the main influence on plant purchasing became the women. As an industry we accepted that the key customer was the 55 to 70-year-old female. Around the millennium this changed. The way the family unit (all forms) made decisions went from set areas of influence to both partners discussing and deciding on all purchases.

Figure 2. Plants need a story, not just new. This plant (*G.* 'Rozanne') was voted plant of the Century in 2003 by RHS.

Ornamental gardening has moved from being a public statement of a person's wealth or design skills. Either way a beautiful garden is satisfying to the gardener, an added value to the property and something that makes the local community happy. An edible and beautiful garden ticks all the boxes.

As discussed above people originally grew plants for use in the home. They needed to grow what they could not buy. The move back to growing plants for the home has come from several drivers including fashion trends. For many it is still about growing what they cannot buy. First is the range on offer in the food retailers compared with the garden centers. In local supermarkets there are usually only four or five types of tomato—none of which were bred for flavor, worldwide there are over 4500 cultivars of tomato available. Varieties bred for flavor, use, size, color and acidity—all characteristics that are important to the end consumer. In our range we have over; 45 tomatoes, 40 chilies, 20 basil, 30 geraniums scented, 50 mint and 40 thyme. The retail food stores are lucky to have more than 3 or 4 of each.

Second when they grow their own edibles, they have complete control over what fertilizer, pesticides, and other chemicals are used on their food. They can decide to be completely organic or just avoid the worst of the pesticides. They can harvest when they want and when the plants are best harvested for flavor rather than storage. The diver behind their plants is usability and taste not storage, transport and profit. Although most of the large retailers have strict controls on the residual level of pesticides, many consumers don't trust them.

Finally, home grown is often cheaper than store bought herbs. Sometimes the store product is cheaper, but it usually has a much shorter shelf life. Sensible buying and growing can make home grown herbs and vegetables very economically. For many varieties continual harvesting over a full season makes them quite cheap to produce.

As mentioned above edible plants are now trendy and some types are in demand and others not. Like clothing this can change at a moment's notice. The varieties in demand will fluctuate as quickly as weekly or as slowly as annually. In some cases, an edible maybc uscd on a popular cooking, renovation or gardening show one day and then sell out the next. This can be frustrating for the industry as it may take several months to rebuild stock levels. As an industry we want to make money, but we also want to make sure our customers get the plants they want.

Food trends are incredibly hard to predict and control as there are a multitude on influencing factors that we cannot control or even modify. The Australian market is pushed by the range of extremely popular cooking shows like MasterChef. An edible featured on an episode will be in demand for the next few weeks. Consumers will go looking in supermarkets at first then to the garden centers. If the industry does well it should be able to persuade buyers to come back in a month or two when the plant is available.

At the same time the growers should be continually searching for "new" and old edibles. They should be constantly trialing these plants to learn how to grow them and to test consumer demand/usage. Indeed, a good grower will not only be aware of current demand but also helping to set future demand.

The cooking shows and food bloggers are always searching for different edibles that they can promote internally. This will often result in them being displayed then/or processed through the administration sector. Like most of the media they are after new material and will respond to new, well presented offerings.

The new direction in edibles will be with us for many years and we just need to understand the drivers as discussed above. The home consumer wants new, tasty and healthy plants with a story. There are still hundreds of edibles out there in industry that are just waiting to be promoted and made available to the masses.

Combined Proceedings IPPS 68:137-139. 2018.

Spotted Lanternfly: Our Latest Threat

Trilby Libhart
Pennsylvania Department of Agriculture, 2301 North Cameron Street, Harrisburg, Pennsylvania 17110 USA

tlibhart@pa.gov

Keywords: Insect, exotic invasive, pest control, *Lycorma*.

INTRODUCTION

In September 2014, a Pennsylvania Game commissioner educator noticed damage to the tree-of-heaven (*Ailanthus altissima*), along with an insect he did not recognize, on his property in eastern Berks County, Pennsylvania. The landowner decided to report this finding to the Pennsylvania Department of Agriculture (PDA) Entomology Program which responded with an immediate site visit. The PDA entomologists were able to collect several hundred specimens which were identified as *Lycorma delicatula* (White), a new pest in Pennsylvania and North America. (Figure 1).

The spotted lanternfly, *Lycorma delicatula*, is a planthopper native to China, India, Vietnam, and introduced to Korea where it has become a major pest. This insect has the potential to greatly impact the grape, hops, and logging industries. When it was first detected in South Korea in 2006 it rapidly spread throughout the country and was reported to be a pest of grapes and peaches. Studies also indicated that the planthopper made use of over 65 different types of plants. In Pennsylvania the spotted lanternfly has been recorded feeding on many different plants including *Ailanthus, Salix, Vitis, Acer, Prunus, Quercus*, and *Juglans*, along with a range of vines, ornamentals, and garden plants. Adult *Lycorma* narrow their host range to *A. altissima,* or tree-of-heaven, before mating and laying eggs.

Figure 1. Adult spotted lanternfly, *Lycorma delicatula*

In Pennsylvania spotted lanternfly completes one generation per year. It begins laying eggs in the fall of the year in clumps of 30–50 eggs which are covered in a foam-covered mass (Figure 2). Egg masses are laid on trees and many other smooth surfaces and are often hidden. Many eggs are laid on surfaces like cinder blocks, rocks, rusty barrels, picnic tables, and other outdoor items. As eggs masses age they begin to look more like dried clay or mud. Egg masses can be removed by scraping the mass off a surface they are attached to and placing them in rubbing alcohol or hand sanitizer.

Figure 2. Spotted lanternfly egg mass.

Starting in mid to late May, the first immature stages begin to hatch from the eggs. The immature stages of the spotted lanternfly will molt three times, getting progressively larger. Each life stage immediately after molting is referred to as an instar. The first three instars are black with white spots can be found from mid-May through August. During these stages the insects crawl up and down host trees and plants each day to feed and to avoid predators. Because the insects move up and down each day, large sticky cards have been used to trap them as they are traveling. This involves wrapping the trees with a band of sticky paper, which catches the insects as they crawl.

Figure 3. Immature spotted lanternflies.

The last nymphal stage of the spotted lanternfly is not just the largest, it is also the most colorful. Fourth instar nymphs start to appear in early July and are red and black with white spots. Starting with the later instar nymphs the spotted lanternfly starts moving from other hosts and begins to concentrate on Tree-of-Heaven almost exclusively.

Adult spotted lanternfly feed for several weeks on tree-of-heaven prior to mating and laying the next generation of eggs. Because they strongly prefer tree-of-heaven, removing most of these invasive trees on infested properties serves to concentrate the spotted lanternfly population so it can be more easily and safely treated with pesticides. Most tree-of-heaven are removed and treated with herbicide to prevent regrowth. A few select males are left standing as "trap trees" which are then treated with a systemic pesticide which kills the spotted lanternfly that have moved into feed on the remaining trees. The pests have little choice but to feed on these treated trees and are eliminated from the population. This has worked well in reducing populations in heavily infested areas but requires a lot of work.

Damage from the spotted lanternfly comes primarily from repeated feeding especially on grapes, hops, and plants in orchards, nurseries and the hardwood industry. Feeding on plant juices results in excretion of a sweet, sticky substance known as honeydew and accumulation of dripping honeydew from heavily infested areas can result in the growth of sooty mold on surfaces (Figure 4). In 2017, heavy feeding was seen on walnut, red oak, maple and hickory trees which resulted in flagging and dieback.

Figure 4. Honeydew and sooty mold.

Currently a quarantine is in place to stop the movement of this pest to new areas and to slow its spread within the quarantine area. The quarantine restricts the movement of the pest and products that may have egg masses present. The quarantine currently covers thirteen counties including Berks, Bucks, Carbon, Chester, Delaware, Lancaster, Lebanon, Lehigh, Monroe, Montgomery, Northampton, Philadelphia, and Schuylkill, and encompasses 6928 square miles. Surveys are currently underway to determine how widespread this pest is in Southeastern Pennsylvania and to ensure the spotted lanternfly is not present in other parts of the commonwealth (Figure 5). Reports from the public aid the Pennsylvania Department of Agriculture in new detections. While the spotted lanternfly can walk, jump, or fly short distances, many of these detections are single specimens whose spread appears to be linked to hitchhiking.

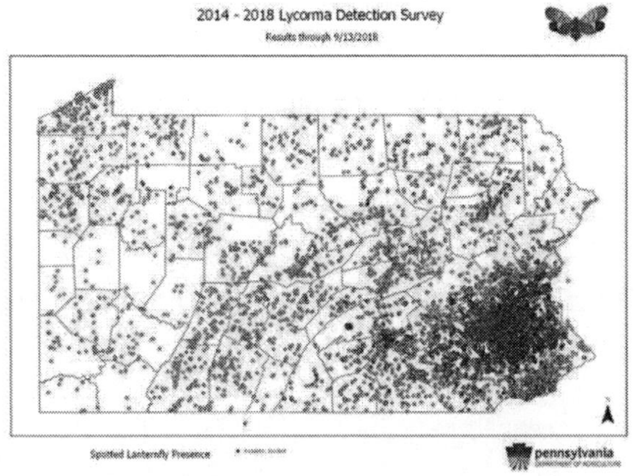

Figure 5. 2014-2018 *Lycoma* detection survey.

The spotted lanternfly program relies on cooperation between local, state, and federal agencies and organizations, along with citizens living in and out of the quarantine area. Local officials and state agencies have been leading the organizational charge and have created a work plan for Pennsylvania.

Currently extension, universities, the USDA and others are researching new and effective ways to deal with this pest. Some work being researched is to the identify the host range of the pest, its impact on grapes, the attractiveness of certain plant volatiles for use in trapping programs, and to analyze its DNA.

The spotted lanternfly is a significant threat to Pennsylvania's vineyard, orchard, nursery, and forest industries. It can also damage landscapes and make recreation unpleasant. Education and outreach are key to future control of this quickly spreading invasive pest. Be informed, aware of your surroundings and report any unusual observations to authorities.

ACKNOWLEDGEMENTS

All images: Pennsylvania Department of Agriculture.

Combined Proceedings IPPS 68:140-144. 2018.

The trial gardens at the University of Delaware Botanic Gardens

Robert E. Lyons

University of Delaware, Newark, Delaware 19716 USA

rlyons@udel.edu

Keywords: volunteers, vines, annuals, tropical, color, cultivar evaluation, All-American selections, Proven Winners.

INTRODUCTION

The University of Delaware Botanic Gardens (UDBG) is generally described as encompassing the grounds surrounding the academic, greenhouse, research, and support buildings on South Campus. This translates into a range of soil types, moisture properties, and sun/shade exposures, which include wetlands, open lands, full shade areas, special collections, a garden of herbaceous perennials, and a Lepidoptera Trail, to name but a few. The Trial Gardens are relatively new to the UDBG complex and are situated just east of Fischer Greenhouse Lab and just north of Roger Martin Drive. They were established approximately 10 years ago by Director John Frett, who converted a transient-purpose space into a traditional arrangement of linear beds patterned after trial gardens found at many universities. A decade of cultivation, which included the installation of turf pathways between trial beds and the annual tilling of mulch and other organic matter, has resulted in an excellent soil medium for plant growth. A very visible and obvious interpretive sign is placed at the forefront of the Trial Gardens and is often consulted by visitors who first enter the Trials (Figure 1).

Figure 1. A very visible and obvious interpretive sign is placed at the forefront of the Trial Gardens.

GENERAL MANAGEMENT

Care and maintenance of Trials Gardens is primarily volunteer-driven, making them quite unique among university peers. A single individual with a professional background in trial gardens matters has willingly volunteered to oversee most aspects of conducting the trials, including bed design, inventory records management, seed sowing and post-germination care, maintaining post-installation aesthetics, and all labeling. Other UDBG volunteers are absolutely critical to the Trials success; they transplant literally hundreds of young plants in spring (mid-May) and help remove all spent vegetation after frost (Figure 2). One more volunteer assists strictly with maintaining the structural integrity and painting of the vine supports and vine fence.

Figure 2. Volunteers working on installing plants in beds along College Ave.

Specific cultivation practices are intentionally low input. Seedling cultivation includes light fertilization and then again only once immediately following transplantation. New transplants are watered upon installation and then once or twice again (weather depending) to insure establishment. That's it. Mulch is applied to all beds (approximately 2 in.) and serves as the primary weed inhibition strategy, augmented by hand-pulling, as needed. Glyphosate is used only infrequently to maintain clean bed edges. Finally, deadheading is done only selectively.

BED COMPOSITION STRATEGY FOR 2018

The overall decisions related to bed composition rest with the Volunteer Curator, who intentionally makes changes each year to create a sense of anticipation and excitement among visitors. In 2018, one bed was reserved for the paid entries (Proven Winners) and another bed had only All-American Selections Winners (AAS) (Fig. 3).

Figure 3. One bed was reserved for the paid entries (Proven Winners) and another bed had only All-American Selections Winners (AAS).

University of Delaware Botanic Gardens is a designated AAS Display Garden, which means that the latest winners (and others) of the AAS program are sent to UDBG for planting somewhere in the public's view. University of Delaware Botanic Gardens chooses to include them in the Trial Garden. In 2018, the Curator acquired additional former AAS winners to accompany the latest winners, providing a rough timeline of AAS winners.

In 2018, the remaining beds were designed following a philosophy of "Curator Choice," whereby plant selection was often based on several loose categories:

- Vines: this growth habit always presents a challenge but should be included due to the sheer number and diversity of garden-worthy possibilities. The UDBG Trials accommodate vines via pyramid-like structures placed within the beds, and a tall painted fence constructed of repurposed materials.

  2018 examples — *Vigna caracalla*/snail vine and *Ipomoea lobata*/Spanish flag.

- Display impact: What will draw the eye, capture the attention, and provide genuine visual interest via habit, size, form, and/or color?

  2018 examples — *Xanthosoma aurea* 'Lime Zinger'/elephant ears and *Leonitis leonurus*/lion flower.

- Curiosity: relying on the unusual and unexpected.

  2018 examples — *Carica papaya*/papaya trees and *Euphorbia cotinifolia*/tropical smokebush.

- Genus diversity: take advantage of the opportunity to include many species of the same genus.

  2018 examples — *Salvia* and *Solanum*.

- Cultivar comparisons: take advantage of the opportunity to include an array of historic and current cultivars of the same species.

  2018 examples — *Catharanthus*/Vinca and profusion zinnias (Figure 4).

Figure 4. Cultivar comparisons: take advantage of the opportunity to include an array of historic and current cultivars of the same species.

FORMAL EVALUATIONS

These evaluations are reserved for only paid entries. In 2018, this included only Proven Winners selections, at $40/entry. Efforts to increase the representation of paid entries have been difficult and often met with a similar response…UDBG is close to many other trial gardens and companies are reluctant to pay for evaluations so close to one another. Efforts will continue to increase the paid entries, despite this perception.

Evaluations commence as soon as plants are established. They are recorded every 2 weeks by the same UDBG staff person and continue through the end of September. A report is produced by the end of the calendar year and is sent to the contributing company and posted to the UDBG Website.
http://canr.udel.edu/udbg/

INFORMAL EVALUATIONS

All America Selections Winners have already received their status through a rigorous program of national evaluation. Hence, they are not evaluated again in the UDBG Trials. However, they are included for general comparative and publicity purposes. Each label indicates the year in which a particular AAS winner received its recognition.

All other plants are evaluated on an empirical basis. Their performance is also conveyed on tours, visitor observations, and through posts on UDBG social media platforms (Facebook & Instagram).

The direct link to the Trials evaluation reports
http://canr.udel.edu/udbg/gardens-plants/garden-areas/landscape-trial/

Combined Proceedings IPPS 68:145-151. 2018.

The Effects of Air-Root Pruning on Seedlings of Species with Taproots

Brandon Miller and Nina Bassuk
School of Integrative Plant Science, Cornell University, Ithaca, New York 14853 USA

bmm244@cornell.edu

Keywords: Nursery production, root architecture, containers, *Aesculus*, *Carya*, *Juglans nigra*, *Nyssa*, *Pinus*, *Quercus*, *Ungnadia*, *Xanthoceras*.

Abstract

Cultural practices that influence the development of the root system shortly after germination can have a lasting impact on the quality of container-grown nursery crops. Standard plastic containers have been shown to promote crooks in primary roots and root circling, potentially leading to girdling roots, and overall decreases in plant quality. Taxa that exhibit a prominent taproot are particularly susceptible. This phenomenon poses a significant issue for species predominantly propagated by seed in container nurseries. Alternative container products that employ air-root pruning techniques may provide a solution, however, these products are often expensive and thorough investigations of their influence on coarsely-rooted taxa are lacking. This study explores the influence of air-root pruning, in comparison to culture in standard plastic containers, on vegetative growth responses and root architecture of seedlings shortly after germination of eight species of woody plants that exhibit taproot development. At the conclusion of the study, all air-pruned taproots lacked crooks or circling, whereas deflection of the taproots was observed in each of the controls. Significant differences in the vegetative growth responses were not observed.

INTRODUCTION

Woody plants that are claimed to be difficult to produce are often associated with the morphology of their roots (Burkhart, 2006). These root systems generally consist of coarse roots or a root system dominated by a taproot, which display minimal fibrous-root branching (Gilman, 1990a). As a result, these plants may exhibit reduced transplant success and a resistance to standard growing methods (Gilman, 1990a; Jacobs et al., 2009). Numerous taxa are noted for this attribute, including species belonging to the genera *Carya* (Dirr, 2009*)*, *Nyssa* (Stephens and Sutton, 2015), *Pinus* (Gilman, 1990a), *Xanthoceras* (Dirr, 2009), and others. Many of these taxa are important for use in urban landscapes as ornamental plants (Dirr, 2009). Interestingly, some common species utilized in managed landscapes, such as *Gymnocladus dioicus*, exhibit taproot development yet are amenable to traditional production and transplant techniques. This contrasting evidence suggests that our understanding of how taproots influence transplant success and the responses of taproots to manipulation and disturbance are poorly understood.

Alternative production practices such as root manipulation through careful root pruning (Carlson, 1974; Geisler and Ferree, 1984; Harris et al., 2001; McGraw and Smith, 1998) or the use of plant growth regulators (Crunkilton et al., 1994; Prager and Lumix, 1983) have been explored for use in modifying the root systems of some species deemed "difficult-to-transplant".

Root pruning has shown to be an effective tool to encourage root branching, create a denser root ball, and enhance shoot growth rates over time (Gilman, 1990b). Undercutting is one such technique employed in the field production of juvenile trees in the nursery. This practice utilizes a blade instrument to sever the taproot at a specific depth in the soil. Root pruning via undercutting has shown to alter root morphology and improve post-transplant survival for some taxa (Schultz and Thompson, 1990; Mullin, 1966). However, the responses of species belonging to different genera range broadly. For example, taproots of species belonging to the genus *Carya* are known to regenerate after pruning (Miller, 2017), whereas taproots of *Quercus* spp. generally do not.

Furthermore, some taxa respond differently to root pruning performed at various stages in development (Zhang et al., 2015). Other factors such as genetic variability within a species can also influence generalizations about growth responses to root manipulation (Kalliokoski et al., 2008). Studies that explore these techniques generally focus on plants past their first year of development. Experiments conducted to examine the effects of root manipulation shortly after germination, during initial root development, are lacking. However, the few studies that exist provide evidence that this area of research merits further study (Harris et al., 2001; Zhang et al., 2015).

While root-pruning techniques have been shown to be effective tools for manipulating the roots of some taxa, there are some disadvantages. Mechanical root pruning, especially in field-production nurseries, can act as a vector for soil-borne pathogens. In addition, these techniques are often not viable for application in container nurseries where the implementation of these practices would be more time consuming and expensive.

As alternatives, numerous container technologies taking advantage of air-root pruning have emerged. Air-root pruning is the use of containers with open, or highly-porous, surfaces. Upon root extension within these regions of the container, root tips are exposed to the air and desiccate. Desiccation of the root meristems is claimed to promote increased, proximal root branching and is also often claimed to eliminate serious root deflection within a container. It is suspected that increasing the number of roots within the container can ease transplant shock in taxa that are often difficult to transplant while simultaneously providing a higher quality plant without girdling roots. We questioned whether air-pruning techniques implemented shortly after germination could be effectively utilized to circumvent root-architecture issues with a diverse array of taxa that exhibit taproot development.

Our objectives were to:

- determine how air-root pruning influences root and shoot development compared to standard plastic containers,
- establish a baseline understanding of how unrelated taxa respond to air-root pruning, and
- quantify the effects of air-root pruning on morphological responses often associated with high-quality nursery stock.

MATERIALS AND METHODS

The species examined in this study include *Aesculus glabra*, *Carya glabra*, *Juglans nigra*, *Nyssa sylvatica*, *Pinus koraiensis*, *Quercus montana*, *Ungnadia speciosa*, and *Xanthoceras sorbifolium*. Seeds of *N. sylvatica* and *P. koraiensis* were purchased from Sheffield's Seed Company (Locke, New York). Seeds of *C. glabra*, *J. nigra*, and *Q. montana* were collected from wild populations in Ithaca, New York. Seeds of *A. glabra* and *X. sorbifolium* were acquired through the North Central Regional Plant Introduction Station (Ames, Iowa) through the National Plant Germplasm System. Seeds of *U. speciosa* were donated by The Huntington Botanical Gardens (San Marino, California).

Seeds were mixed with moistened, shredded peat and sealed in plastic bags. With the exception of *P. koraiensis*, seeds were stratified for 200 days in a cooler maintained at 4 °C. Seeds of *P. koraiensis* were warm stratified at 24 °C for 50 days followed by cold stratification at four degrees Celsius for 150 days.

After stratification, 25 seeds of each species were randomly assigned to a treatment (control or air prune). Seeds assigned to the control treatment were communally sown 2 cm deep into standard #7 nursery containers and seeds assigned to the air prune treatment were communally sown 2 cm deep in custom-made wooden containers measuring 61 cm × 30.5 cm × 15.24 cm (Figure 1).

Containers were constructed using cedar wood frames and the bottom of each container was composed of 0.635cm mesh wire. All containers were filled with 12.7 cm of LM6 potting medium. Containers were placed on a greenhouse bench in a completely randomized design. Plants were grown for 67 days. During this growing period, plants were irrigated twice weekly with a water-soluble fertilizer (21–5–20 plus Epsom) solution at a concentration of 300 ppm and once weekly with clear water. Taproots of seedlings grown in cedar containers were air pruned by growing out of the potting medium through the openings of the mesh bottom. Taproots of seedlings grown in standard #7 nursery containers grew to the bottom of the container and deflected, forming a "j" root.

Figure 1. Air-root pruning container.

At the conclusion of the growing period, seven seedlings were chosen at random for data collection. Data were gathered by washing the soilless media from the roots of each plant in order to measure the following growth responses and inspect the architecture of each root system. The location of the cotyledon scar, or in the case of seedlings of *P. koraiensis* the zone of transition from hypocotyl tissue to root tissue, was used as standardized locations for taking measurements of growth responses. Stem height was measured from the most distal shoot meristem to the standardized location. Stem caliper was measured at 2 cm above the standardized location using a digital caliper tool. Root length was determined for primary roots by measuring the distance from the standardized location to the most distal root meristem. First-order lateral roots were counted by hand. Any lateral roots less than or equal to 2 mm in diameter were considered. Five plants of each treatment combination were destructively harvested in order to measure leaf surface area, root mass, and shoot mass. All leaves were removed from each seedling and measured using a LICOR leaf surface area meter in order to determine leaf surface area (cm^2). Shoots and roots were oven dried and weighed to determine their respective masses. Shoots and roots were separated at the standardized location.

All data were analyzed using JMP Pro® 14 software (JMP® Version 14. SAS Institute, Inc., Cary, North Carolina, USA).

RESULTS AND CONCLUSIONS

All air-pruned taproots lacked crooks or circling (Figure 2), whereas deflection of the taproots was observed in each of the controls (Figure 3). According to Tukey's Honestly Significant Difference test, mean stem height, stem caliper, root length, number of first-order lateral roots, leaf surface area, root mass and shoot mass were not significantly different between controls and air-root pruned seedlings for each taxon (Table 1). Although the differences in first-order lateral root counts and root mass were not significant, it was clear that the development of these root systems were different between treatments. In the case of *C. glabra*, the fleshy taproot was not as long and significant extension of the first-order lateral roots had occurred in the air-pruned seedlings (Figure 4). Similarly, seedlings of *Pinus koraiensis* that were air-root pruned exhibited longer first-order lateral roots, however, the number of first-order lateral roots was not different from controls (Figure 5).

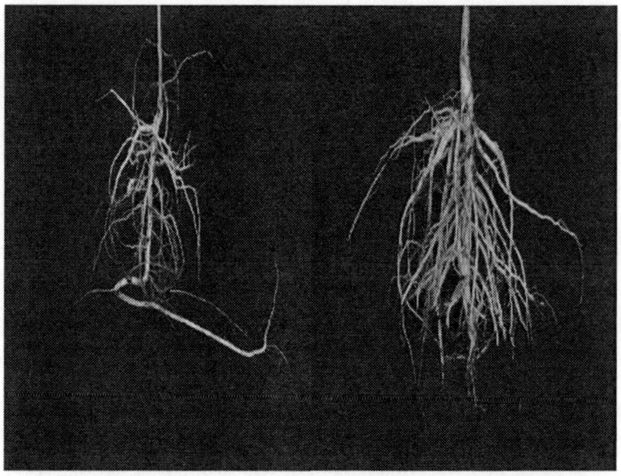

Figure 2. Roots of *Xanthoceras sorbifolium*. Control (left) and air-root pruned (right).

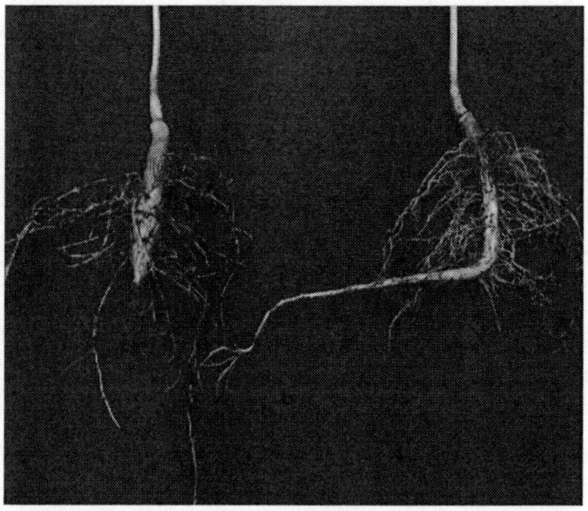

Figure 3. Roots of *Aesculus glabra*. Control (right) and air-root pruned (left).

Table 1. effects of air-root pruning on plant growth.

Taxon	Treatment	Stem Height[z] (cm)	Stem Caliper[y] (mm)	Root Length[x] (cm)	First-order lateral Root Count[w]	Leaf Surface Area (cm^2)	Root Mass (g)	Shoot Mass (g)
Aesculus glabra	Control	13.6	4.9	22.5	50.0	254.9	3.4	2.7
	Air Pruned	8.9	4.8	10.1	49.9	228.2	4.9	2.7
Carya glabra	Control	10.1	2.2	23.6	72.0	111.1	0.9	0.6
	Air Pruned	7.8	2.8	10.1	66.0	94.2	1.3	0.7
Juglans nigra	Control	41.9	6.1	37.4	75.0	1811.6	3.1	10.1
	Air Pruned	38.6	6.6	11.4	54.1	-	-	-
Nyssa sylvatica	Control	19.6	2.1	17.9	24.3	91.3	0.2	0.4
	Air Pruned	14.2	2.4	14.4	23.9	101.5	0.2	0.5
Pinus koraiensis	Control	6.7	1.7	12.8	10.7	11.0	0.1	0.3
	Air Pruned	5.4	2.0	10.4	13.0	12.9	0.1	0.3
Quercus montana	Control	21.1	2.2	30.7	51.3	207.7	0.6	1.6
	Air Pruned	18.6	2.3	9.5	40.4	199.7	0.6	1.3
Ungnadia speciosa	Control	15.8	2.9	28.4	48.9	358.5	0.7	2.6
	Air Pruned	19.5	3.0	11.6	33.0	578.1	1.2	4.5
Xanthoceras sorbifolium	Control	33.9	4.2	28.4	38.4	604.7	2.6	4.6
	Air Pruned	37.6	4.4	11.1	29.3	551.0	3.3	4.9

[z] Cotyledon scar to standardized location.
[y] Measured 2 cm above the standardized location.
[x] Standardized location to most distal root meristem.
[w] Fibrous first-order lateral roots (<2mm diameter)

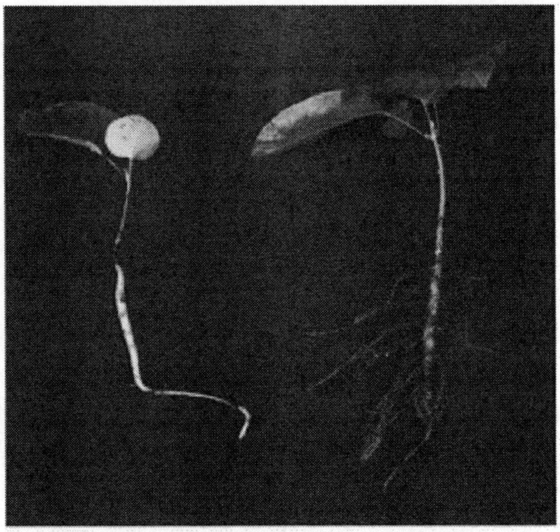

Figure 4. Seedlings of *Carya glabra*. Control (left) and air-root pruned (right).

Figure 5. Seedlings of *Pinus koraiensis*. Control (left) and air-root pruned (right).

Many of the species examined in this study exhibit determinate growth. We question whether this factor influenced the lack of variability in growth responses such as stem height. In a follow up study, we intend to grow replicate seedlings for a second growing season to determine if air-root pruning shortly after germination positively influences vegetative growth within this timeframe.

In this study the effects of air-root pruning on seedlings shortly after germination of eight species of woody plants exhibiting taproot development were examined.

By comparing unrelated taxa of horticultural interest, this study provides a broad baseline of responses for nursery growers to consider when determining if air-pruning techniques should be implemented. Because the air-pruning flats were of simple design, this study may also encourage nursery growers to develop their own air-root pruning implements in order to accommodate their cropping cycle, infrastructure, and budget. Based on these results, we conclude that air-root pruning flats should be considered for germinating taxa with taproots in order to reduce undesirable root architecture and promote the production of high-quality seedlings.

Literature Cited

Bauer, K., Grauvogel-stamm, L., Kustatscher, E., and Krings, M. (2013). Fossil ginkgophyte seedlings from the Triassic of France resemble modern Ginkgo biloba. BMC Evolutionary Biol. *13*:177.

Burkhart, B. (2006). Selecting the right container for revegetation success with tap-rooted and deep-rooted chaparral and oak species. Ecol. Restoration *24*(2):87–92.

Carlson, W.C. (1974). Root initiation induced by root pruning in northern red oak. Forest Res. Rev., Ohio Agr. Res. Develop. Center, Wooster, Ohio, USA

Crunkilton, D.D., Garrett, H.E., and Pallardy, S.G. (1994). Growth and ectomycorrhizal development of northern red oak seedlings treated with IBA. Hortscience *29*:771–773.

Dirr, M.A. (2009). Manual of woody landscape plants: Their identification, ornamental characteristics, culture, propagation and uses. Stipes, Pub. Champaign, Illinois, USA.

Dirr, M.A., and Heuser, C.W. (2006). The reference manual of woody plant propagation: From seed to tissue culture. 2nd ed. Timber Press. Portland, Oregon, USA.

Geisler, D. and Ferree, D.C. (1984). Response of plants to root pruning. Hort. Rev. *6*(67):155–188.

Gilman, E.F. (1990a). Tree root growth and development. I. Form, spread, depth and periodicity. J. Environ. Hort. *8*:215–220.

Gilman, E.F. (1990b). Tree root growth and development. II. Response to culture, management and planting. J. Environ. Hort. *8*:220–227.

Harris, J.R., Niemiera, A., Fanelli, J., and Wright, R. (2001). Root pruning pin oak liners affects growth and root morphology. HortTech. *11*:49–52.

Jacobs, D.F., Salifu, K.F., and Davis, A.S. (2009). Drought susceptibility and recovery of transplanted Quercus rubra seedlings in relation to root system morphology. Annals For. Sci. *66*(5):1–12.

Kalliokoski, T., Nygren, P., and Sievänen, R. (2008). Coarse root architecture of three boreal tree species growing in mixed stands. Silva Fennica *42*(2):189–210.

McCraw, B.D., and Smith, M.W. 1(998). Root pruning and soil type affect pecan root regeneration. HortTech. *8*:573–575.

Miller, B.M. (2017). The horticultural potential of six species of North American hickories. Iowa State University, Ames, Graduate Theses and Dissertations. Paper 15376.

Mullin, R.D. (1966). Root pruning of nursery stock. Forest. Chron. *42*:256–264.

Prager, C.M. and Lumis, G.P. (1983). IBA and some IBA-synergist increases of root regeneration of landscape-size and seedling trees. J. of Arboricult. *9*:117–123.

Schultz, R.C., and Thompson, J.R. 1(990). Nursery practices that improve hardwood seedling root morphology. Tree Planters' Notes *41*(3):21.

Sjöman, H., Hirons, A.D., and Bassuk, N.L. (2015). Urban forest resilience through tree selection—variation in drought tolerance in *Acer*. Urban For. & Urban Greening. *14*(4):858–865.

Stephens, M., and Sutton, M. (2015). Transplanting and a deeper look at "fall hazards". 15 January 2018. http://nysufc.org/transplanting-and-a-deeper-look-at-fall-hazards/2015/10/13/.

Sun, W.Q.1992. Quantifying species diversity of streetside trees in our cities. J. Arboricult. *18*(2):91–93.

Zhang, R., Peng, F.R., Yan, P., Cao, F., Liu, Z.Z., Le, D.L., and Tan, P.P. (2015). Effects of root pruning on germinated pecan seedlings. HortScience *50*:1549–1552.

Combined Proceedings IPPS 68:152-157. 2018.

Growing Plants in the Roaring Forties

Jill Reader
Lincoln University, P O Box 85084, Lincoln 7647, Christchurch, New Zealand

Jill.Reader@lincoln.ac.nz

Keywords: Nurseries, New Zealand, biosecurity, *Xylella*.

Abstract

New Zealand's first nurseries were established in the 1840s, at a time when there was both an influx of immigrants to the young colony and a worldwide thirst for new and exotic plants. By the 1880s many nurseries were importing large numbers of plants as well as producing their own stock, and there was a thriving trade in the export of native plants.

Few of those early nurseries remain. At present there are over 600 nurseries to cater for New Zealand's population of 4.7 million. Import and export of plants has almost halted. There have been a number of ongoing and recent challenges to the industry, including the ever-present threat of pest and disease incursions. Nevertheless, the nursery industry has a positive future, and growers have an important role to play in New Zealand's horticultural legacy.

INTRODUCTION

New Zealand is a country approximately the size of the United Kingdom, tucked into an isolated spot in the south west of the Pacific Ocean with its closest neighbour Australia lying over 1,700 km away. Much of the country lies within the 40s latitude, with the capital Wellington at 41°S. The term 'roaring forties' was coined by sailors in the 1800s who voyaged to the southern hemisphere and sailed into the roaring forties to catch the strong trade winds heading east.

New Zealand was the last habitable land mass to be settled, first by Maori in approximately the 1200s, followed by European colonization from the early 1800s onwards. Immigration by Europeans in earnest started in the 1840s, when advertisements to potential immigrants portrayed the country as "a veritable Garden of Eden, a place teeming with abundance and capable of further abundance under the improving hand of the settler" (Bradbury, 1995).

The 1840s in England became known as the 'Hungry Forties' due to large scale unemployment, so it is not surprising that many people were attracted by the prospect of a new life in New Zealand. Between 1840 and 1855 thirty thousand immigrants boarded sailing ships for a perilous voyage to the other side of the world to make a new life for themselves and their families in New Zealand.

Around the same time as these thirty thousand immigrants were settling into their new lives, an interest in exotic plants was sweeping the world. Plant hunters were introducing large numbers of plants into cultivation from many corners of the globe. This time was one of the greatest floods of plant introductions in history, an international event which carried on well into the twentieth century, and New Zealand was a participant in this event (Bradbury, 1995).

The English had already had a "heads up" many years before about the unfamiliar nature of New Zealand's native plants. When James Cook sailed the Endeavour to New Zealand in 1769 he was accompanied by Joseph Banks, who had paid the astronomical sum of £10,000 (as a comparison, the Endeavour cost under £4,000 to build) to be allowed to join the expedition and collect plant specimens. Banks had his good friend Daniel Solander, a top student of Carl Linnaeus, accompany him to help with the plant collecting. They returned to England and to Kew with hundreds of dried plant specimens, plus many notes and drawings.

NEW ZEALAND'S FLEDGLING NURSERY INDUSTRY

The first nurserymen arriving in New Zealand were part of the early wave of immigrants in the 1840s. Most had developed their skills through apprenticeships as gardeners on large English estates and were keen to try their skills in a new country. At that stage New Zealand was a young colony, and in many areas of the country there was a need for immigrants to establish a vegetable garden or risk going hungry. However, it was the nurserymen who were skilled in grafting fruit trees, and since New Zealand's native flora lacked edible plants, fruit trees were one of the main crops that helped establish nurseries. Shelter belt trees were also grown as soon as seed was available, particularly for the windy Canterbury Plains. Gorse (*Ulex europaeus*) seed was imported and grown in vast numbers for fences. Unfortunately, the gorse grew so well that it is now the worst pest plant in New Zealand.

Until 1880 New Zealand was a struggling colony, but with both a gold rush and the advent of refrigerated shipping its fortunes grew. Once there was more cash flowing through the economy, many businesses grew. This was true for the many of the nurseries, and with faster shipping there was a good trade importing plants – and having them arrive in good condition. Plants were imported from Australia, England, Europe and Asia on a regular basis.

In his book 'Pioneer Nurserymen of New Zealand' Allen Hale describes the huge range of exotic plants the nurseryman Thomas Mason grew in his garden by the 1870s: "250 named rhododendrons, 60 named camellias, hundreds of azaleas, while his collection of Japanese maples and tree peonies is unsurpassed" (Hale, 1955). For people who had the funds to spend, there was a vast range of exotic plants to choose from, and the temperate climate meant that most species thrived. It followed that by default New Zealand's isolation led to it becoming a repository for some rare or endangered exotic plants.

Around 80% of New Zealand's 2,500 native plants are endemic at the species level, increasing to 93% endemism for alpine species (Mark, 2012). Because these plants were a little different, overseas plant collectors were keen to add some to their collections. Many thousands were exported; hebes, pittosporums, dodonaeas, flaxes and tree ferns were especially popular.

Duncan and Davies Nursery in New Plymouth, the biggest nursery in the Southern Hemisphere by the 1940s, exported both native and exotic plants. The rich volcanic soils and good rainfall in Taranaki were ideal conditions for species such as rhododendrons to thrive, and in the 1930s the nursery was exporting species such as magnolias, rhododendrons and viburnums to Australia (Jellyman, 2011). They also grew a huge range of plants for the domestic market. There were many other smaller nurseries, mostly growing exotic plants. However, over time there developed a big demand for native plants. Revegetation contracts, lifestyle block numbers increasing, and farm riparian plantings saw growers specialize in eco-sourced plants, with most sold as liners.

MAJOR CHALLENGES IN THE PAST TWENTY YEARS

Some of the challenges listed below have been around for more than twenty years. In particular, the lack of interest in horticulture as a career has been an ongoing problem both in New Zealand and overseas. Thirty years ago an article published in Commercial Horticulture quoted a visiting University of New Hampshire Professor:

"The greatest concern, however, is the falling numbers of horticultural students at universities. Students are no longer interested in horticulture and agriculture because of poor Press and low perception of the trade. They are more interested in business and management courses and becoming quickly solvent after completing their education. A student can work for McDonald's (hamburgers) and command $6–7 per hour whereas the nursery industry can only offer $3.20. That is the real problem that faces the nursery industry today" (Routley, 1988).

Major challenges:

- Expansion of the two box stores, which now control around 60% of the market. The plant range in these stores is limited, and specialist growers are hurting.
- The spread of cities and towns and high real estate prices means it can be uneconomic to use land for nurseries or garden centers.
- A lack of demand for established nursery businesses when they come up for sale.
- A lack of interest by new generations in established family businesses.
- A lack of interest in horticulture as a career. Few high schools promote horticulture as a career, and there is a limited range of horticultural qualifications available in New Zealand.
- A lack of skilled labour.
- A diminishing pool of expertise in scientific/academic areas. As people are retiring it is becoming increasingly difficult to find young qualified staff to replace them.

In addition, New Zealand has a major challenge to industry growth due to limitations with the Plants Biosecurity Index (PBI), a database that is used as a reference for plant importation. The PBI has two Acts of Parliament governing it, the Biosecurity Act 1993 and the Hazardous Substances and New Organisms Act 1996 and is used as a working index for import standards for 'Seeds for Sowing' and 'Nursery Stock'. Unfortunately, the major challenge is due to there being an estimated 10,000 species present in New Zealand in July 1998 that were omitted from the list. Historically New Zealanders have imported many plants, and there are more exotic species in New Zealand than native species. Unfortunately, many of these historical imports were recorded at the genus level only, and so would not be included in the PBI due to a lack of species information.

As a result, for the past twenty years growers and plant breeders have been unable to import a plant not on the list without an extensive environmental risk assessment costing approximately $30,000 per species. Amendments can be made to the PBI if the plant species is proven beyond doubt to be present in New Zealand but instances of this are rare, and the onus and cost lie with the person or company applying for the addition to the PBI.

NURSERY NUMBERS IN 2018

Few of the early nurseries remain. 2018 figures for established/recognized growers are:

- 245 wholesale nurseries
- 391 wholesale and retail nurseries

These 645 nurseries account for more than 80% of all production (Snell, pers. com.) In addition, many private or open gardens grow their own specialty lines, and there is an increasing trend for landscapers to grow for their own use. Small growers selling through markets and Trade Me (New Zealand's equivalent of EBay) are ubiquitous and seem to be increasing in number and volume.

Nursery size is known for 407 of the established growers. Sixty nurseries are based on less than 1 hectare, with the large majority (259) on between 1 and 5 hectares. Sixty-eight growers have nurseries between 5 and 19 hectares, while 15 have a nursery size between 20 and 50 hectares. Five nurseries cover more than 90 hectares.

INTO THE FUTURE

Biosecurity

In 2016 New Zealand Plant Producers Incorporated (NZPPI) was formed. This industry body incorporates the sectors of horticulture, viticulture, forestry, retail, amenity, landscape and revegetation, providing a platform for R&D plus a united voice for the increasing horticultural biosecurity issues New Zealand is facing. Biosecurity problems at present include:

- Myrtle rust, first found in New Zealand 2016, and despite an initial intensive eradication effort by MPI (Ministry of Primary Industries) measures have since been moved to a control plan rather than eradication. New Zealand's iconic Christmas tree, the pohutukawa, is in the Myrtaceae family and is a host. Department of Conservation staff have been collecting seeds and seedbanking from all native plants in the Myrtaceae family as a precaution.
- *Xylella fastidiosa.* The escalated threat of this disease has led to increased border security measures. It has a wide host range of over 300 species, and apart from potentially devastating many ornamentals here it is a big threat to New Zealand's thriving wine industry.
- Pests such as the brown marmorated stink bug and fruit flies. These are not directly an issue with ornamentals; however New Zealand's economy is reliant on its primary industries and clean, green image. Tighter border security has meant increased difficulty importing plant material.

A positive outlook

While the challenges and biosecurity threats outlined above paint a negative picture, there is a positive spirit within the industry here which bodes well for the future. The box stores may control much of the market, but their garden centres are busy. Nurseries growing native plants for riparian plantings and lifestyle blocks find it hard to keep up with the demand. Many nurseries offering online sales to the public are finding that this portion of their business is increasing each year. Nurseries that have a wide customer base and a diverse range of plants are generally doing well.

Figure 1. The retail area at Southern Woods Nursery.

Southern Woods Nursery, just south of Christchurch, is an example of a business that has found its niche market. They produce most of their own stock, supply plants and information to local lifestyle block owners, have a retail area with a large selection of native and exotic plants, and have an online sales option. While online sales have equated to 15% of their business this year, this figure is expected to rise to about 20% next year (Mannall, pers. com.)

Finally, a discussion about New Zealand's nursery industry would not be complete without mentioning the important role the IPPS has contributed. New Zealand became a Region almost 50 years ago (the 50th celebration is coming up in 2021). Its 220 members are enthusiastic, conferences and field days are well attended, and the society's motto of 'Seek and Share' is evident at every event. It has truly been an integral part of the industry here.

Literature Cited

Bradbury, M (ed.) (1995). A history of the garden in New Zealand. Penguin Books New Zealand.

Hale, A.M. (1955). Pioneer nurserymen of New Zealand. A.H. & A.W. Reed Ltd, Wellington, New Zealand.

Jellyman, A. (2011). The growing world of Duncan and Davies. A Horticultural History 1899–2010. The Sir Victor Davies Foundation, New Zealand.

Mannall, R. Business Development & Marketing Manager, Southern Woods Nursery, 1002 Robinsons Road, Christchurch, New Zealand.

Mark, A.F. (2012). Above the treeline, a nature guide to alpine New Zealand. Craig Potton Publishing, Nelson, New Zealand.

Routley, D. (1988). There are lessons to be learned for NZ in US cut flower experience, says professor. In Commercial Horticulture, April, 1988, pp 14–15. Reference Publishing, The Auckland Printing Co.

Snell, D. Editor, Commercial Horticulture, Reference Publishing, The Auckland Printing Co., Auckland, New Zealand

Combined Proceedings IPPS 68:158. 2018.

Inducing Mutations In Vitro in Chaste Tree (*Vitex Agnus-Castus*) with the Herbicide Oryzalin (Surflan)

N.K.A. Nor Hisham Shah and Mark Bridgen
Cornell University, School of Integrative Plant Science, 3059 Sound Ave., Riverhead, New York 11901 USA

mpb27@cornell.edu

Keywords: Breeding, mutation breeding, woody plants, tissue culture.

INTRODUCTION

Plant mutations are very useful in a breeding program because they are sources of novel phenotypes that could lead to the development of new cultivars. There are several chemical, physical, and molecular techniques that are available to mutate plant materials *in vitro*. This research documents the effects of the mutagen, oryzalin [4-(dipropylamino)-3,5-dinitrobenzenesulfon-amide], available as the herbicide Surflan, on tissue cultured plants of *Vitex agnus-castus*.

MATERIALS AND METHODS

In this study, in vitro cuttings were exposed to oryzalin at three different concentrations, 0.3 mM, 1.3 mM and 2.8 mM, with three different exposure times, 30 minutes, 60 minutes and 90 minutes. Observations of the treated plants were taken for 16 weeks. Morphological changes and the number of cuttings survived were recorded to determine the lethal dose of oryzalin to *Vitex*.

RESULTS

It was found that the lethal dose of oryzalin that causes 50% of death to *Vitex* in tissue culture to be at 0.3 millimolar of oryzalin at 30 minutes of exposure time. Higher concentration or longer exposure time increased the death percentage of the cuttings. After 16 weeks, the surviving explants exposed to oryzalin were maintained in tissue culture for further growth and propagation. Plants that were treated are being field tested to evaluate plant growth and flowering.

This experiment demonstrated the effects of concentration and exposure time of oryzalin on *V. agnus-castus*. The results were as expected: the increase in concentration or the increase in exposure time increased the mortality of plants. The result of this experiment will enable the use of oryzalin for chemical mutagenesis in future research by providing the predicted mortality associated with the dosage of oryzalin. This chemical, when exposed to plants in vitro, can be used to create mutations in plants.

IPPS Vol. 68 – 2018

Combined Proceedings IPPS 68:159-162. 2018.

Pit Greenhouse Construction and Operation

Harold E. Sweetman
Executive Director, Jenkins Arboretum and Gardens, 631 Berwyn Baptist Rd, Devon, Pennsylvania 19333 USA and Owner, Octoraro Farm Nursery LLC, 698 Lees Bridge Rd, Nottingham, Pennsylvania 19362 USA

hesweet@verizon.net

Keywords: Propagation structure, history, propagation.

INTRODUCTION

Older greenhouse concepts from many decades ago or even past centuries seem new today and the benefits are many. Previous generations of nurserymen and estate gardeners embraced the concept of earth shelter structures and used them with great success (Figure 1). With increasingly common and complex energy crises in recent memory, there should be greater energy awareness.

In addition to the environmental stewardship, savings on energy costs can translate into greater profitability. Using new double-wall glazing materials, the energy benefits of being up to 6-ft below grade can be quite significant. This project was not a scientific study of energy efficiency, but an example of how common-sense principles can be applied simply and economically, resulting in a successful propagation program that is scalable to any size nursery.

Figure 1. Earth shelter structure at Summer Hill Nursery.

MATERIALS AND METHODS

Timing

If the timing of construction is in the fall, there may be an opportunity to effectively use nursery staff as nursery activities decrease and the skill levels required are modest if the plans are well conceived and there is competent construction leadership. It is important to close in the structure before the onset of winter. Much of the interior work can continue regardless of weather and at very comfortable indoor temperatures. It is likely that there can be active propagation the first winter of construction with the first crop after only 4 months.

Depth of the pit excavation

A pit greenhouse is a structure that is earth sheltered either by digging a deep pit (Figure 2A) or a shallower pit with soil used for berms. If the site is nearly level, then the shallower pit is quite practical because the excavated soil does not need to be removed but can be used to berm the walls thus creating the benefits of a deeper pit.

Wall construction

Once level footings are in place, 8-in. concrete blocks are stacked dry without mortar joints (Figure 2B). A surface bonding cement specified for this method of wall construction is applied to both sides of the wall and used for bedding the first course of concrete block. This type of cement builds a very strong wall and is also quite effective at waterproofing. The concrete blocks can be strengthened further by filling some of the cores with cement and rebar though this was not done in this construction project. Polystyrene panels 2-in. thick were applied exteriorly for insulation to a depth of 4-ft from sill to below grade.

Figure 2. (A) Pit greenhouse is a structure that is earth sheltered, (B) Once level footings are in place, 8-in. concrete blocks are stacked dry without mortar joints.

Subfloor stone heat storage bed for increased efficiency

Efficiency can be increased further if air is circulated through perforated 4-inch ducts in a subfloor stone bed of 3-in. stones (Figure 3A). Even on cloudy days there can be a "greenhouse effect" and the extra energy can be stored. A thermostat turns on a fan when there is extra heat and stores it under the floor and the same fan releases the heat at night. Because of the constant heat from being below grade, it may even be possible to skip an alarm system as freezing temperatures are not likely to occur with good construction. General guidelines from the passive solar building industry for subterranean heating recommends a total length of perforated pipe ducting to be half of the floor area. The sizing of the fan should allow about five air exchanges per hour. For this project, a cage fan used for inflating double-wall poly film greenhouses has worked well as they were designed for continuous operation.

Supplemental heating

This project included 4 independent zones of hydronic heating pipe from a manually controlled manifold supplying warm water for radiant heat in the floor and under the propagation bench (Figure 3B). A small 30-gal electric hot water heater with a thermostatically controlled circulating pump provides adequate root zone heating at any level desired. Heating only the plant root zone and not the air adds to energy efficiency.

Figure 3. (A) Efficiency can be increased further if air is circulated through perforated ducts in a subfloor (left), (B) project included 4 independent zones of hydronic heating pipe (right).

DISCUSSION

Because of excavation and greater material expense, pit greenhouses are more expensive than simple hoop-house construction. It comes down to: "You can pay a little more now or a lot more later." Energy costs are always increasing and only a few feet below ground the earth remains a constant temperature year-round of 55-58 °F in the Mid-Atlantic region. If the structure is used for propagation of seed and cuttings and not growing, the greater cost of construction is offset by the high value of the propagules. The density of plants per square foot is very high. As an example, standard Anderson deep flats may hold over 100 cuttings. The calculation to estimate the number of years to recoup the cost of construction can be quite simple based on unit cost to purchase propagules versus on site pit propagation. In the project used in this article; the payback is estimated to be less than 3 years.

This modest 12 ft ×28 ft pit greenhouse example has about 300 sq. ft. of growing space (Figure 4A). Results from the first year of production has shown that the number of cuttings per sq. ft. can range from 25–50 resulting in the potential of 7,500–15,000 propagules (Figure 4B). The cost of construction in this example was under $10,000 due to the free labor associated with owner equity. In this modest example, the cost of construction may be recovered in 2–3 years.

In addition to the cost savings of not having to purchase plants propagated from other sources, nursery production can be nimbler with availability and sales. Having control of propagation can allow nurseries to specialize in a niche markets by producing plants that are often unavailable.

Figure 4. (A) Finished pit greenhouse, (B) propagules rooting.

SUMMARY

Although this was not a scientific study, the success of the construction design can be determined by successful plant propagation. Every situation is unique and success in propagation is largely dependent on the propagator and attentive administration. The improved economy of energy efficiency is universal and quite remarkable. There is not a single month throughout the year that plants cannot be propagated. Working in the pit greenhouse in the depths of winter sheltered by the warmth of the earth and sun is a special pleasure for the propagator.

CONCLUSION

Older greenhouse designs have stood the test of time. If the pit greenhouse is used primarily for seed and cutting propagation rather than production growing, the return or profitability of construction costs can be recouped in a very short period of time. A simple calculation of the density of cuttings per unit area times the cost per propagule if purchased and the math become apparent. Pit greenhouses are scalable and may be ideal for many different types of nurseries. Due to the low-tech concepts, pit greenhouses may be a good fit for small or startup nurseries.

Combined Proceedings IPPS 68:163-173. 2018.

New Plant Forum – Eastern Region IPPS

Charles Tubesing, Moderator
Holden Arboretum, Ohio USA

ctubesing@holdenfg.org

PRESENTER

Brent Horvath
Intrinsic Perennial Gardens, Inc.,
Hebron Illinois, 60034
BrentH@intrinsicperennialgardens.com

Prunella vulgaris 'Magdalena' PPAF

Heavy blooming lavender purple bi-color flower clusters begin in July on 6 in. plus stems above notched dark green foliage (Figure 9). The spreading clumps reach 1 ft wide plus. In August and September some re-bloom keeps this larger form interesting. Semi-evergreen foliage has some purple highlights in the cool temperatures of fall and spring. The purple seed heads also remain attractive once the flower fade. These adaptable plants prefer some moisture in sun but are tolerant of dry conditions once established. Easy to propagate from cuttings.

Pennisetum alopecuroides 'Yellow Ribbons' PPAF

A golden yellow pennisetum with thin foliage on compact plants growing 24 in. tall and wide. The color mellows to yellow going into fall. Easy from division.

Sedum 'Lime Joy' PPAF

Vigorous hybrid plants have a lime colored 4 in. buds starting in July, by September the 8 in. flower heads combine to create a large dome of bi-color wine pink flowers (Figure 10) held on strong stems reaching 15 in. Easy from cuttings and division but also propagates from leaf cuttings. One seedling in 2017 yielded 300 plants in 2018.

PRESENTER

James R. Ault, Ph.D.
Director of Ornamental Plant Research, Chicago Botanic Garden, Glencoe, Illinois 60022
jault@chicagobotanic.org

Panicum 'Bad Hair Day' PP29,313

Bad Hair Day switchgrass is an intermediate to large-sized grass growing to 71 in. tall by 88 in. wide in full inflorescence after 3 years in trial in northern Illinois (USDA Zone 5). Its stems are strongly upright, giving the plant a narrow silhouette until midsummer. The grayish, yellow-green foliage leaves are mostly pendulous, creating an attractive waterfall effect. The late-summer inflorescences are the most unique feature of this selection, as they are much denser than other switchgrasses, and also cascade like the foliage. The weeping inflorescences sway in the wind, creating a whimsical mop head appearance that gave it its "bad hair day" name. The cascading foliage and inflorescences persist well into autumn, turning an attractive light yellowish brown after going dormant. Plants have been strongly upright and persistent all winter, even with repeated snow loads. To our knowledge, this is the first switchgrass cultivar in the marketplace that is a hybrid of the bitter switchgrass (*Panicum amarum* 'Dewey Blue') and of common switchgrass (*P. virgatum*). *Panicum amarum* is native to seashore sand dunes along the East Coast and across the Gulf states; combined with the attributes of *P. virgatum*, our selection is not only heat and cold tolerant, but may be more drought and salt tolerant than most switchgrasses. Readily propagated by division of the culms in spring. Likely hardy to USDA Zones 3-9.

Vernonia 'Summer's Surrender' PP28,475

'Summer's Surrender' ironweed was selected in 2013 from a 2010 cross of *Vernonia lettermannii* and *V. arkansana*. It inherited the bushy habit of *V. lettermannii* and the more robust plant size, larger foliage and larger capitula of *V. arkansana* (Figure 12). 'Summer's Surrender' is no shrinking-violet perennial, as it matures into a robust yet uniform and dense broad mound. Three-year-old plants measured 48 in. tall and 83 in. wide and a 5-year-old plant measured 48 in. tall and 74 in. wide. With excellent resistance to both powdery mildew and rust, the foliage remains clean and attractive all summer and fall. The olive-green leaves grow to 5 in. long and half-an-inch wide. Dark purple florets are packed into nearly 1-inch-wide flower heads which are borne in profusion on airy inflorescences from early September to early October in northern Illinois (USDA Zone 5). A diversity of butterflies, moths, and bees are attracted to the flowers. Plant in full sun and reasonably well drained soil. It has proven adaptable to both moist and to drier soil. Use this selection in the back of larger perennial or mixed borders, near lake edges and other moist sites, in pollinator gardens, and in any situation that calls for a larger yet attractive and uniform perennial plant. Readily propagated by shoot tip cuttings taken in early June in USDA Zone 5 and treated with 1,500 ppm KIBA. Likely hardy to USDA Zones 4 to 8.

Vernonia 'Summer's Swan Song' PP28,556

'Summer's Swan Song' ironweed was selected in 2012 from a cross made in 2010 of *Vernonia lettermannii* and *V. angustifolia* 'Plum Peachy'. This was the first of our hybrid seedlings to catch our attention, and it never failed to impress in the subsequent trials. Similar in adaptability, bushiness and fine texture to the popular selection *V. lettermannii* 'Iron Butterfly', 'Summer's Swan Song' grows somewhat larger, has larger capitula, and never lodges due to the elongated and interlocking inflorescence branches (Figure 13). The stems, leaves, and inflorescence branches also take on a dark red cast in full sunlight, adding to its display. The third-inch-wide capitula with their dark purple florets are born in profusion for 5 to 6 weeks from early September to mid-October in northern Illinois (USDA Zone 5). A diversity of butterflies, moths, and bees are attracted to the flowers. The fine foliage is a moderate to dark olive green and measures upwards of 5 in. long and less than a ¼ inch wide. Three-year-old plants of 'Summer's Swan Song' measured 28 in. tall by 35 in. wide and 5-year-old plants measured 31 in5 tall by 43 in5 wide. It has proven adaptable to both moist and to dry soil and has also exhibited excellent resistance to rust and powdery mildew. Use this selection individually or massed in perennial or mixed borders, in pollinator gardens, near lake edges and other moist sites as long as well drained, and in any situation that calls for a midsized attractive and uniform perennial plant. Readily propagated by shoot tip cuttings taken in early June in USDA Zone 5 and treated with 1,500 ppm KIBA. Likely hardy to USDA Zones 4–8.

PRESENTER

Brigitte Crawford
North Creek Nurseries, Inc.,
Landenberg, Pennsylvania USA
brigitte@northcreeknurseries.com

Earlibeauty® Series

The Earlibeauty® series of phlox was introduced by The Primrose Path (Charles and Martha Oliver). The selections are all hybrids of *Phlox* species that are native to the eastern United States. They have been free of powdery mildew in North Creek Nurseries' trial gardens, range from about 16–36 in. tall, and are hardy to USDA Zone 4. There are five cultivars in the Earlibeauty® series and more coming. North Creek is highlighting our three favorites.

Phlox Earlibeauty® 'Daughter of Pearl' PP27267

Phlox Earlibeauty® 'Daughter of Pearl' has white flowers with a small purple blush in the center (Figure 6); flowers form in a conical shape. Its name comes from the fact that 'Minnie Pearl' is in its parentage. It blooms in mid or late June into July, gets about 30–36 in. tall, and has an upright habit similar to that of *Phlox paniculata*.

Phlox Earlibeauty® 'Rose Bouquet' PPAF

Phlox Earlibeauty® 'Rose Bouquet' has pink flowers (Figure 7) and has its main flush of blooms in May, then sporadically blooms through the summer. It gets 16–20 in. tall and spreads into a large dense mass.

Phlox Earlibeauty® 'Solar Flare' PP27265

Phlox Earlibeauty® 'Solar Flare' has bicolor flowers (Figure 8) that are white with pink centers and form in rounded panicles. 'Solar Flare' flowers from late May to June, reaches a height of 24 in., and has a round, mounding, clump-forming habit.

Chrysogonum virginianum
'Quinn's Gold' PP19306

Chrysogonum virginianum is a low growing, shade loving groundcover. It has a long bloom period of yellow flowers that stand out against its dark green foliage. 'Quinn's Gold' was introduced to the trade via Sunny Border Nurseries and is unique in that its flowers exhibit shades of dark yellow, light yellow, and creamy white. It is hardy to USDA Zone 5.

PRESENTER

Tim Wood
Spring Meadow Nursery, Inc.,
Grand Haven, Michigan, 49417 USA
Tim@SpringMeadowNursery.com

Aronia melanocarpa 'UCONNAM012', GROUND HOG™ black chokeberry ppaf, cbraf

The perfect landscape plant for covering ground or filling in difficult spaces such as parking lot beds or areas you don't want to mow (Figure 1). This dwarf aronia grows as a thick, dense mat, eliminating the chores of weeding and mulching. Glossy foliage, white spring flowers, dark purple berries in summer, and autumn foliage hues of orangish-red make this native plant as showy as it is adaptable. Developed by Dr. Mark Brand at University of Connecticut. USDA 3, AHS 9, height: 8-10 in. tall and 3 ft wide. Spring blooming.

***Philadelphus coronarius* 'SMNPVG', ILLUMINATI TINY TOWER™ mockorange ppaf, cbraf**

We've been breeding and evaluating *Philadelphus* for over 15 years but have not discovered anything remarkable—until now. A truly unique and striking shrub with dense, upright branching (Figure 5), and notable dark green foliage. Stems are punctuated in late spring with vertical rows of highly fragrant flowers, creating a four-sided tower effect. USDA 4, AHS 7, height: 3–4 ft tall and 1.5 ft wide, late spring bloom.

***Taxus ×media* 'SMNTHDC', STONEHENGE™ yew ppaf, cbraf**

Think of this evergreen as a thin, fast growing 'Hicksii' (Figure 11). Perfect for hedging or foundation plantings where a narrow plant works best. It has been in our field trial for over five years and has never needed pruning, shearing, or shown signs of winter burn. This shrub is narrow enough to be useful and interesting, thick enough to look good in a container, and fast-growing enough to be profitable. USDA 5, AHS 7, hieght: 8–10 ft tall and 2–4 ft wide.

***Weigela florida* 'Verweig8', MY MONONET PURPLE EFFECT™ weigela ppaf, cbraf**

Hues of cream, green, and pink, layered with a cast of burgundy-purple, make the foliage of this new dwarf, variegated selection noteworthy (Figure 14). From the renowned plantsman Bert VerHoef, it is exceptionally strong flowering for a dwarf, delivering purple-pink flowers in mid-late spring. A companion to My Monet, it is a tad faster growing and more heat tolerant. The foliage looks good right up until frost, giving new energy to this long-time favorite. USDA 4, AHS 6, height: 1.5-2.5 ft, spring blooming.

Combined Proceedings IPPS 68:174-178. 2018.

Engaging Elementary Students into Horticulture with Cooperation of Master Gardeners Through Multidisciplinary Approaches in Rural Kentucky

[1]Zenaida Viloria, [2]Winston Dunwell, [3]Ric Bessin, [4]Edwin Ritchey, [1]Raul Villanueva, [2]Daniel Becker and [5]Amanda Martin

[1]Department of Entomology; [2]Department of Horticulture; [4]Department of Plant and Soil Sciences and [5]Regualtory Services, University of Kentucky, Research and Education Center-Princeton, Kentucky; and [3]Department of Entomology, University of Kentucky-Lexington, Kentucky USA.

zenaida.viloria@uky.edu

Keywords: Education, botanic garden, pollinators, educators.

INTRODUCTION

The UKREC Botanical Garden is a 5-acre setting located at the University of Kentucky-Research and Education Center, Princeton. It was created in 1980 to evaluate and select superior environmentally sustainable plants for enhancing Kentucky's environments and landscapes. The garden is visited by master gardeners, extension county agents, students and residents of nearby communities. Being an enclave in a rural region, with limited resources; the garden has the potential as a learning center to teach science-based knowledge and outreach. This project has offered hands-on activities for fourth and fifth grade students of Caldwell County and Lyon County Schools. Topics included plant diversity in different categories (natives, invasive, ornamentals, small-fruit crops, and vegetables), insect-plant and plant-soil interactions, and the importance of environmental protection for a sustainable future. Science teachers and University of Kentucky Faculty and Staff have prepared, organized and delivered planned activities. These events scheduled in March-May and August-October coincide with the school year. The students will be aware of how plants affect their lives and *vice versa*. An extension agent and master gardeners collaborated during the students' visits. Training classes offered to enhance Master Gardeners' knowledge. The botanical garden scenery offers students an open classroom to learn through a direct contact with the different stages of plant development and interactions with its surrounding.

OBJECTIVES

The objectives are to foster scientific interest and curiosity in elementary students' minds to explore and protect their environments and be active life-long supporters of horticulture demonstration gardens in their schools for horticulture education and healthier nutrition.

SPRING VISITS TO THE UKREC BOTANICAL GARDEN

Caldwell County Elementary School (CCES) has six classes per grade level, fifth graders visited the UKREC Botanical Garden on April 11 and May 14, and fourth graders on April 12 and May 15.

Lyon County Elementary School (LCES) has three classes per grade level, all classes visited the garden on April 17 and May 21. Students were grouped such as each visit had about 75 students, two visits were held a day: 9:00–11:00 AM and 12:00–2:00 PM for CCES.

ACTIVITIES

Instructors and Master Gardeners received students and their science teachers and divided each group of 75 students into five subgroups. Each subgroup was assigned to a working station and every 20 min, they rotated until finished visiting the five stations.

Hands-on activities were delivery on diverse agricultural topics:

- April: Soil types and conservation; soil biology: compost and vermicomposting, carbon footprint calculation; plant hunter (plant morphology); planting vegetable seeds and seedlings (Figure 1).

Figure 1. Earthworms for vermicomposting.

- May: Plant diseases: pathogen and symptom description; flowers and pollinators; mimicry and camouflage; insects: pests vs natural enemies; raised bed vegetable garden (students observed crop development after the first month) (Figure 2).

Figure 2. Pollinators are more than Monarch butterflies and honeybees.

School on-site Activities

- Caldwell and Lyon County Elementary schools were provided with four raised beds; 4 ft (1.22 m) wide by 8 ft (2.44 m) long by 12 in. high (30.48 cm).
- Lyon County Elementary school added 8 raised beds to provide the gardening experience for all classes (Figures 3 and 4).

Figure 3. Students planting raised at beds LCES.

Figure 4. Multiple raised beds at the LCES.

SURVEY

At the end of the second spring visit, teachers (17 CCES, 6 LCES) and students (300 CCES, 150 LCES) were surveyed to assess their experiences and receive their input to plan fall activities. Evaluation sheets were presented to students and teachers and the results are in Figure 5. A majority of the students, up to 86.2%, have gardened. Plant hunter morphology was the least favorite among the science teachers. Fourth grade teachers from both schools considered that activities on pollinators and soil fit their curriculum in large extend. Carbon Foot Print calculator activity did not match CCES curricula for fourth and fifth grades, besides this session evidenced that a very low number of students recycle at home, and recycling is barely done at the schools. It might have been the reason for those students evaluated it as the least favorite of all the learning sessions except for the Lyon County 4th grade. The insect mimicry and camouflage were highly ranked by students and teachers. In general, students enjoyed learning most of the activities; particularly LCES 4th graders. LCES science teacher is very active and committed to use gardening as a tool to complement her lectures and teach students to work and maintain their garden.

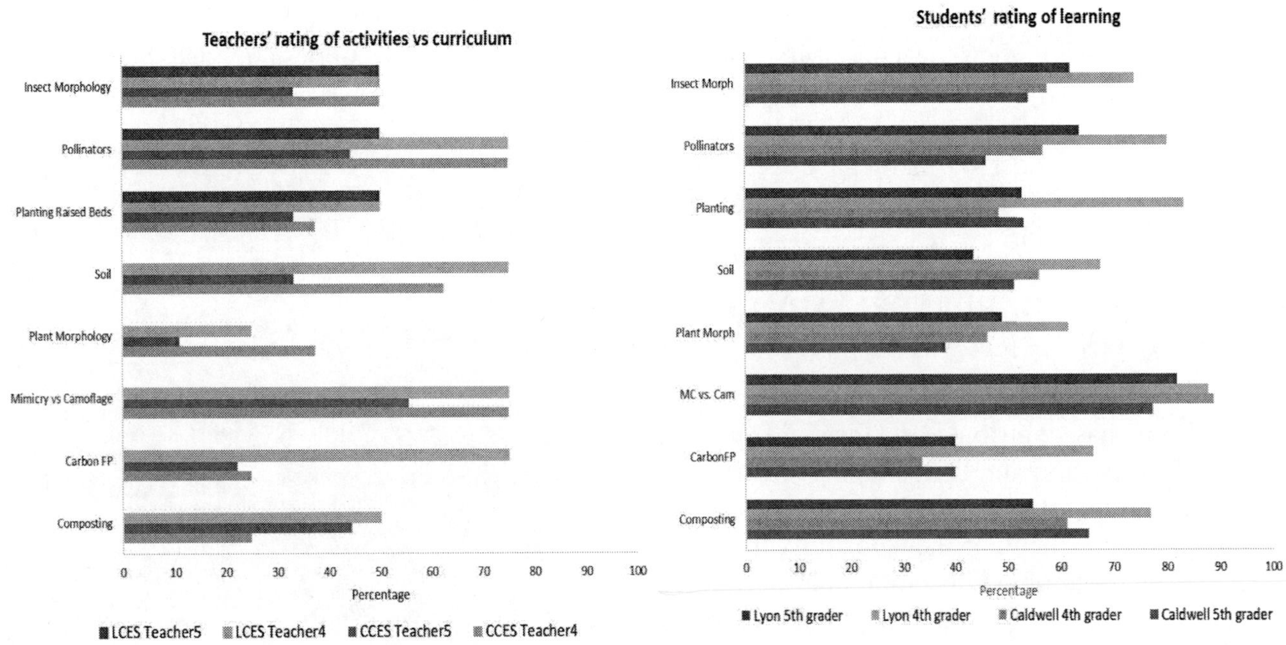

Figure 5. Teacher's and students' evaluations of activities offered in 2018 spring.

The quality of the events varied according to the schools, all LCES teachers judged the events as excellent (Figure 6). The majority of CCES fourth grade teachers considered that the activities were excellent. However, CCES fifth grade teachers thought that the activities were just good for their students. Students' evaluation was similar regardless the school, fourth graders rated the events as excellent. Based on this information, and considering that the fourth graders would be our fifth graders in the fall program, we planned totally different approach. The main goal of fall program was to provide students the opportunity to learn how to evaluate fair tests, use laboratory equipment, write a report and give a presentation. Different elements of their curriculum will be integrated, such math, analysis and writing.

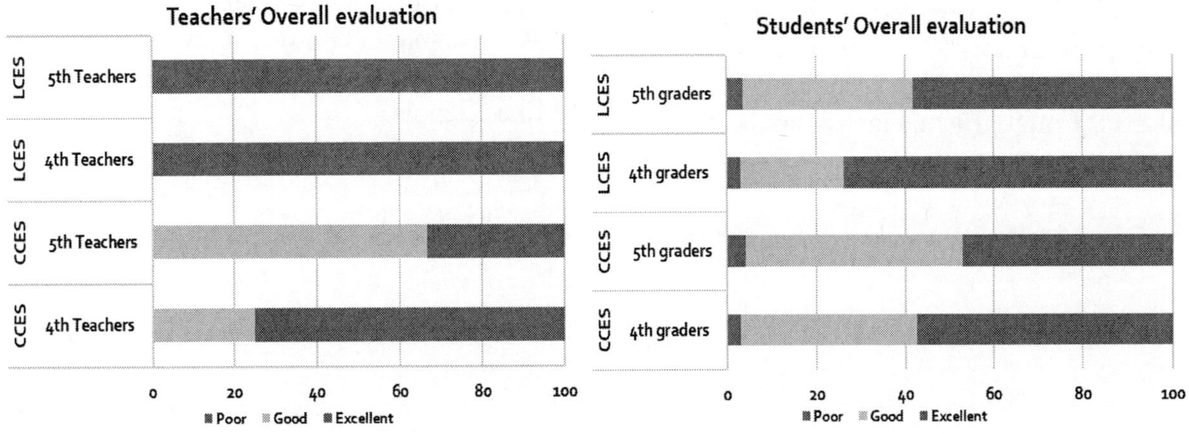

Figure 6. Overall evaluation of the 2018 spring events at the UKREC Botanical Garden.

Fifth graders had an excellent experience according to their teacher Mrs. Megan Radivonyk, who said "This has been quite a learning experience for our students.... I don't think most of them have ever done something like this before. They have all really enjoyed it".

MASTER GARDENER ADVANCED TRAINING

Lyon County Master Gardener Association has contributed to the field days of this program through helping instructors delivering information, keeping students in groups and helping them to move between working stations. This program is also aimed to cooperate in the professional developed of master gardeners. Advanced Master Gardener Training included hands-on and interactive sessions on pruning techniques (Figure 7), natural enemies of garden pests, plant disease identification, and UKREC Soil Laboratory Tour. A total of 24 master gardeners from 5 counties attended the event that was classified as very good for most of them.

Figure 7. Master Gardeners advanced training pruning workshop.

ACKNOWLEDGEMENTS

This project was funded by The University of Kentucky 2018 Sustainability Challenge Grant Program. We thank the collaboration of Ag. Extension Ag agent Susan Fox, Lyon County Master Gardener Association, Mrs. Amy Perdue and Ms. Abby Smith (Lyon County Elementary School) and Mr. Ronnie Martin and Ms. Laura White (Caldwell County Elementary School).

The participation of volunteers has contributed to a large extent to the success of this program, our gratitude to Rocio Davila visitor scientist, and Yaziri Gonzalez and Izabella (graduate students), Kaleb Tamez, Alex Teutsch (undergraduate students), and UK personnel Ginny Travis, Christine Bradley, Dwight Wolfe, Ryan Scott and Mike Nichols.

Combined Proceedings IPPS 68:179-183. 2018.

Practical Approaches to IPM in Propagation

Jeremy Webber
Koppert Biological Systems, Inc., Howell, Michigan 48843 USA

jwebber@koppert.com

Keywords: Insect control, best management practices, fungus gnats, nematodes, predator mites, thrips, spider mites, aphids.

INTRODUCTION

Fungus gnats are the single greatest insect pest in propagation facilities globally-excessive moisture, conducive media temperatures, and exposed rooting media provide ample opportunity for reproduction. Pressure from this pest along with shore fly is nearly ubiquitous in any propagation facility that Koppert Biological Systems works with. Since fungus gnat larvae directly feed on actively growing roots, severe populations can consume new roots nearly as fast as young plants can produce them and economic damage is quickly an issue.

Shore fly on the other hand cause no direct damage to plants as they feed on algae that grows on a medium's surface. Their unsightly presence in the finished crop and frass deposited on the crop's foliage are the principal issue with their development.

Beyond this, pest pressures vary wildly depending on crop type, region, season, and facility. One grower's issue may not be the same at a similar facility just down the road, so custom programs utilizing both biocontrols and compatible pesticides are frequently required for the seemingly endless combinations of pest pressures and individual goals that are found from facility to facility.

FUNGUS GNAT AND SHORE FLY—THE COMMON DENOMINATOR

Likely the most adopted biocontrol in ornamental crops on the planet — *Steinernema feltiea* is relatively easy to work with and is extremely effective at controlling the larvae of fungus gnats, as well as the pupae of western flower thrips (*Frankliniella occidentalis*) when applied correctly (Figure 1).

Part of its success is its similarity to conventional products during application—there aren't any creepy crawlies here, just a paste of some sort that's diluted in a stock or spray tank and applied to the crop. The process feels like a pesticide application, so there's familiarity and ultimately this helps wary growers dip their toes in the IPM pool so to speak.

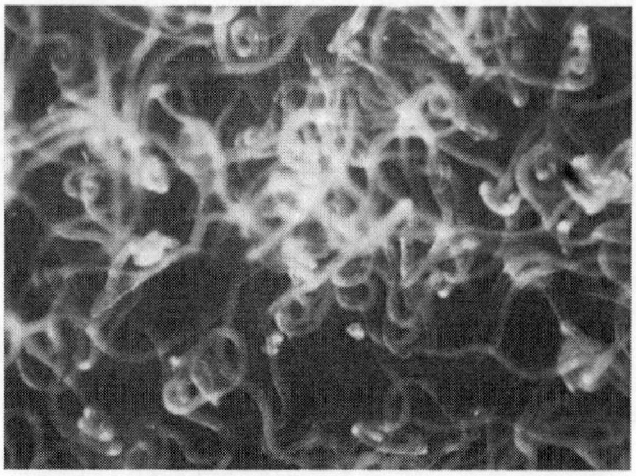

Figure 1. Nematodes.

To keep the metaphor going—what actually gets them swimming around in the deep end of the pool are the effects. When applied properly, *S. feltiae* will kill nearly every larval fungus gnat in a greenhouse, with no concern of resistance. For such a ubiquitous pest, this is the key takeaway—no resistance and consistent, predictable efficacy.

Application is relatively easy—nematodes from the two largest manufacturers in the world are both recommended for application in more or less the same manner. The product should be diluted in a few gallons of lukewarm water to activate it, then injected into your irrigation system with a portable injector and sprenched into your crop or mixed into a finished spray solution and applied over the top of the crop with a sprayer. The rate is generally accepted to be 500,000/m², usually applied weekly. At this rate, a container of 250 million will treat 500 m² or approximately 5,000 ft².

The addition of a surfactant is highly recommended for best efficacy. While the typical application frequency to fungus gnat susceptible crops is 1/week, for the most predictable results Koppert recommends tweaking the application interval versus the application rate for lower or higher pressure. For instance, poinsettia should receive intros every 5-7 days as they are highly sensitive to damage, but preventative applications in crops with no sensitivity to the pest and low pressure could receive intros every 10-14 days. Koppert recommends speaking with an IPM consultant to discuss the particulars of the application process in greater detail.

Unfortunately, this nematode has no significant effect on shore fly. While *Steinernema carpocapsae* has shown sporadic results versus this pest, the most consistent results in IPM programs has been seen from occasional usage of compatible insect growth regulators (IGRs) at times when shore fly are most prone to development (high heat, excessive moisture, algae growth). Ultimately—this pest's larvae grow in and feed on algae—any steps taken to maintain or eliminate algae where possible will ultimately go towards control. Most labeled IGRs are tank mix compatible with *Steinernema*—so control of both species is possible without adding an additional application. Koppert and most other biocontrol companies have apps and/or databases on the web to check the specific side effects of a chemical with a biocontrol agent—always check there or with an IPM consultant if you're trying a tankmix just to be safe.

Stratiolaelaps scimitus is another option for drier situations that has a similar effect to nematodes, but with a completely different mode of action so to speak. This soil-based generalist predatory mite navigates through the upper ranges of the rooting media and is highly effective at feeding on any number of pests, and perhaps even more importantly—it reliably reproduces on them too. Populations establish willingly when introduced early, and in many scenarios, can last through the length of the crop without reintroduction. Since rates and application methods vary greatly depending on crop type, media, pressure, etc.—Koppert recommends seeking the advice of an IPM consultant for the most effective introduction strategy. Since "strats" aren't fond of swimming, application early in a propagation scenario can be a situation to avoid. Nematodes typically function best in high humidity and misting scenarios whereas "strats" tend to do much better when the media is no longer continually saturated, and the crop is hardening off.

In addition to nematodes, predatory mites (Figure 2), and IGRs, there is likely no better usage of the pest control dollar than with mass trapping (Figure 3). So far—all controls discussed attack the larval phase of development. While it's hardly the most user friendly or visually attractive option—physically removing the adult population from the crop with numerous sticky cards or tape has a significant curative and preventative effect. Fungus gnat adults are quite attracted to yellow—so it stands to reason that if 30 can be killed by a small monitoring card in a week—many, many more can be taken out in the same amount of time by simply increasing the surface area of the sticky card. Flexible yellow pest control tape is the best option here. Exact implementation depends on how your facility is designed, your workflow, and what type of pest you're going after specifically—so advice from a consultant is recommended for best implementation. Bottom line though—every adult controlled is a major number of larvae taken out of the equation as well, so the more area in yellow sticky tape, the better.

Figure 2. A predatory mite, *Stratiolaelaps miles*.

Figure 3. Sticky tape! Mass trapping is king when pressure is high.

OTHER PESTS—WESTERN FLOWER THRIPS, SPIDER MITES, AND APHIDS

While fungus gnats and shore fly are the most common pests in propagation, relatively speaking, they are also very easily dealt with. Western flower thrips and spider mites however can be much more troublesome. Their incidence overall in propagation facilities is lower than that of fungus gnat, but not by much. Because of this, no two programs for their control are exactly alike as there are numerous variables from facility to facility. Aphids also fall into this nebulous category—there are as many different approaches to their control as there are propagation businesses. Following a few of the more common approaches.

Western flower thrips

As discussed previously—*S. feltiea* is an excellent control for the pupal phase of development, so it's likely that a well-executed program for fungus gnat control will keep western flower thrips (WFT) from easily developing in a crop, and vice versa. Our clients typically use a combination of nematodes, regularly applied predatory mites, and mass-trapping in scenarios where this pest is a concern. This method targets larvae with the predatory mites, the pupa with the nematodes, and the adults with the sticky traps. All three of these controls combined make it very difficult for a WFT population to reproduce from within the crop. Blow-ins frequently occur however when outdoor weather is warm enough, this is where the mass-trapping approach really shines. Since flowers are typically rare in propagation, the cards act as beacons to the easily fooled adults and the problem can be dealt with before it even starts. When the pressure is higher however, compatible chemistry can be employed as well. There are numerous options for predatory mites as well as chemistry depending on your region, crop, and goals, so please seek out an IPM consultant for a detailed recommendation on what works best.

Spider mite

A fairly common pest, *Tetranychus*, most commonly enters propagation facilities on cuttings, either in-house or from off shore. They can be dealt with relatively easily using a combination of predatory mites (Figure 4) and/or compatible chemistry. If pressure is only occasional in hotspots or limited to specific plants, programs can be executed on an as-needed basis so long as scouting is thorough, and populations are detected early. Several different predatory mites are out there, some are specific just to spider mite, and others work for many different pests simultaneously.

For example, *Amblyseius swirskii* or *Amblydromalus limonicus* can have considerable efficacy versus spider mite even though they are primarily being used for WFT. Other mites specifically for augmenting the spider mite control of your program can be added to this cocktail for higher pressure scenarios. Getting the pieces put together effectively will require a good relationship with a trusted advisor.

Figure 4. Specialist predators like *Phytoseiulus persimilis* come into the picture when spider mite pressure increases.

Aphids

Biological control of aphids in ornamentals is still in its infancy. As opposed to other pests like WFT and spider mite, it is quite difficult to find pesticide resistant populations of aphids here in the USA. They are certainly reported on occasion—however populations generally respond well to applications of any number of pesticide categories. Many labeled products are new and relatively safe to biologicals as well as the applicators and therefore biocontrols haven't been as needed in this arena as in the ones mentioned previously. It's completely possible to have a biocontrol program for fungus gnat, WFT, and spider mite, and to use conventional products for aphids. That said, there are numerous ornamental growers working to minimize their usage of these newer products, as well as edible growers that have a much smaller list of labeled products to work with. In these cases, Koppert tends to work with broad spectrum predators like green lacewing larvae (*Chrysoperla carnea*) (Figure 5) or the gall midge (*Aphidoletes aphidimyza*). Parasitic wasps seem to get the most press for aphid control, but they typically have narrow host ranges and are not a one-size fits all solution. Typically, when a grower is working with a more broad-spectrum predatory bio approach, if a population is discovered through weekly scouting, either a corrective small-scale pesticide app can then be made to bring the population back in line or additional predators can be applied to the hotspot. Growers vary in how they're comfortable handling this scenario, and fortunately, there are numerous options at their disposal.

Figure 5. For aphid control Koppert recommends generalist predators such as green lacewing larvae to keep it as simple as possible.

Combined Proceedings IPPS 68:184-193. 2018.

Ecological All Stars

Steven A. Wright
Director of Horticulture/Curator of Plant Collections, Jenkins Arboretum & Gardens, 631 Berwyn Baptist Road, Devon, Pennsylvania 19333 USA

steve@jenkinsarboretum.org

Keywords: Pollinators, butterflies, caterpillars, bird habitat, carbon balance, ecology, *Liriodendron*, *Aralia*, *Populus*, oak, *Quercus*, holly, *Ilex*, willow, Salix, *Vaccinium*, *Tilia*, maple, *Acer*, *Prunus*.

INTRODUCTION

For more than a century, baseball has been considered America's favorite pastime. References to playing baseball in America go back to the late 1700s and professional leagues were started in the late 1800s (baseball-reference.com). Since then, thousands of games have been played and player performance (statistics) has been recorded. The most celebrated of those statistics have traditionally been home runs, runs batted in, and batting average—the three prongs of the lauded triple crown. Others like hits, walks, on base percentage and so on were recorded but often not valued to the same extent.

That way of thinking began to change in the early 1980s with the growing usefulness and availability of computers and the introduction of sabermetrics (Schell, 2011). Developed by Bill James and introduced to the wide public in 1982, sabermetrics is, in short, calculations that consider various contributions a player has made to their team (Schell, 2011). One of the most commonly used sabermetric statistic is called Wins Above Replacement (WAR). WAR considers numerous aspects of a player's performance, not just those that are most celebrated. The "power" numbers that once might have solidified a player as a Most Valuable Player candidate are now simply parts of the calculation. In recent years, players with a higher WAR have been given more consideration than those that might lead the categories like home runs, runs batted in, and batting average.

By now you're probably asking how this relates to ecology or plants at all for that matter. In truth, it doesn't, but it is what inspired me to take a closer look at plants and think more critically about all the ecosystem services that they provide, not just those that typically come to mind. Typically, when we discuss the ecological or environmental value of plants, our mind jumps to those that provide nectar for pollinators, or fruits and seeds for birds, or serve as host plants for a range of moths and butterflies. Yes, these are all important and deserve consideration, but these are not the only considerations in determining the overall ecological value of a plant. That determination is complicated, convoluted, and likely incalculable. After a significant amount of reading, research, observation, and reflection, there is no doubt that some plants deserve much more celebration than they typically receive. Though I am sure there are more, I have identified seven factors that influence the environmental/ecological value of a plant.

ECOLOGICAL VALUE FACTORS

Factor 1: Pollinator value

When discussing the pollinator value of plants, we typically think of those that provide abundant nectar. Species like *Pycnanthemum muticum*, *Clethra alnifolia*, and *Phlox paniculata* are among the best nectar sources and their status as superior pollinator plants is well-known, as it should be.

Often, however, we forget about the importance of pollen. Jenkins Arboretum's long-time apiarist compared the two with the analogy that pollen is like cheeseburgers and nectar is like candy bars. Pollen, which provides mainly fats and proteins, is truly sustaining and vital to the survival of bees. Nectar on the other hand, provides mainly carbohydrate sugars which provide energy. Both are important, but is one more so than the other? Could it be argued that pollen is more valuable than nectar? Regardless of the answers to these questions, we should be including pollen plants in our lists of plants to attract bees. For the purposes of evaluating a plant's ecological value, species that produce both pollen and nectar are given more consideration than those that provide only one of the two.

Figure 1. The picture on left is American hazelnut (a great pollen source) Photo credit: Will Cook; and the picture on the right is summersweet (a great nectar source) Photo Credit: Willowbend Nurseries.

Factor 2: Lepidopteran larval hosts

Doug Tallamy's book, *Bringing Nature Home*, was the first time many of us have seen a quantification of the value of plants to butterflies and moths (Tallamy, 2009). Knowing that oak trees support 534 species of lepidopteran larvae (Tallamy, 2009), far more than all others, helps us to better understand the huge role these trees play in our environment. There are a couple of different ways to think about host plants though. First, we now know that oaks, cherries, and willows support hundreds of different moth and butterfly species and we label them as most valuable (Tallamy, 2009). This may be true, but we cannot discount the value of plants that support the specialists. That is, we must remember that without milk-weeds (*Asclepias* spp.), we would not have monarchs; without pipevines (*Aristolochia* spp.), we would not have pipevine swallow-tails; and without senna (*Senna* spp.) we would not have cloudless sulphurs. There are dozens of similar examples and this needs to be considered for overall ecological value.

Factor 3: Food for birds

Aside from backyard birdfeeders, when we think about feeding birds, most of us immediately picture plants such as hollies (*Ilex* spp.) and elderberries (*Sambucus* spp.) that produce fleshy berries. With a little more thought, we then realize we need to include plants such as pines (*Pinus* spp.), birches (*Betula* spp.), and coneflowers (*Echinacea* spp.) that provide high quality seeds. Fruits and seeds are both highly valuable, that can-not be denied, but these are only the things that the plants are producing. What is equally valuable, but often overlooked, is what the plants are supporting.

As noted in the previous section, there are countless plant species that support hundreds of different moth and butterfly larvae (and many other insects as well). These larvae are an important source of fats and proteins for various bird species (Figure 2). The plants that support insects are also supporting birds. That means that a black willow (*Salix nigra*) is just as important for songbirds as a black elderberry (*Sambucus canadensis*), if not more so, but I have never seen black willow listed as a plant to attract birds (Soren, 2018).

Factor 4: Nesting and cover

In the increasingly urban world we live in, with ever decreasing wildlife habitat, animals of all kinds need a place to escape—a place to den, nest, and raise offspring. Some plants provide better nesting and cover than others and they deserve some attention as well. Plants with dense foliage or branches and those that form thickets provide great nesting and cover opportunities. Evergreen trees and shrubs also provide protection from winter winds and snow and are perhaps the best for nesting and cover. Species like inkberry (*Ilex glabra*) which combines all of these traits are ideal.

Factor 5: Erosion control

With increasing human population, we also see increasing areas of impervious surface which increases stormwater runoff into our waterways and leads to severe erosion along streambanks. We also see soil erosion in areas that are only lightly vegetated. It could be argued that erosion is the most significant threat to our ecosystem because the land cannot support life if there is no soil on which to grow plants. Species like gray dogwood (*Cornus racemosa*) that form dense colonies or thickets, grow vigorously, and form a dense root system are highly valuable in helping to control erosion.

Figure 2. We typically think of berries and seeds. However, plants that host a lot of insects are just as valuable (if not more) than those that produce fleshy fruits or high-quality seeds. Photo Credits: Top left: Gypsy Flores, Top right: Ryan Sanderson, Bottom left: Fred Ortlip, Right: Johnny Wee.

Factor 6: Stormwater mitigation

The impervious surface mentioned above, combined with the more frequent and more extreme weather events we've seen over the past few years, have led to serious stormwater concerns including soil erosion and flooding. Stormwater management is a complex issue with many contributing factors. I will not go into those details here, but there are things we can do with plants that will help reduce stormwater. Trees that have large leaves and crowns will intercept significant amounts of rain water and, in some cases, will hold it, never letting it hit the ground. Next, trees that grow quickly and to a large size will pull water from the soil through evapotranspiration. Additionally, if we remember that water always runs downhill, bottomland species will likely remove more water than upland species. With all of that said, sycamores (*Platanus* spp.) and cottonwood (*Populus deltoides*) are among the most valuable plants at mitigating storm water.

Factor 7: Carbon sequestration

Climate change is a topic for a different time, but one that is hard to deny. We are well aware of the effects that carbon dioxide and other greenhouse gasses are having on our environment. Understanding this, we should be doing all we can to help offset the enormous emissions of these gasses. As horticulturists, we can help with plants—specifically, trees that are long-lived, fast-growing, and large in size (girth and height) that will store large amounts of carbon. There are several species that fit this description, but two of the best are tulip poplar (*Liriodendron tulipifera*) and baldcypress (*Taxodium distichum*).

My Ecological All-Star Team

After considering all of the above factors, I have selected 10 species as "ecological all-stars". Each of these possesses a combination of characteristics that make them highly ecologically or environmentally valuable. There is no doubt that there are dozens more that deserve consideration, but I also wanted to promote diversity and show that there is a wide range of plant options. Some groups of plants, like the white oaks for example, would have several representatives in the list if allowed. The list that follows is in no particular order, that is, it is not a ranking. It is also important to keep in mind that these plants have been selected for their ecological value, not necessarily their horticultural value.

In fact, some have relatively little ornamental value or have other characteristics that might make them less desirable in a garden setting. They do; however, all have enormous value for restoration purposes. I should also mention that the explanations below are not a comprehensive evaluation of each of the plants, but rather an explanation of the factors that set them apart and led to their addition to the team.

Species 1: Devil's walking stick (*Aralia spinosa*)

Considering I am promoting plants for their ecological value, devil's walking stick deserves significant attention. Its spreading, thicket-forming habit makes it valuable for erosion control as well as cover for wildlife. It produces enormous, 3 ft long flower panicles that provide both nectar and pollen rewards to pollinators (Trees, Shrubs, and Woody Vines of Illinois). These flowers are then followed by huge clusters of dark purple berries that are cherished by songbirds of all kinds (Trees, Shrubs, and Woody Vines of Illinois) (Figure 3).

Figure 3. *Aralia spinosa*. Photo credits—left: Ellen Honeycut, right: Bill Hubick.

Species 2: Tulip poplar (*Liriodendron tulipifera*)

Though often dismissed by gardeners for its weedy nature, the tulip poplar provides numerous ecological benefits. Perhaps the most obvious trait of this species is its enormous size (Figure 4). In fact, it is among the largest species in the eastern USA in both girth and height. It is also a very fast grower. Combined, these traits make tulip poplar very valuable for both stormwater mitigation and carbon sequestration. In addition, though they often go unnoticed because they are usually held so high on the tree, the flowers of tulip poplar produce high quality pollen (Eierman, 2013) as well as some of the highest volume of nectar (Angel, 2018).

Figure 4. Tulip poplar (*Liriodendron tulipifera*).

Species 3: Eastern cottonwood (*Populus deltoides*)

A fast-growing bottomland species with large leaves and crowns, the cottonwood is a great choice for both carbon sequestration and stormwater mitigation. It is the flowers though that really set this species apart. The male flowers of cottonwood, and other *Populus* spp., contain very high-quality protein (Collison, 2016). Not only is it high in nutritional value, but bees that consume this pollen realize significant health benefits, not the least of which is a significant increase in lifespan (Collison, 2016). This is an important consideration as beekeepers continue battling against colony collapse disorder and we're now seeing certain bumblebee species being listed, or considered for listing, on the Endangered Species list.

Species 4: White oak (*Quercus alba*)

Though it is not fast-growing like others on this list, the white oak is very long lived (Figure 5). The combination of slow growth, which makes for denser wood, and longevity make it another species that should be considered for carbon sequestration. Its broad crown and large leaves make it valuable for stormwater mitigation. In addition, *Quercus* spp. support 534 species of butterfly and moth larvae—more than any other plant (Tallamy, 2009). The sweet acorns of white oak are highly prized by all kinds of wildlife including certain birds, like blue jays and some woodpeckers. Overall, this species is an ecological workhorse.

Figure 5. White oak (*Quercus alba*).

Species 5: American holly (*Ilex opaca*)

The only evergreen on my list, American holly has several ecologically valuable traits (Figure 6). Its dense, evergreen branches provide great nesting and cover sites for birds. The flowers provide both nectar (female flowers) and pollen (male flowers) to pollinators (Encyclopedia of Life). In addition, the berries, though not highly nutritious (Encyclopedia of Life), are a valuable source of food for overwintering birds.

Figure 6. American holly (*Ilex opaca*).

Species 6: pussy willow (*Salix discolor*)

Most people, regardless of profession, are familiar with the fuzzy spring buds of the pussy willow. Many, however, do not realize that those buds open to become highly valuable flowers. The flowers of plants in the genus *Salix* produce both high quality nectar and pollen (Eierman, 2013; Forcone et al, 2011). This nutrition is valuable and important, but the very early bloom time boosts this plant onto the list as there are very few other foraging options for early season bees. In addition to this, pussy willow tends to sucker and form colonies which provides good cover for wildlife and helps to reduce erosion.

Species 7: Farkleberry (*Vaccinium arboreum*)

Farkleberry makes the list mainly because of its size (Figure 7). *Vaccinium* spp. rank #7 in the number of lepidopteran larvae they support, at 288 (Tallamy, 2009). As the largest member of the genus, growing 15–30 ft, it follows that it would support the most lepidopterans. Also, because it is the largest, it follows that there are more flowers for pollinators, like blueberry bees, whose visits are rewarded with both pollen and nectar (Trees, Shrubs, and Woody Vines of Illinois).

The abundance of flowers is followed by an abundance of fleshy fruits that are consumed by various songbirds.

Figure 7. Farkleberry (*Vaccinium arboretum*), Photo credit: Will Cook.

Species 8: American basswood (*Tilia americana*)

Another long-lived, large tree, American basswood is among the top performers in sequestering carbon and mitigating storm water. Its flowers, however, are its real claim to fame. Perfuming the air with a sweet fragrance in late spring/early summer, basswood flowers are bee magnets. The enormous volume of nectar produced by this species makes it among the most valuable plant for bees. Honeybees are especially attracted to basswood and they can produce 800–1,100 pounds of honey per acre when nectaring on the species. This is rivaled only by black locust which can result in honey volumes of 800–1,200 pounds (Wikipedia) per acre.

Species 9: Silver maple (*Acer saccharinum*)

Though I debated adding the silver maple to the list for various reasons, there is no doubting this species' ecological value. Like many others on the list, the silver maple grows very quickly and very large which makes it valuable for carbon sequestration and storm water mitigation. It is also a valuable host plant for various moths and butterflies and maples as a genus support 285 different species, ranking them 8th highest (Tallamy, 2009). Being the largest of the maples, silver maple will support more than other species. Finally, like pussy willow, there is enormous value in the flowers, specifically the timing. The very early bloom and high nectar value make the silver maple highly valuable to early-season bees. Again, its large size means there are more flowers from which to forage.

Species 10: Chokecherry (*Prunus virginiana*)

Though often written off as a weedy tree, chokecherry combines numerous factors that might well make it the most valuable plant on my list (Figure 8). In spring, chokecherry's long panicles of white flowers attract pollinators of all kinds. It has been found that *Prunus* spp. are among the best producers of both high-quality pollen and nectar (Collison, 2016). Those flowers are followed by fleshy berries prized by various wildlife, especially songbirds and, containing 30% to 50% fat, they rank among the most nutritious of all berries (Wikipedia). In addition, as a host to 456 different species (Tallamy, 2009), *Prunus* spp. rank 2nd only to *Quercus* spp. in supporting lepidopteran larvae. What sets chokecherry apart from others in the genus though, is that it tends to form thickets. These thickets provide good nesting and cover sites for various wildlife and help to reduce erosion.

Figure 8. Black cherry (*Prunus virginiana*), Photo credit: Moonshine Designs Nursery.

CONCLUSION

Throughout this article I have focused on the environmental and ecological value of plants with little regard to their aesthetic or economic value. I am aware that the latter are also important, especially for an industry responsible for supplying landscape architects and installers with plant material. In a capitalist economy, the product that sells is the product that gets produced, but I am proposing a shift in mindset. That is, we need to encourage a culture that prizes plants for their overall value— ecological as well as aesthetic. This shift needs to start at the producer level. Those of us who develop, propagate, produce, promote, and ultimately sell plants are the ones responsible for what ends up in the landscape.

I am not necessarily suggesting that we start producing more devil's walking stick or silver maple for our landscapes. What I am suggesting though, is that we start growing more of the plants that combine numerous aesthetic and ecologically beneficial traits. There are dozens of examples, but I will close with one that might help to clarify the point.

Burning bush, *Euonymus alatus*, is one of the most common plants in the landscape. They are tough, heat and drought tolerant, perform well in full sun to part shade, and have outstanding red fall color. They are great landscape plants, there is no doubt. Unfortunately, though, this species has escaped cultivation and become extremely invasive displacing native plants and reducing biodiversity in our natural areas. It provides very little ecological benefit and despite this ecological disaster, it is still a plant frequently propagated, sold, and planted in both home and commercial landscapes. The shift in mindset that I am proposing would see a phasing out of plants like burning bush and a phasing in of other species that would fill the same role in the landscape. These new, alternative plants would also provide numerous ecosystem services.

One possible alternative to burning bush would be winged sumac, *Rhus copallinum*, especially the cultivar 'Morton' (sold under the trade name *R. copallinum* var. *latifolia* 'Morton', Prairie Flame™ shining sumac). This species combines all of the same attributes as those listed for burning bush above. In addition, it is a native species that is much less aggressive, has flowers that are highly attractive to bees and other pollinators, and has nutritious fruits that are a valuable source of food for birds through fall and into winter when food is relatively scarce. In all, *Rhus copallinum* would be a much better addition to the landscape.

There are many other examples of plant alternatives. Though we are now starting to see more of those alternatives in the landscape, I hope that, in time, they become the rule rather than the exception. Again, it starts with us; let's see if we can make that happen.

Literature Cited

Angel, P.N. (2018). Trees for honeybees. eastern Kentucky beekeeping school. January 20, 2018. https://perry.ca.uky.edu/files/trees_for_honey_bees_eky_bee_school_hazard_ky_1-20-18pp.pdf. Accessed 9/21/2018

Baseball Reference. https://www.baseball-reference.com/ Accessed 9/12/2018

Collison, C. (2016). Pollen quality. bee culture. The Magazine of American Beekeeping. August 2016

Eierman, K. (2013). High value pollen sources for honey bees: get planting! EcoBeneficial Blog. https://www.ecobeneficial.com/2013/03/high-value-pollen-sources-for-honey-bees-get-planting-2/ Accessed 9/10/2018.

Encyclopedia of Life. *Ilex opaca*. http://eol.org/pages/582641/details Accessed 9/21/2018

Forcone, A., Aloisi, P.V., Rubbel, S., and Munoz, M. (2011). Botanical composition and protein content of pollen collected by *Apis mellifera* L. in northwest of Santa Cruz (Argentinea Patagonia). Grana *50*:30-39, DOI: 10.1080/00173134.2011.552191.

Schell, R. (2011). SABR, baseball statistics, and computing: the last forty years. Baseball Research J., Fall 2011.

Soren, S. (2018). Planting native to attract birds to your backyard. Stackhole Books.

Tallamy, D.W. (2009). Bringing nature home: how you can sustain wildlife with native plants. Timber Press, Portland, Oregon 97204 USA.

Trees, Shrubs, and Woody Vines of Illinois. *Aralia spinosa* (devil's walking stick). https://www.illinoiswildflowers.info/trees/plants/devil_ws.html Accessed 9/10/2018.

Trees, Shrubs, and Woody Vines of Illinois. *Vaccinium arboreum* (farkleberry). http://www.illinoiswildflowers.info/trees/plants/farkleberry.html. Accessed 9/20/2018

Wikipedia. List of Northern American nectar sources for honey bees. http://en.wikipedia.org/w/index.php?oldid=542320470

PROCEEDING'S PAPERS

WESTERN REGION OF

NORTH AMERICA

Dr. Gene K. Blythe, Regional Editor

Fifty-ninth Annual Meeting - 2018

Honolulu, Hawaii USA

Combined Proceedings IPPS 68:194. 2018.

Morphological Characterization of Six Accessions of Pa'uohi'iaka (*Jacquemontia ovalifolia* subs. *sandwicensis*)

Darel Kenth S. Antesco and Orville C. Baldos
Department of Tropical Plant and Soil Sciences, University of Hawaii at Manoa, Honolulu, Hawaii 96822, USA

obaldos@hawaii.edu

Keywords: Convolvulaceae, ground covers, Hawaiian plants, native plants.

Abstract

Pa'uohi'iaka *(Jacquemontia ovalifolia* subsp. *sandwicensis*, Convolvulaceae*)* is a native Hawaiian plant that is commonly used as an ornamental ground cover and has potential use as a hanging basket plant. Despite variations in flower color and leaf pubescence, little effort has gone into varietal selection. This study aimed to document easily identifiable morphological characteristics to distinguish between accessions and use these characteristics to select accessions with hanging basket potential. Six accessions were collected from Maui (Puhala Bay, McGregor Point, and Ahihi-Kinau), Hawaii (Mahana Bay), and Oahu (Shidler Business College and Leeward Community College [cultivated]).

Ten sample plants of each accession were grown and pruned after 3 months to encourage lateral branching. Qualitative and quantitative characteristics were recorded at one month after pruning. Analysis of variance (ANOVA) and Tukey's pairwise comparisons were conducted on each quantitative characteristic to identify differences among accessions. Analysis revealed significant differences ($p \leq 0.05$) in average stem diameter, leaf length, leaf width, leaf thickness, petiole diameter, length of internodes, number and length of lateral branches, flower diameter, and number of open flowers.

No significant differences were observed in plant canopy and number of preformed roots. Morphological characteristics, such as number of lateral branches and length of lateral branches, are important factors for selection and potential use of pa'uohi'iaka as a hanging basket plant. Based on these characteristics, selections from Leeward Community College, Puhala Bay, and Mahana Bay show promise for further evaluation.

Combined Proceedings IPPS 68:195. 2018.

Morphological Characterization and Diversity Analysis in Ilima (*Sida fallax*)

Darel Kenth S. Antesco and Orville C. Baldos
Department of Tropical Plant and Soil Sciences, University of Hawaii at Manoa, Honolulu, Hawaii 96822, USA

obaldos@hawaii.edu

Keywords: Malvaceae, Hawaiian plants, herbaceous plants, native plants.

Abstract

The use of native Hawaiian plants as ornamentals has grown in the last two decades due to state laws aimed at promoting the use of natives in landscaping. Despite increased awareness and use of native Hawaiian plants, species and varieties available in nurseries are limited. Ilima (*Sida fallax*, Malvaceae) is a culturally important species commonly used in leis and in landscaping. Although various forms exist, ranging from the prostrate coastal ecotype to the upright mountain ecotype, cultivar development has been limited. Collection, characterization, and evaluation of plants from different islands and habitats is essential for developing horticultural selections. In this study, 19 accessions of ilima were collected from wild and cultivated sources on four islands (Kauai, Maui, Molokai, and Hawaii) and grown in a common garden.

Morphological characterization was done using a descriptor list and assessed using the Shannon-Weaver Diversity Index. High (> 0.67) to medium (0.34-0.66) levels of diversity in 15 quantitative and qualitative characteristics were observed. Characteristics exhibiting a high diversity index were leaf length (0.81), leaf width (0.71), plant canopy (0.72), stem diameter (0.79), and number of plant samples with open flowers (0.69). Morphological characteristics with medium level of diversity included position of the lamina (0.39), mature leaves (0.53), leaflets (0.52), plant height (0.41), number of branches (0.63), floral diameter (0.55) and number of open flowers (0.60). A low level of diversity (< 0.33) was observed in petiole color, petiole orientation, growth habit, and floral type.

Combined Proceedings IPPS 68:196-201. 2018.

Developing Propagation Protocols for Native Hawaiian Plants with Potential Landscape and Indoor Uses

Orville C. Baldos[a], Aleta K. Corpuz, and Darel Kenth S. Antesco
Department of Tropical Plant and Soil Sciences, University of Hawaii at Manoa, 3190 Maile Way, St. John 102, Honolulu, Hawaii USA 96822, USA

obaldos@hawaii.edu

Keywords: *Heteropogon contortus*, *Chenopodium oahuense*, *Melanthera integrifolia*, *Eragrostis deflexa*, *Jacquemontia ovaliifolia*, *Peperomia* spp.

Abstract

Native plants as ornamentals are gaining popularity across the United States, including Hawaii where native plants have been promoted since the 1990s and are required in state-funded landscaping projects. Due to limited information and to increase availability and variety of native Hawaiian plant selections, a research program was established at the University of Hawaii at Manoa to collect, select, and evaluate non-endangered and underutilized species for ornamental uses.
Research has provided useful information on propagation of *Heteropogon contortus* (pili grass), *Chenopodium oahuense* (aweoweo), *Melanthera integrifolia* (nehe), *Eragrostis deflexa* (Pacific lovegrass), *Jacquemontia ovaliifolia* (pau o hiiaka), and *Peperomia* spp. (ala ala wai nui).

INTRODUCTION

The use of native plants as ornamentals is gaining popularity in the United States. In recent years people have become more aware that plants in the landscape not only provide benefits to human needs (e.g. aesthetics and functionality), but also provide ecosystem services. The composition of plants in a landscape can have effects on wildlife, water quality, climate, soil microbiology, and other factors on the surrounding area. Due to issues such as water use/conservation, biodiversity conservation and invasive species spread, interest in the use of native plants has increased. Besides environmental issues, the use of native plants is also encouraged because it also provides a sense of place (e.g. saguaro of the Sonoran Desert, redwoods of Northern California). A recent survey by the American Society of Landscape Architects indicate that native plants and drought tolerant plants were the top landscape/garden elements demanded by residential consumers in 2018.

NATIVE HAWAIIAN PLANTS

In Hawaii, the use of native plants in landscapes has been promoted since the 1990s. This is due to a state law requiring the use of native plants in state-funded landscaping projects. Demand for native Hawaiian plants is expected to increase in the next 20 years due to amendments to the previous state law. In the revised legislation (Act 233 (15) Relating to Hawaiian Plants), the percentage footprint of native plants in state-funded landscaping projects will be increased from 10% by 2019 to 35% by 2030. Despite active promotion to use native Hawaiian plants in landscaping, several bottlenecks still exist. Tamimi (1999) and Ricordi et al. (2014) found that availability and variety of plants as well as lack of horticultural information (e.g. care and use in landscape design) are the key limitations to maximizing the use of native Hawaiian plants in landscaping. In addition, there has been little to no research done to develop selections or nativars suitable for landscape and other ornamental uses.

SUSTAINABLE ORNAMENTAL HORTICULTURE LABORATORY AT THE UNIVERSITY OF HAWAII AT MANOA

To increase availability and variety of native Hawaiian plant selections, I (first author) established a research program to collect, select, and evaluate non-endangered and underutilized species for several ornamental uses (i.e. landscape, potted ornamental, indoor/houseplant, and floriculture use). Research on propagation comprises a big portion of our program since successful release of a native plant selection requires a well-developed propagation technique. Propagation methods that we are currently exploring are the following: 1) seed, 2) cuttings/division, and 3) tissue culture. For seed propagation, we are interested in exploring dormancy and the pretreatments necessary to relieve it. For research on cuttings/division, we would like to know if accessions vary in rooting and if rooting hormones can improve root morphology. We are also interested in developing tissue culture protocols for more difficult-to-propagate plants.

Currently, we are conducting research on native plants for landscape/potted plant use and indoor use. Species that we are studying for landscape/potted plant use are aweoweo (*Chenopodium oahuense*), ilima (*Sida fallax*), nehe (*Melanthera integrifolia*), Pacific lovegrass (*Eragrostis deflexa*), pili grass (*Heteropogon contortus*) and pau o hiiaka (*Jacquemontia ovaliifolia*). Species we are currently evaluating for indoor use are sedges (*Carex* spp.) and ala ala wai nui (*Peperomia* spp.). I would like to share some propagation experiments that we have conducted on some of these potential species.

Heteropogon contortus (pili grass)

Pili grass is an indigenous, drought tolerant, perennial grass found on all main Hawaiian Islands. It is a culturally important species since the ancient Hawaiians have used it as a thatching material for the traditional 'hale' (houses). In recent years, demand for pili grass has increased due to interest in building these structures with authentic plant materials. There is also interest in using pili grass for erosion control, roadside revegetation, green roofs, and urban landscaping. Upright and prostrate forms of pili grass occur in Hawaii and can be selected for various uses. Pili grass is usually propagated from seed but can also be propagated with some success through division. Freshly harvested seeds of pili grass possess dormancy and require an after-ripening period of 1 year. Conditions to hasten or optimize the after-ripening period have not been studied before, so we looked at the influence of storage humidity (12%, 50% and 75% equilibrium RH), temperature (10°C, 20°C and 30°C) and storage period (0, 1, 3, 6, 9 and 12 months) on dormancy loss of pili grass. Results from this study indicate that the optimum conditions were storage of the seeds at 12% equilibrium RH and 30°C for 1 year (Baldos et al., 2014). With this treatment we can achieve germination percentages greater than 90%.

Aside from after-ripening, we also found that smoke water application can stimulate germination of 1-month-old seeds. I tested two known components of smoke (KAR_1 and cyanide) and found that cyanide (at the concentrations found in smoke) can break dormancy of seeds (up to 30% germination). If making smoke water is a hassle, a 1% dilution of food grade liquid smoke (tested free of cyanide) can also break seed dormancy (Baldos et al., 2015).

Chenopodium oahuense (aweoweo)

Aweoweo is an endemic, drought tolerant shrub or tree found on all main Hawaiian Islands. It has been used by ancient Hawaiians for medicinal purposes and as a famine food. Aweoweo growth forms are diverse, ranging from prostrate (occurring on the coast) to shrub/tree form. This diversity in growth forms can allow selections for various ornamental uses. Aweoweo can be propagated from seeds or through stem cuttings. We currently have two selections of prostrate aweoweo that we are evaluating for potted and landscape use. One question that we wanted to know about these selections was its rooting response with or without application of rooting hormone. We did a rooting experiment comparing the two selections with or without 3,000 ppm of indolebutyric acid (Hormex 3). Results of the experiment indicated that the two accessions are easy and fast to root (>90% rooting within 23 days). Rooting hormone application did not improve percent rooting, rooting index (visual scale of root density), or length of the longest root. There was a significant difference in longest root length between selections with the compact selection having longer roots than the prostrate selection.

Melanthera integrifolia (nehe)

Nehe is an endemic coastal perennial herb found on all main Hawaiian Islands. Flowers and fruits of nehe are used in lei making. It has been promoted as a groundcover for landscapes but is still not used as widely. A recent collection trip has shown that there is substantial variation in leaf shapes and growth of nehe. This provides the potential to select plants suitable for potted/hanging baskets and landscape use. Propagation of nehe can be easily done through stem cuttings with no rooting hormone required. Due to the variations found in our collection, we wanted to characterize rooting of the different accessions. We also wanted to find out if accessions will benefit from rooting hormone application (despite what the literature mentions).

We studied rooting behavior of three accessions (Koko Head, Makapuu, and Southpoint) treated with or without 3,000 ppm indolebutyric acid. Results of the study indicated that nehe is easy to root (>90% rooting). There were differences in root surface area of accessions: Makapuu exhibited the highest root surface area, followed by Koko Head (intermediate) and Southpoint (lowest root surface area). Contrary to what the literature mentions, rooting hormone application improved root morphology of cuttings (i.e., root surface area, number of root tips, and total root length), regardless of accession. This new information may be beneficial for improving root morphology of accessions such as Southpoint (with low root surface area).

Eragrostis deflexa (Pacific lovegrass)

Pacific lovegrass is an endemic bunchgrass found in the dry, leeward forests of Lanai and Hawaii Islands. Its upright growth and linear leaves make it a potential landscape accent or groundcover. It is a rare grass with very limited information about its propagation. Typically, it is propagated from seeds. Since success in propagation through division is not known, we did some experiments to evaluate survival and rooting of bare rooted clumps combined with the application of rooting hormone (as a soak). For our soak treatments, we compared non-soaked (control) clumps with soaked clumps in tap water, 1:20 Dip 'N Grow® (500 ppm indolebutyric acid + 250 ppm naphthalene-acetic acid) and 1:10 Dip 'N Grow® (1,000 ppm indolebutyric acid + 500 ppm naphthaleneacetic acid). Results of the soaking study indicated that soaking was detrimental to the plant (100% mortality). Non-soaked, control plants displayed moderate success (70% with new roots and shoots). Based on the results, seeds are still the best way to propagate this species.

Jacquemontia ovaliifolia (pau o hiiaka)

Pau o hiiaka is an endemic, coastal vine found on all main Hawaiian Islands. It has been used by the early Hawaiians for medicinal purposes. It has been promoted for use as a native groundcover in landscaping and has potential use as a container/hanging basket plant. Our germplasm collection of pau o hiiaka shows differences in flower color, leaf size, hairiness, and branching habit. This diversity in characteristics allow selection for various ornamental purposes. Propagation of pau o hiiaka can be easily done from cuttings. We wanted to characterize rooting of each accession, so we compared rooting of 1-node and 4-node cuttings with preformed roots. Results of the study indicate that certain accessions exhibited differences in rooting character-istics. The Southpoint and McGregor Point (hairy) accessions exhibited shorter roots than the other accessions (Puhala Bay, Ahihi-Kinau, Lyon Arboretum, Shidler Business School) tested. The use of 4-node cuttings was superior to 1-node cuttings as the former had more roots and higher rooting percentages (97% for 4-node vs. 73% for 1-node cuttings). Based on the results of the study, a minimum of 4-nodes on cuttings is recommend for propagating pau o hiiaka.

Peperomia spp. (ala ala wai nui)

Ala ala wai nui are succulent herbs found growing in dry to wet forests on most main islands (except Kahoolawe). Of approximately 1000 species found worldwide, about 25 species (2 native and 23 endemic) are found in Hawaii. Our collections of species show varying leaf shapes and colors. This allows selection for houseplant use. We have trialed four species (*P. blanda*, *P. cookiana*, *P. oahuense*, and *P. sandwicense*) for tolerance under low light levels. Results of the study indicate *P. sandwicensis* as the species with the most potential for indoor use.

To complement our indoor trials, we also conducted studies to explore the potential of whole-leaf cuttings as a plant material for propagation. Leaf cuttings of *P. blanda*, *P. cookiana*, *P. oahuensis*, and *P. sandwicense* were treated with or without 1,000 ppm indolebutyric acid and allowed to root under mist (*P. cookiana*, *P. oahuense*, and *P. sandwicense*) or sprinkler irrigation (*P. blanda*). Results of the study indicate that all species except *P. sandwicensis* were fairly successful (>84% rooting) with propagation from leaf cuttings. Application of rooting hormone did not improve rooting, except for *P. oahuense* (rooting hormone application increased root length). The low rooting percentage of *P. sandwicensis* may be due to excessively wet conditions. Testing under dry conditions is recommended.

TISSUE CULTURE

Aside from seed and cutting propagation, our lab is also looking at tissue culture as a means to mass propagate select species, such as ala ala wai nui. One of the initial hurdles to successful tissue culture is finding the right sterilization technique. It may be difficult for ala ala wai nui since most of the species have fine hairs. A preliminary study on sterilization and tissue culture of *P. sandwicensis* has shown some success. However, more detailed studies are required to refine the technique.

SUMMARY

Based on our experiences, we can say that exploration of propagation protocols is an essential step to introducing new ornamental native plant selections. Our studies have shown that there is no one-size-fits-all method. Different species have different propagation challenges. Ease of propagation can be dependent on species and vary among selections. Hormone application may be beneficial, even for easy-to-root species.

Literature Cited

Baldos, O.C., DeFrank, J., Kramer, M., and Sakamoto, G.S. 2014. Storage humidity and temperature affect dormancy loss and viability of tanglehead (*Heteropogon contortus*) seeds. HortScience *49*:1328-1334.

Baldos, O.C., DeFrank, J., and Sakamoto, G.S. 2015. Germination response of dormant tanglehead (*Heteropogon contortus*) seeds to smoke-infused water and the smoke associated stimulatory compounds, karrikinolide and cyanide. HortScience *50*:421-429.

Ricordi, A.H., Kaufman, A.J., Cox, L.J., Criley, R., and Cheah, K.T. 2014. Going native in Hawai‘i. Landscape J. *33*(2):127-139.

Tamimi, L.N. 1999. The use of native Hawaiian plants by landscape architects in Hawaii [MLA thesis]. Blacksburg (VA): Virginia Polytechnic Institute and State University.

Additional resources

Abbott, I.A. 1992. La'au Hawai'i: Traditional Hawaiian Uses of Plants. Honolulu (HI): Bishop Museum Press., p. 99-102.

Bornhorst, H.L. 2005. Growing Native Hawaiian Plants: A How-to Guide for the Gardener. Honolulu (HI): Bess Press, p. 25-26.

Elliot, D.D. and Tamashiro, S.Y. 2009. Native Plants Hawai'i Project. URL: <http://nativeplants.hawaii.edu/> (accessed 25 June 2018). Honolulu (HI): University of Hawaii.

Herring, E. 2001. Hawaiian Native Plant Propagation Database. https://www.ctahr.hawaii.edu/hawnprop/plants/pep-lept.htm (accessed 25 June 2018). Honolulu (HI): College of Tropical Agriculture and Human Resources, University of Hawaii at Manoa.

Lilleeng-Rosenberger, K.E. 2005. Growing Hawai'i's Native Plants: A Simple Step-by-Step Approach for Every Species. Honolulu (HI): Mutual Publishing, p. 290-291.

Office of Hawaiian Affairs. 2015. OHA-5 HB 206/SB435 relating to Hawaiian plants. http://www.oha.org/wp-content/uploads/2015/01/OHA-5-Hawaiian-Plants-External-White-Paper-Final.pdf (accessed 25 June 2018). Honolulu (HI): Office of Hawaiian Affairs.

Wagner, W.L., Herbst, D.R., and Somer, S.H. 1999. Manual of Flowering Plants of Hawaii, revised ed. Honolulu (HI): University of Hawaii Press, p. 1029.

Combined Proceedings IPPS 68:202-204. 2018.

All-America Selections Winners For 2018: The Newest and Best Ornamentals and Edibles for North America and Producers

Diane Blazek[1,] and Eugene K. Blythe[2]
[1]All-America Selections, 1311 Butterfield Road, Suite 310, Downers Grove, Illinois 60515, USA; [2]Coastal Research and Extension Center, Mississippi State University, South Mississippi Branch Experiment Station, Poplarville, Mississippi 39470, USA

dblazek@aaswinners.com ; blythe@pss.msstate.edu

Keywords: annuals, plant trials, new plants, vegetables.

Abstract

Twelve varieties became All-America Selections (AAS) National Award Winners for 2018. AAS includes a network of over 80 trial grounds across the United States and Canada where new, never-before-sold varieties are "Tested Nationally and Proven Locally®" by skilled, impartial AAS Judges. Only the best performers are declared AAS Winners. Once these new varieties are announced as AAS Winners, they are available for immediate sale and distribution. An additional three varieties were selected as AAS Regional Award Winners for 2018. Regional winners undergo the same trialing process as national winners, but are recognized as varieties that exhibit outstanding performance in specific regional climates.

AAS NATIONAL WINNERS FOR 2018

Brassica rapa 'Asian Delight' (F_1 pak choi)

'Asian Delight' forms small to medium-sized (5- to 7-inch) heads that have a tasty, tender white rib and dark green, textured leaves. Plants are slower to bolt than other varieties. Bred by Seed Solutions of America LLC.

Canna generalis 'South Pacific Orange' (F_1 canna)

Attractive, vivid, bright orange flowers contrast nicely with bright green foliage on compact, seed-grown plants. More vigorous, uniform, and more basal branching than comparison varieties. Bred by Takii & Co., Ltd.

Capsicum annuum 'Onyx Red' (ornamental pepper)

Compact, well-branched plants are adorned with eye-catching, dark black foliage. The contrast between the diminutive black foliage and many shiny red peppers is striking and makes a bold statement in the garden. Bred by Takii & Co., Ltd.

Capsicum annuum 'Red Ember' (F_1 cayenne pepper)

This new introduction is earlier to mature than similar varieties, an important characteristic in areas with shorter growing seasons. 'Red Ember' produces many rounded-end fruits with just enough pungency for interest on durable, medium-sized plants. Bred by Johnny's Selected Seeds.

Capsicum chinense 'Roulette' (F_1 habanero pepper)

'Roulette' resembles a traditional habanero pepper in every way (fruit shape, size and color, and plant type) with one exception – no heat! When mature, the one-ounce, thick-walled fruits are red with a pleasant citrusy habanero flavor. Bred by Seminis Vegetable Seeds.

Cuphea FloriGlory® 'Diana' (Mexican heather)

Compact plants with dark green foliage bloom over a long period and are heat tolerant, making them ideal for borders, mass plantings, and containers. The intensely colored, magenta flowers are slightly larger and more numerous than on current varieties. Plants are vegetatively propagated. Bred by Westhoff.

Gypsophila muralis 'Gypsy White Improved' (gypsophila)

Semi-double blossoms on this new, improved variety exhibit better branching, growth habit, and heat tolerance than the older 'Gypsy White', making it perfect for containers, small spaces, and garden beds. Flowers are semi-double, larger in size, and more numerous than its predecessor, resulting in attractive mounds of white flowers. Bred by Sakata Seed Corp.

Solanum lycopersicum 'Red Racer' (F_1 cocktail tomato)

Cocktail tomatoes have a good sweet/acid balance and are a smaller variety tomato (although larger than cherry or grape tomatoes). Fruits are uniform in size and mature in clusters on determinate plants. Plants may be grown in small spaces and in containers. Bred by EarthWork Seeds (distributed by Garden Trends Wholesale).

Solanum lycopersicum 'Valentine' (F_1 tomato)

'Valentine' has an appetizing, deep-red color with a very sweet (Brix of 7 to 9) taste and will hold longer on the vine without cracking or losing the excellent eating quality. Gardeners should plan on staking the indeterminate vines for best results. Plants tolerate heat and keep on producing through the summer. Bred by Johnny's Selected Seeds.

Tagetes patula Super Hero™ Spry (crested French marigold)

This compact French marigold produces dark maroon lower petals and golden yellow upper petals on top of the dark green foliage. Flowers show a more uniform and stable color pattern, plants are earlier to bloom, and deadheading is not required. Bred by Ernst Benary of America, Inc.

***Zea mays* 'American Dream' (sweet corn)**

With excellent seed germination and very tender, super sweet kernels, this new sweet corn makes a great addition to the home garden. 'American Dream' matures slightly earlier than comparison varieties and produces vigorous, healthy plants with cobs that have good tip fill of bicolored kernels. Bred by Illinois Foundation Seeds, Inc.

***Zinnia elegans* 'Queeny Lime Orange' (zinnia)**

Sporting large, dahlia-like blooms on sturdy, compact plants, this variety provides cut flower gardeners and growers with an attractive hue for today's floral trends. The unique flower color evolves from dark coral/peach/orange to a light peach with a dark center as the flowers age. Bred by Hugo Dittmar (available through Floragran).

AAS REGIONAL WINNERS FOR 2018

***Capsicum annuum* 'Mexican Sunrise' (F_1 Hungarian pepper) (Regions: Southeast, Mountain/Southwest)**

These earlier maturing, conical, pendant-shaped peppers feature a thick-walled fruit that can be eaten at any stage. The semi-hot fruits exhibit a full spectrum of colors from lime green to yellow, then orange and red as they mature. Bred by Seeds by Design.

***Capsicum annuum* 'Mexican Sunset' (F_1 Hungarian pepper) (Regions: Southeast)**

The high heat, conically pointed peppers develop as a thick-walled fruit that can be eaten at any stage, but the taste tends to improve as the fruits ripen. Fruits set early and prolifically, and the plants may be grown in the ground or in containers. Bred by Seeds by Design.

***Solanum lycopersicum* 'Chef's Choice Red' (F_1 tomato) (Regions: West/Northwest)**

This fifth addition to the popular Chef's Choice tomato series is produces globe-shaped, tomato-red, beefsteak-type tomatoes with just the right balance of acid to sugar. Plants will produce 30 or more scar-free fruits throughout the season on disease-resistant, indeterminate vines with dark green foliage. Bred by Seeds by Design.

In summer 2018, the first two AAS Regional Winners for 2019 were announced:

***Citrullus lanatus* 'Cal Sweet Bush' (F_1 watermelon)**

This short-node watermelon produces two to three fruits weighing 10 to 12 pounds each on bushy vines with enough foliage cover to protect the fruits. Bred by Seeds by Design and Enza Zaden.

***Cucumis melo* 'Orange SilverWave' (F_1 melon)**

Plants produce attractive, 5-inch oval melons with an extremely sweet, orange flesh and unique rind color. Bred by Asia Seed Co., Ltd.

More information on AAS and AAS winners is available at: www.all-americaselections.org.

Combined Proceedings IPPS 68:205-209. 2018.

Forty Years of Trying to Containerize Everything: Successes and Failures

David Cliffe
Narromine Transplants, 120 Eumungerie Road, Narromine, New South Wales 2821, Australia.

david@transplants.com.au

Keywords: Cherry, *Eucalyptus*, grafting, pecan, polystyrene containers.

Abstract

There have been major advances in containers used for production of nursery stock during the past decades, particularly with the advent of plastics in the manufacturing of nursery containers. Most seedlings were being grown in seed beds in the early 1970s, with peat block and plastic bags started to make their appearance. Availability of polystyrene led to the development of Speedling® trays for vegetable seedling production. Plug trays gained popularity in forestry around the world, along with the development of mechanical transplanters. Containerization of woody crops still presents some challenges, with systems developed over the past 20 years for nursery production of grape, tea tree, cherry, and pecan, and citrus.

INTRODUCTION

During forty years of involvement in the nursery trade and before, I have endeavoured to be as innovative as possible in growing as many genera in containers as I could. During that time, there have been significant advances in the use of different materials for forming containers, rather than the recycling of kerosene drums and jam tins used when crops first came out of in-ground nurseries. Container growing has been further assisted by the availability of high-quality growing media, controlled-release fertilizers, and a better understanding of the physical properties substrates need to possess to produce high quality plants. This paper attempts to cover as many different containers that I have used or have seen being used in other nurseries, describe some of the successes we have had, and highlight some of the failures we have encountered along the way.

THE EARLY DAYS OF CONTAINER-IZATION

I remember my mother taking me to a nursery almost 70 years ago to buy bedding plants. Some were grown in-ground with the seedlings dug and wrapped in newspaper and some were grown in wooden flats. As I recall, the wooden flats had a division in the centre, used a very sandy medium, were returnable after use, and probably held about 100 seedlings.

There were also 4-gallon and 5-gallon kerosene drums in use, mainly with a heavy soil-based medium for growing citrus and some other fruit trees. Jam tins with holes punched in the bottom were quite commonplace and used for a wide range of genera. *Eucalyptus*, *Callistemon*, and other native plants were grown in metal tubes that could be unclipped and the tree extracted from them for planting, with this type of container was used extensively in the forestry and revegetation industries during the 1960s and 1970s. From this time forward, the use of plastics for container manufacture started to become commonplace and revolutionized the way we grew plants.

FROM BARE-ROOT SEEDLINGS TO THE SPEEDLING® FLAT AND BEYOND

In the early 1970s, most seedlings were being grown in seed beds, either by nurseries or individual growers of vegetables, tobacco, forestry trees, and bedding plants. Changes to this system probably started with the use of peat blocks, particularly in Europe, along with large grow blocks. Plastic grow bags also made an appearance during this time. In Australia, much of the citrus tree production was in 5-litre and 6-litre bags. Plastic grow bags are low cost and still in use today.

When polystyrene was made available for general manufacturing, it lent itself to being used for a huge range of molded products. George K. Todd, the son of an upstate New York vegetable grower who had trained as an electrical engineer, put his mind to coming up with a simpler method of growing seedlings than by using seed beds and wooden flats. He developed a 200-cell, polystyrene tray that was 76 mm deep with the same outer dimensions as traditional wooden flats and became known as the Speedling® 100A. This invention, which George Todd patented as the Todd Planter Flat and then branded the name Speedling®, revolutionized the seedling nursery industry globally. Introduced into Australia in the late 1970s by celery growers under license to Todd, growers kept the inverted pyramid shape but changed the cell numbers, depth, and outer dimensions of the tray to suit their benching that was originally set up to hold wooden flats.

In 1982, Narromine Transplants, at the time a subsidiary of Yates Seeds Ltd., entered into an arrangement with the Speedling Corporation of America (by then based in Florida) to use Speedling® flats, but with the original American sizes. This licensing arrangement also spread to South Africa and Israel around the same time. As an example of the adoption of the Speedling® concept in Australia, the entire vegetable seedling crop changed from in-ground production to containers within about an 8-year period.

Speedling Corporation went on to develop a range seedling flats of various cell sizes, including its popular 080A with 338 cells. This flat, with a tray depth of 57 mm, for the first time allowed a high density of seedlings per square metre and, consequently, a much better utilization of greenhouse space. We used this flat at Narromine Transplants to propagate a range of traditional vegetable seedlings, including processing tomatoes, but the cell size proved to be too small for the establishment of vegetable crops in harsh Australian conditions. We moved back to a 200-cell flat (but with only a 50-mm depth) and used the 080A for in-house propagation of tree seedlings that were then transferred into forestry tubes.

The 100A flat lent itself to the propagation of high-quality seedlings such as brassicas, peppers, tomatoes, tobacco, and some early work with forestry seedlings. Australian asparagus growers were keen to try growing green asparagus hybrids, such as UC157, from seed rather than using traditional crowns. The 100A flat lent itself to this but we soon realized there was a problem with the aggressive root system of this plant which penetrated the polystyrene foam flats, thus rendering both the flat and the seedling unusable. Through our Speedling nursery contacts in South Africa we were able to obtain a solution using copper oxide and water-based paint. Flats were dipped into the solution prior to use with asparagus and this stopped root penetration with a high degree of success. This dipping solution was also used with other crops that had the ability to penetrate the polystyrene, particularly *Eucalyptus*. This need disappeared once injection-moulded plastic flats became available.

From the original Todd Planter Flat, the patent of which was fiercely guarded by George Todd, there was a proliferation of other cell type flats around the world, all with different cell shapes designed to circumvent the patent. Today there are thousands of examples of cell-growing systems.

CONTAINERS IN FORESTRY

At a seedling growing refresher course around 1985, I raised the possibility of using Speedling®-type propagating flats for growing forestry seedlings and was laughed off the conference floor. Five years later we heard that the South Africans were using a 127-mm-deep polystyrene Speedling® flat to grow *Acacia* and *Eucalyptus* for the forestry industry there, so in 1992, after a visit to that country, we imported a few hundred trays to use as a trial and grew our first *Eucalyptus* seedlings for sale the following year. We had great success with this early production, but transport over large distances, coupled with the fragility of the flat in the field and the problems of handling by human transplanters, led us to look at the possibility of using a more user-friendly flat.

We also had problems with the germination of some *Eucalyptus*, such that direct-seeding into the final container was not an option without the use of hand labour to "block-up" empty cells. The Swedes had developed a 40-cell flat known as a Hiko V93 Tray with cylindrical, 93-cm^3 cells with root trainers and a tapered base. This tray was much more manageable in the field and it just happened that three trays set side-by-side were the same width as the original 100A Speedling® tray so that, with a bit of ingenuity, we were able to use our existing benching.

Our next challenge was to overcome the problems of low and intermittent germination of some of the species we were dealing with and the need to reduce hand labour. To this end we experimented with 512-cell plug flats, initially designed for the bedding plant industry, but we thought we should be able to single-seed them and eventually mechanically transplant them into the final container. Our competitors thought we were crazy and some of our clients exhibited a healthy amount of skepticism as well until we were able to prove that there was no detrimental effect on root systems. The degree of uniformity that we finished up with was exceptional. This practice is currently the norm in Australia, along with the use of mechanical transplanters being the accepted process of final transfer.

The forestry industry in Australia, South Africa, Chile, Brazil, Uruguay, and Argentina turned to the use of hybrid *Eucalyptus* for pulp production. This move required the production of these hybrids as clones via cutting production. The system of stock plant creation and maintenance, developed mainly in Brazil and South Africa, was adopted by two Australian nurseries in cooperation with one another, with Narromine Transplants being one of the participants. The process required was adapted by us using coir fibre slabs for stock plant maintenance, 512-cell plug trays for cutting propagation, and Hiko V93 flats as the final container. We have continued to use this production system for a number of other Australian native plants.

GRAPEVINES, ROOTSTOCKS, *MELLEUCA*, AND *LEPTOSPERMUM* SEEDLINGS AND CUTTINGS

During the early 2000s there was a rush to plant grapevines for wine grape production around the world, and Australia was no exception. We chose to grow our vines, both own-rooted and grafted, using waxed paper containers based on a system developed in the USA. This system required a higher price than for open-rooted vines grown in-ground. There was initial resistance to the use of the paper container, which was meant to be planted in its entirety in the vineyard. The system was clean, quick, and successful, and it appealed to customers that were in a hurry to plant and could afford the additional price. We have more recently grown a series of cherry rootstocks from cuttings in a paper container system.

Propagation of *Melaleuca alternifolia* for the production of tea tree oil is a challenge. Seed cannot be sown directly in the field and plant populations in production plantations require around 30,000 seedlings per acre. We have successfully propagated this crop by suspending seed (typically 14,000 seeds per gram) in a gel and sowing the seed in small clumps into seedling flats. Currently, we are also propagating a range of *Leptospermum* species for honey production and clones of selected *Melaleuca alternifolia* from cuttings using 512-cell plug trays and Hiko V93 flats as the final containers.

CHERRIES, CITRUS, AND PECANS

Several years ago, we were approached to grow cherries in a container size of 0.93 litres. We had never attempted any fruit tree work previously and started off growing rootstocks in coir bags for cutting production. Cuttings were stuck into a 240-cell tray, grown under mist, and then transplanted into the final containers. The trees grew well but there was a great deal of grower resistance to such small plants. Most growers were used to open root whips and after one season they reverted to open-root plants. We continue to produce citrus with the rootstock seedlings grown in Hiko V93 flats and then transferred to a 3.3-litre air-pruning container for budding and growing-on. This container grows excellent trees and there has been no buyer resistance with these plants.

During the past two years, we have been attempting to grow pecans, *Carya illinoinensis*, in containers. Traditionally grown in-ground from seedling rootstocks and usually patch-budded, we were approached by Australia's biggest grower to attempt containerization. We started with 0.93-litre containers planted with a seedling rootstock grown in V93 Hiko flats during summer and then grafted using terminal side and top wedge grafts during late winter; it was a dismal failure. We grew-on the failed grafts and patch-budded in late January using new wood, but with only limited success.

I had a conversation with another IPPS member about our problems, he told me that he had tried using the callus system that we used for grapevine propagation, but with containerized pecan seedling rootstocks that were bare-rooted, bench grafted, put through a callus box system, and then grown-on in a larger container. We tried this with a large number of rootstocks and had varying success, sometimes up to 93% but in general 40% to 60% strike rates. We initially used a

3.7-litre round container used mainly for citrus, but this was mistake as pecans have very strong, fast growing root systems and we experienced root curling at the base of the container. We now use our 3.3-litre air-pruner citrus container for pecans which allows a much more acceptable root system to develop.

We are still struggling to increase our grafting take rate and, in the coming season, will be more selective with grafting wood, seedling readiness, and general post-callusing handling in the nursery. One of the main problems we think pecans have in the nursery is that there are no clonal rootstocks available and seedling rootstocks are quite variable in height and vigour, leading to grafting and budding failures.

We will continue to try adapting as many plants as we can to container production.

Combined Proceedings IPPS 68:210-216. 2018.

Propagation of Ornamental Figs (*Ficus*)

Richard A. Criley
Department of Tropical Plant and Soil Sciences, University of Hawaii at Manoa, Honolulu, Hawaii 96822, USA

criley@hawaii.edu

Keywords: *Ficus carica, Ficus benjamina, Ficus elastica, Ficus pumila,* figs, grafting, layering, micropropagation.

Abstract

Many species of *Ficus* are used for interiorscaping and landscaping. Their growth habits range from prostrate groundcovers to shrubs, trees, and vines. Nearly all are easily propagated by the usual vegetative techniques of cutting, layering, grafting, and micropropagation. Fewer are propagated by seed as pollination requires specialized wasps, but some of those that do produce seed have become invasive. *Ficus* species are suitable subjects for plant propagation classes as many, such as *F. elastica*, *F. rubiginosa*, and *F. benghalensis*, can be propagated by single-node cuttings as well as by stem cuttings and air layers. Aerial roots are characteristic of some species, which also signal ease of rooting. Many of the selections used in interiorscapes have been tissue cultured, giving rise to diverse forms, some of which are more compact and better branched.

INTRODUCTION

Among the most celebrated *Ficus* species are the edible fig (*F. carica*), the Bo (Bohdi) tree (*F. religiosa*, associated with Buddha), and the Indian Banyan (*F. benghalensis*, associated with three Hindu gods). All have been propagated vegetatively to maintain lines known for their eating qualities or religious connections. Another with religious connections is *F. sycamorus*, associated with the Egyptian goddesses Isis, Hathor, and Nut. Many citations on *Ficus* propagation can be found in an online search, but only a small selection is cited in this review.

CUTTINGS

Citing the ancient Greek philosopher and botanist Theophrastus, Condit (1947) related that "the fig is propagated by cuttings more readily than any other fruit tree." Branches up to 4 feet in length would be planted in deep holes directly in the field where they would be grown, or shorter cuttings would be produced in a nursery row, then transplanted to the final site. According to Condit (1947), any wood up to 2 or 3 years of age could be used, with a diameter of ⅜ to ¾ inches and about 9 inches in length. These would be handled as hardwood cuttings and heeled into the ground or nursery beds until callused, then lined out in the field and grown for a season, and dug, heeled in, and planted out in the second season. Such practices would probably work for ornamental *Ficus*, but since the target use is not as an orchard, there would be little need to follow such protocols. Condit noted that rooting occurred so readily that rooting hormones were not necessary.

Among the early foliage plants produced in south Florida were members of the genus *Ficus* (Neel, 1975, 1979). Many cultivars of *F. elastica* were propagated by air layering as was the large-leaved *F. lyrata* (Henley, 1979). Henley also noted that *F. benjamina* was produced from semi-hardwood cuttings. Neel (1979) provided tables reporting common methods of tropical plants in which cuttings were most used for *F. benjamina* and *F. microcarpa* and air layering was used for *F. elastica* and *F. lyrata*.

Conditions for rapid multiplication of leaf-bud cuttings of *F. elastica* 'Robusta' were evaluated by Morgan and Lawlor (1976). Rate of rooting improved as bottom heat increased from 18°C to 30°C; a pH between 4.5 and 5.5 for a peat-based medium yielded the longest roots; and the rooting hormones NAA and IBA enhanced rooting in comparison with controls, with a powdered formulation of 0.8% IBA generally being most effective. In the best treatments, root initiation was evident in 15 days and, at 3 weeks, roots of 3 to 5 cm length were recorded. Gooch and Criley (1980) reported that node position influenced extent and rooting percentage of *F. elastica* 'Decora' with node positions 6 through 11 having the highest rooting potential (node position 1 was at the fourth expanded leaf as nodes of the terminal three leaves were too immature). Wang (1988) reported that bottom heat (28°C) enhanced root development of *F. benjamina* as did higher light levels (290 $\mu mol \cdot s^{-1} \cdot m^{-2}$ versus 90 $\mu mol \cdot s^{-1} \cdot m^{-2}$) in the propagation bench.

Propagation of *F. elastica* by single-node cuttings consisting of a section of stem, a healthy bud, and a leaf trimmed to one-third or one-half its length is also noted by Henley (1979). Single-node cutting propagation produces more plants than air layering, but development of a sizable plant takes longer. During the glory days of the foliage plant trade in Florida, Poole and Conover (1984, 1990) described the use of single-node cuttings of the large-leaved *Ficus* (five cultivars of *F. elastica* and *F. lyrata*) to produce more plants from a stock plant than the then common method of air layering. Whole leaves with a portion of stem and the axillary bud could be used, but the leaf had to be rolled and held with a rubber band to reduce the space used in the mist propagation bench. They reported that a leaf could be reduced to 50% of its area and still achieve a satisfactory root grade and shoot production. Treatment of the base of the stem with 0.8% IBA was beneficial. The leaf-and-eye method used for *F. elastica* is often coupled with a foliar spray of 10 ppm IBA (Griffith, 2006). High nutrition levels for the stock plant also yielded single node cuttings with better root systems (Poole and Conover, 1979). Dole and Gibson (2006) also advocated single-node (leaf-bud) cuttings for large leaved *Ficus*, noting that leaf size can be reduced to accommodate more cuttings in the propagation bench.

Among the more recent and accessible reports on propagation of *Ficus* are those noted by Griffith (2006). For the smaller-leafed varieties of *F. benjamina* and *F. microcarpa*, cuttings prepared with or without rooting hormone root readily under mist or high humidity with bottom heat (28°C). Griffith (2006) notes that *F. lyrata* is mostly propagated by tissue culture, which may have produced new compact forms. *F. maclellandii* 'Alii' and the newer 'Amstel King' are propagated by air layering to obtain larger sized trees quickly, as cutting propagation is more difficult.

Dole and Gibson (2006) provided a table suggesting that cuttings are preferred for vining type *Ficus* species such as *F. pumila*, whereas upright types should be propagated by tip cuttings or layers. Single-eye and leaf-bud cuttings are less preferred techniques. Seed propagation is noted but is seldom available. Stock plant management and the physiological status of the plant influence ease of rooting, with *F. benjamina* rooting readily from tip or leaf bud cuttings in any medium, whereas *F. maclellandii* 'Amstel King' and 'Alii' are considered difficult to root. *F. benjamina* tip or leaf-bud cuttings from the more distal part of a shoot root more readily than cuttings made from the stem base. Aging also influences ease of rooting as Davies and Joiner (1980) and Davies (1983) found juvenile *F. pumila* cuttings rooted more readily than cuttings from mature wood.

Although *F. pumila* normally produces adventitious roots during its vining (juvenile) phase, the mature tissues are more difficult to root. Davies and Joiner (1980) reported that 1000-1500 ppm IBA was best for juvenile leaf bud cuttings, whereas mature cuttings required treatment with 2000-3000 ppm IBA to achieve the best root number, quality, and length, and took longer to root than did the juvenile cuttings. Juvenile stem cuttings rooted in 40 days or less with IBA treatments, whereas mature cuttings were still not rooted and required up to 90 days to achieve 60% to 80% rooting (Davies and Joiner, 1978). Juvenile leaf-bud cuttings were likewise faster to root and had higher rooting percentages than mature cuttings, even when both types were treated with 1000 ppm IBA. Interestingly, a small percentage of leaf cuttings (blade and petiole only) from mature plants rooted with a treatment of 1000 ppm IBA/NAA, but none did so from juvenile plants, likely because of insufficient petiole (juvenile leaves are sessile).

LAYERING

According to Condit (1947), Thomas Jefferson propagated his edible fig by simple layers in which a slant cut is made in recently matured wood, and the shoot is bent into the ground and covered with soil leaving the tip exposed and in an upright position. Condit (1947) also describes trench layering, stooling (mound layering), and air layering as useful practices for edible fig. For ornamental figs, however, air layering has been a more common practice.

Hartmann and Kester (1983), in the fourth edition of their plant propagation book, provided a photo series of air layer practices for *Ficus elastica*. Air layering of *F. elastica* was also the focus of a paper by Broschat and Donselman (1981) who recommended two lengthwise slits through the stem, rather than complete removal of bark by girdling, to minimize leaf spotting due to water stress.

Both *F. benjamina* and *F. elastica* have been propagated by air layers, rooting in about 6 weeks with application of an auxin (Wadewitz, 1981). A length of 1.5 to 2 feet (48 to 61 cm) was recommended; also, the terminal end of *F. elastica* shoots should have about 6 or 7 leaves distal to the girdled zone.

GRAFTING

As cuttings have been so widely used to propagate *Ficus*, little attention has been given to grafting or budding, except with respect to the edible fig. Condit (1947) noted that rootstock selection for tolerance to diseases, nematodes, and various soil conditions or for improved growth of weak-growing varieties led to evaluation of other species on which *F. carica* could be grafted. These sometimes were only successful in the short run (e.g., *F. carica* grafted onto species native to Florida), whereas other rootstocks [*F. cocculifolia*, *F. gnaphalocarpa*, *F. glomerata* (syn. *F. racemosa*), and *F. palmata*] showed longer-term success with no incompatibility for chip-budded or grafted *F. carica* (Krezdorn and Glasgow, 1970).

Budding and grafting of edible fig trees were done in the 3rd century B.C. (Cato, cited by Condit, 1947). Figs were topworked using shield buds and patch buds. Cleft grafts and bark grafts have also been used on branches up to 4 or 5 inches in diameter. The author recently observed an ornamental wax ficus (*Ficus microcarpa* var. *crassifolia*) bark-grafted onto a rootstock to produce bonsai-like ornamental plants. The technique might be a useful addition to plant propagation laboratories or for student plant sales.

The practice of stenting (in which a scion is grafted onto a non-rooted rootstock and the formation of the union and adventitious roots on the rootstock occurs simultaneously) was successful for the production of the variegated *F. benjamina* 'Starlight' onto the green-leaved cultivar 'Green Leaf' (Babaie et al., 2014). An omega cut was used for the graft and the rootstock was treated with 2000 to 6000 ppm IBA. Scion and rootstock were 5 to 10 cm in length and 5 to 10 mm in diameter. After 7 weeks, the best rooting occurred using the 4000 and 6000 ppm IBA treatments and grafting take ranged from 83% to 90%.

MICROPROPAGATION

The edible fig, *F. carica*, was initially propagated aseptically using shoot tips with 3 or 4 leaf primordia to free plants of a fig mosaic virus (Murithi et al., 1982; Demiralay et al., 1998; Gunver and Ertan, 1998), but later work was initiated to produce plants in quantity for high density fig orchards (Pontikis and Melas, 1986). A key to their success was the incorporation of phloroglucinol into the multiplication medium. The highest rooting percentage (80%) was achieved with 4.9 µM IBA and no BA in the rooting medium. Nobre and Romana (1998) excised 5-6 mm shoot tips with 2 or 3 leaf primordia and achieved shoots at 15 weeks.

The antioxidant polyvinylpyrrolidone (0.05%) was required for differentiation and growth of these shoot tips, which were then subcultured from single nodes. Benzylaminopurine (BAP at 1 $mg \cdot L^{-1}$) was essential for proliferation of multiple shoots from axillary buds on stem segments of the variegated *F. benjamina* 'Starlight' (del Amo and Picazo, 1992) as was phloroglucinol (80 $mg \cdot L^{-1}$). Rooting was superior using IBA (0.5 to 1.0 $mg \cdot L^{-1}$) versus NAA with only a 3- to 5-day induction period and subsequent transfer to PGR-free medium. On the other hand, NAA (1 $mg \cdot L^{-1}$) was effective for rooting microshoots of *F. religiosa* derived after stimulation with BAP (1 $mg \cdot L^{-1}$) from callus of stem segments of mature trees (Jaiswal and Narayan, 1985).

Prior to the development of tissue culture, conventional propagation techniques were used throughout the foliage industry. Debergh and Wael (1977) used 1-cm^2 leaf sections containing the main vein and achieved plantlet development with rootable shoots in 8 weeks. In the 1980s, Twyford Plant Laboratories in California developed tissue culture systems for several cultivars of *F. benjamina*, *F. lyrata*, and *F. elastica* (Lloyd, 1990). An Australian nursery has had great success with microcuttings of *F. lyrata* taken from micropropagated plants that had been established ex vitro (Bunker, 1981). Such cuttings were found to root quickly, and a large inventory could be built up quickly.

Makino et al. (1977) described in detail the medium and phytohormone additions, explants, subculturing, conditions of culture and rooting, and establishment that led to successful micropropagation of *F. benjamina*, *F. elastica*, and *F. pandurata* (*F. lyrata*). Initially using 3- to 5-mm shoot tips, Jona and Gribaudo (1987) were later successful in inducing adventitious buds from fragments of young leaves of *F. lyrata*. To their modification of a Nitsch and Nitsch medium, various cytokinins were added singly and in combination in increasing combinations of 24, 48, or 72 μM. Adventitious shoot production increased with cytokinin concentration, but rooting was inhibited at the higher concentrations. GA3 at 4 μM improved shoot elongation before rooting, which was induced on individual shoots with a 7-day exposure to 5 μM IBA or NAA in the medium.

SEED

According to Condit (1947), Theophrastus (Father of Botany) tells of growing figs from seed in the 3rd century B.C. Fertile seeds were separated from sterile seeds by immersion in water, with the former sinking and the latter floating. Germination occurs as readily as for tomato seeds. Except for breeding efforts with the edible fig (*Ficus carica*), there has been little use of seed to produce *Ficus* for ornamental uses.

For the most part, this is due to a lack of seed set with no natural pollinators and a complex floral structure that does not easily lend itself to artificial pollination. In Hawaii and a few other areas, seedlings of *F. microcarpa* and *F. religiosa* are invasive plants spread by birds that consume the fruits. Germination is rapid and young plants are found on trees and in sidewalk cracks and rain gutters and dispersed throughout the landscape. Were these desirable ornamental species, such seedlings could be used in the landscape trade, but their invasiveness, ultimate tree size of both species, and insect damage occurring on *F. microcarpa* render such usage unlikely.

Propagation and planting techniques for six Asian dioecious *Ficus* species were evaluated for their inclusion in forest restoration plantings. *F. auriculata*, *F. fulva*, *F. hispida*, *F. oligodon*, *F. semicordata*, and *F. variegata* were efficiently and more economically produced from seed than by cuttings (Kuaraksa and Elliott, 2013). Matthew et al. (2011) reported germination percentages of seeds of *F. racemosa*, *F. microcarpa*, *F. religiosa*, and *F. benghalensis* were 5.0, 2.3, 27.7, and 82.0%, respectively, but seeds of the first two were short-lived (< 6 months), whereas seeds of *F. religiosa* began declining in viability after 12 months and seeds of *F. benghalensis* were still viable at 18 months.

Literature Cited

Babaie, H., Zarei, H., and Hemmati, K. 2014. Propagation of *Ficus benjamina* var. Starlight by stenting technique under different concentrations of IBA in various times of taking cutting. J. Orn. Plants *4*:75-79.

Broschat, T.K., and Donselman, H.M. 1981. Effects of light intensity, air-layering and water stress on leaf diffusive resistance and incidence of leaf spotting on *Ficus elastica*. HortScience *16*:211-212.

Bunker, E.J. 1981. Growing certain Australian native shrubs and trees from softwood cuttings. Comb. Proc. Intl. Plant Prop. Soc. *31*:130-133.

Condit, I.J. 1947. The Fig. Chronica Botanica Co., Waltham, MA.

Davies, F.T., Jr. 1983. Influence of juvenility and maturity in propagation. Comb. Proc. Intl. Plant Prop. Soc. *33*:559-564.

Davies, F.T., Jr. and Joiner, J.N. 1978. Adventitious root formation in three cutting types of *Ficus pumila* L. Proc. Comb. Proc. Intl. Plant Prop. Soc. *28*:306-313.

Davies, F.T. Jr. and Joiner, J.N. 1980. Growth regulator effects on adventitious root formation in leaf bud cuttings of juvenile and mature *Ficus pumila*. J. Amer. Soc. Hort. Sci. *105*:91-95.

Debergh, P. and DeWael, J. 1977. Mass propagation of *Ficus lyrata*. Acta Hort. *78*:361-364.

del Amo, J. B. and Picazo, I. 1992. In vitro propagation of *Ficus benjamina* cv. Starlight from axillary buds with BAP and phloroglucinol. Gartenbauwiss *57*:29-32.

Demiralay, A., Yalçin-Mendi, Y., Aka-Kaçar, Y., and Çetiner, S. 1998. In vitro propagation of *Ficus carica* L. var. Bursa Siyahi through meristem culture. Acta Hort. *480*:165-167.

Dole, J.M. and Gibson, J.L. 2006. Cutting Propagation. Ball Publishing, Batavia, IL.

Gooch, M. and Criley, R.A. 1980. Effect of node position on the rooting of single node cuttings of *Ficus elastica* 'Decora'. Univ. Hawaii Coop. Ext. Serv. Hort. Digest *55*:5-6. June 1980.

Griffith, L.P., Jr. 2006. Tropical Foliage Plants. 2nd edition. Ball Publishing, Batavia, IL.

Günver, G. and Ertan, E. 1998. A study on the propagation of figs by the tissue culture techniques. Acta Hort. *480*:169-171.

Hartmann, H.T. and Kester, D.E. 1983. Plant Propagation Principles and Practices. 4th edition. Prentice-Hall, Inc., Englewood Cliffs, NJ.

Henley, R.W. 1979. Tropical foliage plants for propagation. Comb. Proc. Intl. Plant Prop. Soc. *29*:454-467.

Jaiswal, V.S. and Narayan, P. 1985. Regeneration of plantlets from the callus of stem segments of adult plants of *Ficus religiosa* L. Plant Cell Rpt. *4*:256-258.

Jona, R. and Gribaudo, I. 1987. Adventitious bud formation from leaf explants of *Ficus lyrata*. HortScience *22*:651-653.

Krezdorn, A.H. and Glasgow, S.K. 1970. Propagation of *Ficus carica* on tropical species of *Ficus*. Proc. Trop. Reg., Amer. Soc. Hort. Sci. *14*:156-164

Kuaraksa, C. and Elliott, S. 2013. The use of Asian *Ficus* species for restoring tropical forest ecosystems. Restor. Ecol. *21*:86-95.

Lloyd, G. 1990. The impact of tissue culture on *Ficus* spp.: Propagation and production. Comb. Proc. Intl. Plant Prop. Soc. *40*:163-165.

Makino, R.K., Nakano, R.T., Makino, P.J., and Murashige, T. 1977. Rapid cloning of *Ficus* cultivars through application of in vitro methodology. In Vitro *13*:169. (abstr.).

Mathew, G., Skaria, B.P., and Joseph, A. 2011. Standardization of conventional propagation techniques for four medicinal species of genus *Ficus* Linn. Indian J. Nat. Prod. Resources *2*: 88-96.

Morgan, J.V. and Lawlor, H.W. 1976. Influence of external factors on the rooting of leaf-bud cuttings of *Ficus*. Acta Hort. *64*:39-46.

Muriithi, M., Rangan, T.S., and Waite, B.H. 1982. In vitro propagation of fig through shoot tip culture. HortScience *17*:86-87.

Neel, P.L. 1975. Production of selected ornamental species in south Florida. Comb. Proc. Intl. Plant Prop. Soc. *25*:368-373.

Neel, P.L. 1979. Macropropagation of tropical plants as practiced in Florida. Comb. Proc. Intl. Plant Prop. Soc. *29*:468-480.

Nobre, J. and Romano, A. 1998. In vitro cloning of *Ficus carica* L. adult trees. Acta. Hort. *480*:161-164.

Pontikis, C.A. and Melas, P. 1986. Micropropagation of *Ficus carica* L. HortScience *21*:153.

Poole, R.T. and Conover, C.A. 1979. Propagation and growth characteristics of *Ficus elastica* 'Decora'. Foliage Digest *2*:14-15.

Poole, R.T. and Conover, C.A. 1984. Propagation of ornamental *Ficus* by cuttings. HortScience *19*:120-121.

Poole, R.T. and Conover, C.A. 1990. Propagation of *Ficus elastic* and *Ficus lyrata* by cuttings. https://mrec.ifas.ufl.edu/foliage/resrpts/rh_90_6.htm

Wadewitz, D.W. 1981. Propagation of *Ficus* species by air layering. Comb. Proc. Intl. Plant Prop. Soc. *31*:257-258.

Wang, Y.T. 1988. Influence of light and heated medium on rooting and shoot growth of two foliage plant species. HortScience *23*:346-347.

Combined Proceedings IPPS 68:217-218. 2018.

Two Australian Propagators on Assignment in Oman

David Hancock[1] and Dermot Molloy[2]
[1]Natural Area Nursery, 99C Lord Street, Whiteman, Western Australia 6068, Australia
[2]Royal Botanic Gardens Victoria, Birdwood Avenue, South Yarra, Victoria 3141, Australia

david@naturalarea.com.au

Keywords: Arabian plants, wetland plants, wetland purification.

Abstract

The authors report on two trips to the Sultanate of Oman to assist with horticultural projects. The first trip (November 2015) focused on assisting the management and staff of the Oman Botanic Garden on issues of seed propagation, propagation mixes, and nursery practice. The second trip (May 2018) focused on advising on requirements for propagation of wetland species at the Bauer Nimr nursery.

PROJECT 1: OMAN BOTANIC GARDEN NURSERY

The Sultanate of Oman has commenced a long-term project to develop a world class facility to display the plants of Oman and the Arabian Peninsula and to document, conserve, and research the ethnobotany of the region. It is proposed that the garden area cover 420 hectares and showcase the plants in their climatic settings via large temperature- and humidity-controlled biome domes, as well as display traditional Omani horticulture, such as date groves and Aflaj irrigation systems.

Through an IPPS Melbourne, Australia conference, the authors met two visiting propagators from the Oman Botanic Garden (OBG) nursery, which is situated approximately 20 km from Muscat. After the Melbourne conference, the Oman team travelled to Perth and visited Natural Area Nursery and other venues.

The authors then travelled to Oman in November 2015 to assist OBG management and staff with issues regarding seed management, germination, propagation mixes, and nursery processes. Also, a trip to the mountains and the OBG satellite altitude nursery (Jebel Al Akhdhar) was undertaken and seed and propagation material was collected.

IPPS Vol. 68 – 2018

The OBG nursery has produced over 100,000 plants and 350 varieties. Many of large plants (in containers of sizes up to 3,000 litres) have been prepared for installation in the OBG complex when established. The nursery covers 30,000 square metres and has large, air-conditioned propagation houses and a herbarium that holds the largest documented collection of Arabian plants in the world.

Hancock & Molloy have maintained regular contact with the team at OBG and made a short, follow-up visit in May 2018.

PROJECT 2: BAUER NUMR, OMAN NURSERY PROJECT

In May 2018, Bauer Nimr LLC, a subsidiary of the German engineering and environmental consultancy Bauer Resources, appointed Perth Plant Propagation Pty. Ltd. (via its Director and consultant David Hancock and sub-consultant Dermot Molloy) to provide advice on the strategy and operational requirements for propagation of up to 800,000 wetland plants at the Nimr nursery in the Sultanate of Oman for supply by April 2019. The Nimr oilfield in Oman produces 9 litres of water for every litre of oil extracted. Following extraction of the oil from the oil water mix, past practice was for the oily water to be returned to a deep aquifer by pressure pumping to as deep as 1.5 km. This process incurred substantial energy use. Bauer Resources entered into arrangements with Petroleum Development Oman to treat and purify the oil-laden water via a constructed wetland. Bauer subsequently established the largest constructed wetland in the world and daily treats 110,000 cubic metres of oil-contaminated water per day and thereafter passes the water into evaporation ponds.

The success of the wetland purification system has led to a new project for an additional 250 hectares of wetland to absorb additional water and oil from expansion of the Nimr oilfield. The plan for the new wetland involves plant species not previously propagated before in Oman, and the Nimr nursery was not equipped to grow the necessary species from seed.

The requirements of the plant production, being on a large scale and a short timeframe with no nursery to speak of at commencement, was a significant challenge. Given that this plant supply was critical to the additional stage of the Nimr constructed wetland, the authors provided advice on plant propagation and on the planning and completion of upgrades to the existing nursery.

Critical success factors for the project were identified in a report and included:

1. Adequate seed and seed viability.
2. Early provision of necessary equipment & infrastructure.
3. Diligence in seeding and maintenance of seed trays.
4. Reliable power and water systems.
5. Experienced and capable nursery management.
6. Early detection and treatment of plant pathogens.

Combined Proceedings IPPS 68:219-220. 2018.

Auxin drench treatment improves rooting of *Salvia* 'Blue Hills' and *Scabiosa* 'Pink Mist'

Benjamin K. Hoover
Department of Horticulture and Crop Science, California Polytechnic State University, San Luis Obispo, California 93407, USA

bkhoover@calpoly.edu

Keywords: Biochar, herbaceous cuttings, K-IBA.

Abstract

Auxin is frequently used to improve rooting uniformity of herbaceous cuttings. Indole-3-butyric acid (IBA) is normally applied in a powder to the base of the stem or in solution as a dip or foliar spray. The objectives of this study were to: (1) assess the rooting of herbaceous cuttings after treatment with auxin in powder form or as a liquid drench applied to the substrates and (2) assess if biochar incorporation in the substrate affected auxin drench efficacy.

The base substrate was 1:2 (v/v) sphagnum peat and coarse perlite, and amended with biochar (0%, 10%, 20%, 40%, or 80% by vol.). The biochar was a coconut-based product obtained from a commercial supplier (Bay Area Biochar, Concord, California, USA) and screened to a maximum particle size of 2 mm prior to incorporation. Cuttings of two herbaceous perennial plants (*Salvia* ×*sylvestris* 'Blue Hills' and *Scabiosa columbaria* 'Pink Mist') were obtained from a commercial supplier.

Auxin was supplied as a powder [Rhizopon (Phytotronics, Earth City, Missouri, USA) at 1,000 ppm and 3,000 ppm] by dipping the stem and knocking off excess, or as a substrate drench (Hortus Water Soluble Salts at 1,000 and 3,000 ppm K-IBA) applied as 10 ml solution per rose pot (top width: 2.25 inches, depth: 3.25 inches, vol: 185 cm^3) prior to sticking cuttings. One cutting was stuck in each rose pot. A separate randomized complete block design was used for each species. There were 4 replicates per treatment for *Scabiosa* and 9 replicates per treatment for *Salvia*.

The cuttings were placed in an enclosure with high humidity (~95% RH), with bottom heat supplied at 70°F. Data were collected after 28 days for *Salvia* and 29 days for *Scabiosa*. Cuttings were removed from the rose pots, the roots were washed, and each cutting was assigned an adventitious rooting rating. Roots were excised from the stem and scanned with a flatbed scanner (Epson Perfection V19). Scans were analyzed for

IPPS Vol. 68 – 2018

first-order lateral root counts and two-dimensional root area with ImageJ software.

Neither species showed an interaction effect between auxin treatment and biochar on rooting responses. Substrates with the 80% rate of biochar produced plants with lower ratings and lower root area than the control ($P < 0.05$). Drenching substrates with 1,000 ppm K-IBA produced increased rooting in all metrics compared to the control. Drenching substrates with 3,000 ppm K-IBA did not improve the rooting rating of cuttings compared to the control but did increase primary root counts ($P < 0.05$). Cuttings receiving the 3,000 ppm K-IBA drench frequently initiated many short roots from the petioles of leaves near the substrate surface. This study suggests that drenching of substrates prior to sticking may be an effective method of auxin application.

Combined Proceedings IPPS 68:221-222. 2018.

Biochar and Sand-Amended Cutting Substrates: Particle Size Effects

Benjamin K. Hoover
Department of Horticulture and Crop Science, California Polytechnic State University, San Luis Obispo, California 93407, USA

bkhoover@calpoly.edu

Keywords: Biochar, herbaceous cuttings, K-IBA.

Abstract

Auxin is frequently used to improve rooting uniformity of herbaceous cuttings. Indole-3-butyric acid (IBA) is normally applied in a powder to the base of the stem or in solution as a dip or foliar spray. The objectives of this study were to: (1) assess the rooting of herbaceous cuttings after treatment with auxin in powder form or as a liquid drench applied to the substrates and (2) assess if biochar incorporation in the substrate affected auxin drench efficacy.

The base substrate was 1:2 (v/v) sphagnum peat and coarse perlite, and amended with biochar (0%, 10%, 20%, 40%, or 80% by vol.). The biochar was a coconut-based product obtained from a commercial supplier (Bay Area Biochar, Concord, California, USA) and screened to a maximum particle size of 2 mm prior to incorporation. Cuttings of two herbaceous perennial plants (*Salvia* ×*sylvestris* 'Blue Hills' and *Scabiosa columbaria* 'Pink Mist') were obtained from a commercial supplier.

Auxin was supplied as a powder [Rhizopon (Phytotronics, Earth City, Missouri, USA) at 1,000 ppm and 3,000 ppm] by dipping the stem and knocking off excess, or as a substrate drench (Hortus Water Soluble Salts at 1,000 and 3,000 ppm K-IBA) applied as 10 ml solution per rose pot (top width: 2.25 inches, depth: 3.25 inches, vol: 185 cm^3) prior to sticking cuttings. One cutting was stuck in each rose pot. A separate randomized complete block design was used for each species. There were 4 replicates per treatment for *Scabiosa* and 9 replicates per treatment for *Salvia*.

The cuttings were placed in an enclosure with high humidity (~95% RH), with bottom heat supplied at 70°F. Data were collected after 28 days for *Salvia* and 29 days for *Scabiosa*. Cuttings were removed from the rose pots, the roots were washed, and each cutting was assigned an adventitious rooting rating. Roots were excised from the stem and scanned with a flatbed scanner (Epson Perfection V19). Scans were analyzed for first-order lateral root counts and two-dimensional root area with ImageJ software.

Neither species showed an interaction effect between auxin treatment and biochar on rooting responses. Substrates with the 80% rate of biochar produced plants with lower ratings and lower root area than the control ($P < 0.05$). Drenching substrates with 1,000 ppm K-IBA produced increased rooting in all metrics compared to the control. Drenching substrates with 3,000 ppm K-IBA did not improve the rooting rating of cuttings compared to the control but did increase primary root counts ($P < 0.05$). Cuttings receiving the 3,000 ppm K-IBA drench frequently initiated many short roots from the petioles of leaves near the substrate surface. This study suggests that drenching of substrates prior to sticking may be an effective method of auxin application.

Combined Proceedings IPPS 68:223-227. 2018.

Gardens by the Bay and Applied Research: Bringing a World of Plants to Singapore

Andrea Kee
Gardens by the Bay, 18 Marina Gardens Drive, Singapore 018953

andrea.kee@gardensbythebay.com.sg

Keywords: Conservatories, plant trials, public gardens, waterfront gardens.

Abstract

Gardens by the Bay is a national garden covering 101 hectares in the Central Region of Singapore, adjacent to the Marina Reservoir. The park consists of three waterfront gardens: Bay South Garden, Bay East Garden and Bay Central Garden. Cooled conservatories display plants that cannot be grown outdoors in the climate of Singapore. The Research & Horticulture Department conducts trials for plants to be displayed at the Flower Field in Flower Dome and the Orchid Corner in Cloud Forest.

INTRODUCTION

A national garden and premier horticultural attraction for local and international visitors, Gardens by the Bay is a showpiece of horticulture and garden artistry that presents the plant kingdom in a whole new way, entertaining while educating visitors with plants seldom seen in this part of the world, ranging from species in cool, temperate climates to tropical forests and habitats.

WATERFRONT GARDENS

Gardens by the Bay comprises three distinctive waterfront gardens: Bay South, Bay East and Bay Central (Figure 1).

Figure 1. The three waterfront gardens at Gardens by the Bay in Singapore: Bay South, Bay East, and Bay Central.

Bay South, the largest of the three gardens with an area of 54 hectares, opened in 2012. With its award-winning, cooled conservatories and iconic Supertrees, Bay South has placed Singapore squarely on the international map and is a source of national pride. Within Flower Dome, the ever-changing floral displays including crowd favourites such as Tulipmania, Orchid Extravaganza, and Cherry Blossoms bring the beauty and diversity of the plant kingdom to life for all to enjoy.

Bay East, the second largest garden, offers a tranquil respite from the bustling city and a stunning view of the Singapore skyline even as it unfolds over the next century. This green space is open to the public and has immense potential for future development as a waterfront garden.

Bay Central is the garden which will serve as a link between Bay South to Bay East when developed. The garden features a 3km waterfront promenade boasting stunning views of the city.

While plant displays remain the focal point of Gardens by the Bay, engaging programmes and excellent service form key pillars in enhancing the Gardens' overall visitor experience.

OUTDOOR GARDENS

The outdoor gardens comprise different themed gardens: Heritage Gardens, World of Plants, Sun Pavilion, Silver Garden, Dragonfly Lake, Kingfisher Lake, and Supertree Grove. A large representation of the plant kingdom comes from different parts of the tropical and subtropical regions. The main plant collections are orchids, palms (Figure 2), aroids, and tropical trees. The outdoor gardens are meant for both serious and not-so-serious plant lovers.

Figure 2. Rare and unusual palms in the outdoor gardens at Gardens by the Bay in Singapore.

COOLED CONSERVATORIES

The two conservatories (Figure 3) are of great interest and visitors can be expected to be wowed by seeing plants that cannot be grown outside in the tropics. Flower Dome offers a Mediterranean feel while the Cloud Forest exudes a sense of misty, montane forest in the highlands.

Figure 3. View of the two conservatories at Gardens by the Bay in Singapore: Flower Dome and Cloud Forest.

FLOWER TRIALS TO PRODUCTION TO FLOWER DISPLAY

Currently, there are seven changing plant-centric displays in Flower Dome and two curated orchid displays in Cloud Forest. Since the two conservatories are ticketed, the shows have to be attractive to draw in local and foreign visitors. In 2017, visitor ship was 3 million.

In the initial years since the garden's inception, the displays were largely centered on the readily available plants that could be procured from local or overseas nurseries. Over time, there have been many challenges. The displays need to be innovative and novel. Visitors have become more sophisticated and their expectations are higher.

Many criteria must be fulfilled. While design is an important aspect of the show, so are the plant materials. Each display is focused on the plant materials to bring out the central theme of the show. The ratio of flower colours to green foliage has to be high, otherwise the display is deemed uninteresting or unappealing.

The role of the Research & Horticulture Department is to support the flower display team by conducting trials for plants to be displayed at the Flower Field in Flower Dome (Figure 4) and the Orchid Corner in Cloud Forest. Most of the trials are carried out to assess flowering duration and to prolong the postharvest lifespan of the flowers. Besides trials, the growers at Gardens by the Bay are tasked with cultivating specialty crops like dahlias and begonias, as these plants do not travel well if they are imported from overseas.

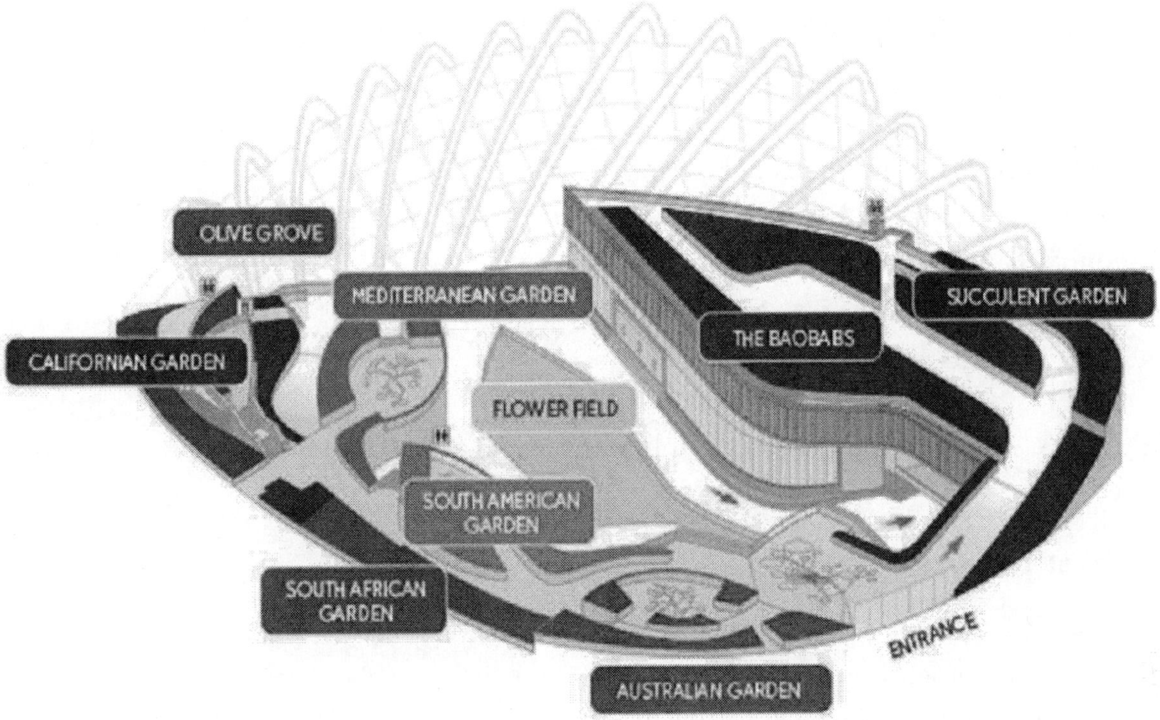

Figure 4. The gardens inside the Flower Dome at Gardens by the Bay in Singapore.

Combined Proceedings IPPS 68:228-229. 2018.

LED and Fluorescent Lighting Effects on Hydroponically Grown 'Tom Thumb' Lettuce

Kent D. Kobayashi and Teresita D. Amore
Department of Tropical Plant and Soil Sciences, University of Hawaii at Manoa, Honolulu, Hawaii 96822, USA

kentko@hawaii.edu

Keywords: Artificial light, sequenced lighting, vegetable production.

Abstract

Food safety, environmental impact, and efficient energy usage are growing concerns for horticultural production systems. Producing lettuce under artificial lighting could be a solution to address these concerns. Light-emitting diodes (LEDs) offer the advantages of a narrow light spectrum, low power consumption, and low heat production. The objective of this study was to compare the effects of LED lighting and fluorescent lighting and the sequence of the lighting on the growth of compact 'Tom Thumb' lettuce (*Lactuca sativa* L.) in a noncirculating hydroponic system.

'Tom Thumb' lettuce seeds were started in Oasis cubes under T5 high output fluorescent lighting in a laboratory. Seedlings were then transferred to 5.1-cm net pots, which were placed in 1.9- liter containers containing a hydroponic nutrient solution of Hydro-Gardens Chem-Gro lettuce formula 8-15-36 hydroponic fertilizer with added calcium nitrate (19% Ca and 15.5% N) and magnesium sulfate (9.8% Mg and 12.9% SO_4). One-half of the seedlings were grown under red+blue+white LEDs (110 μmol/m^2/s) and one-half were grown under T5 high output fluorescent lighting (111 μmol/m^2/s). The photoperiod was 12 hours. After 12 days, one-half of the plants under the LED lighting were moved under the fluorescent lighting, and one-half of the plants under the fluorescent lighting were moved under the LED lighting for 16 more days.

The four treatments were LED lighting (LL), fluorescent lighting (FF), initial LED lighting followed by fluorescent lighting (LF), and initial fluorescent lighting followed by LED lighting (FL).

At the end of the study, differences among treatments were significant for plant height, leaf chlorophyll content, root dry weight, total plant dry weight, shoot dry weight produced per amount of nutrient solution used, and the pH and electrical conductivity of the nutrient solution. There were no significant differences among treatments for shoot dry weight, shoot-root

IPPS Vol. 68 – 2018

ratio, percent dry weight partitioned to the shoots, percent dry weight partitioned to the roots, and the amount of nutrient solution that was used by each lettuce plant. In summary, the sequence of LED and fluorescent lighting could be an alternative to using only LED lighting or fluorescent lighting for growing lettuce plants.

Combined Proceedings IPPS 68:230-231. 2018.

Topworking Mature *Pinus edulis* Trees with *Pinus monophyla*

Kylie Lawson and Larry Rupp
Plants, Soils, and Climate Department, Utah State University, Logan, Utah 84322, USA

larry.rupp@usu.edu

Keywords: Grafting, pine grafting, pinyon pines, single-leaf pinyon pine.

Abstract

Pine nuts are commonly collected from native stands of single-leaf pinyon pine (*Pinus monophylla*) throughout the Great Basin. The goal of this research was to test the feasibility of topworking wild pinyon pines (*Pinus edulis*) to improve pine nut production.

During 2017, both spring and fall season grafting was attempted. For spring grafting, scion wood of *P. monophylla* was collected near Austin, Nevada on March 7 and grafted to mature, wild-grown *P. edulis* trees in Kaysville, Utah on April 19. For fall grafting, scion wood was collected near Eureka, Utah on August 23 and grafted to additional Kaysville rootstocks between August 25 and August 28.

These trials compared different scion accessions, scion types, and graft type. Scions were prepared as buds only (B) or buds with needles (BN). The B treatment involved removing all needles from the scion wood and dipping the bud in wax (April) or latex grafting sealant (August) to prevent desiccation. Grafts were tied with 5-inch × 3/8-inch grafting rubbers, wrapped with Parafilm®, and coated with latex grafting sealant (Doc Farwell® grafting seal). The BN treatment involved tying with grafting rubbers and then enclosing the completed graft in a 1-mil clear plastic sleeve to prevent desiccation. Also, one layer of 1.5-mil, opaque, white plastic tied with twist ties (April) or 6-mil white plastic was tented over the graft and stapled (August) to reduce heat.

Graft types included bark, side-wedge, and side-stub grafts. Bark grafts were performed by cutting off the end of a rootstock branch and slicing the bark with a vertical cut about 1 inch in length down from the cut end. A sloping cut of similar length was made on one side of the B scions which were then slid inside the bark with the cut surface against the wood. Side-wedge grafts were completed on 1- to 2-year-old rootstock wood. The scions (B and BN) were cut into a V-shape about ½ to 1 inch long. A corresponding cut was made at a downward angle into the rootstock about one-third the width of the stem. Side-stub grafts consisted of BN scions and were almost identical to side-wedge grafts, with the exception that the graft was performed on the main trunk of the tree rather than on newer wood. Three scion accessions were randomly applied to each rootstock tree. For each scion accession, the four graft types were completed, resulting in 120 grafts in April and 108 grafts in August. When evaluated on May 18, 2018, the main effects of scion treatment and season were statistically significant, showing greater success with BN scions (43% as compared to 12% with B) and spring grafting (27% as compared to 6% with fall). Bark grafts were the least successful of all the graft types, with only 1 surviving among the April grafts (0.8%) and none surviving among the August grafts. The BN side-wedge grafts were the most successful among the April grafts (16%) and the BN side-stub grafts were the only successful grafts among the August grafts (6%).

The experiment was repeated on April 18, 2018, using 10 *P. edulis* stock trees at the Blue Creek Experimental Farm with scion wood collected from 3 trees near Eureka, Utah, on March 5. Scions were handled as in the BN treatment, and 6-mil white plastic was tented and stapled over the graft. Graft types were side-wedge and side-veneer grafts. Side-veneer grafts consisted of scion wood cut shallowly along one side and notched at the bottom. A corresponding shallow cut was made in the rootstock and the scion set into the rootstock with cut sides facing each other. Grafts were tied with grafting rubbers as above and wrapped with Parafilm®.

For each rootstock tree, 3 scion accessions were used. For each accession, one side-wedge and one side-veneer graft were completed, resulting in 6 grafts per rootstock and 60 grafts total. After four weeks, the plastic covers were slit and then removed 1 week later. Grafts were evaluated after 10 weeks (June 29). Assessment of the grafts showed 78% with elongated candles, 8% alive but not growing, and 13% dead. Grafting success for side-wedge grafts and side-veneer grafts was 82% and 83%, respectively, with no significant difference. The average scion growth of accession #1 (1.1 cm) was significantly less than that of scions #2 or #3 (2.8 and 2.3 cm, respectively). These preliminary studies indicate that wild *P. monophylla* scions can be established in the canopy of uncultivated *P. edulis* trees.

Combined Proceedings IPPS 68:232-233. 2018.

Selecting and Grafting Wild *Pinus monophylla* on Containerized *Pinus edulis* Rootstocks

Kylie Lawson and Larry Rupp
Plants, Soils, and Climate Department, Utah State University, Logan, Utah 84322, USA

larry.rupp@usu.edu

Keywords: Grafting, pine grafting, pinyon pines, single-leaf pinyon pine.

Abstract

Single-leaf pinyon pine (*Pinus monophylla*) is a source of wild-collected, edible, soft-shelled pine nuts that are in great demand. This project identified high yielding, wild trees and evaluated two means of grafting the selections to *Pinus edulis* rootstocks to establish a stock block.

Trees were selected and sampled from productive wild stands near Hamlin Valley (February 23, 2018), Eureka (March 5, 2018), and the Raft River Mountains (February 12, 2018), UT and Austin, NV (February 2, 2018). These stands were identified by pine nut collectors and represent a broad sample from the Great Basin. From each stand, six trees were selected based on visual observations of the number of cones in the tree and on the ground.

Cone production was quantified by collecting six upper-canopy shoots from the top half of each tree at six compass points (N, NE, SE, S, SW, NW) and counting abscission scars from the past 10 years. Analysis of the cone scar data showed that some trees were more productive than others both within and between stands. The three most productive trees with the highest quality scion wood in each stand were selected for use in grafting. Scion wood was collected concurrent with shoots for cone scar analysis and from the same branches.

Second year seedlings of *P. edulis* (Pitkin Forest Nursery, University of Idaho) were used as rootstocks. Seedlings were delivered in November 2017, potted into 4-inch × 4-inch × 12-inch pots with a bark-based mix (Metro-Mix 900) and kept above freezing temperatures in a cold frame to prevent dormancy. Scion wood was stored for one week after collection at 4°C before grafting. A total of 24 scions from each mother tree were grafted, with 12 side-wedge grafts and 12 side-veneer grafts with one graft per rootstock.

Side-wedge grafts consisted of scion wood cut in a V-shape and inserted into a corresponding cut extending into the stem about one-third of the stem diameter. Side-veneer grafts consisted of scion wood cut shallowly along one side and notched at the bottom. A corresponding shallow cut was made into the rootstock and the scion set into the rootstock with cut sides facing each other. Both grafting styles were tied with 3/8-inch × 5-inch grafting rubber bands. After grafting, the seedling trees were covered with 1 to 2 mm of plastic and row cover material in a cold frame for 6 weeks. The covers were removed once every week to ventilate the trees and to monitor the health of the grafts. After six weeks, the plastic was removed, followed by the row cover after one more week. These treatments were applied to each group of collected scion accessions sequentially based on collection date.

The trees were moved to a shaded greenhouse for 3 weeks before being moved to a full sun greenhouse on May 2, 2018, at which point all treatments were applied uniformly to the entire group. On June 4, 2018 (12 weeks from the last grafting) ,97% of the grafts were healthy and growing, with side-wedge grafts being more developed than side-veneer grafts. On July 2, 2018 (16 weeks after the last grafting), 92% of the grafts were alive, 77% of the grafts had produced significant needle growth, and the tops of the rootstocks were removed at 1 cm above the graft. An analysis of the grafts indicated side-wedge grafts were more successful than side-veneer grafts (95% and 89%, respectively), had longer average candle lengths (3.6 cm and 3.3 cm, respectively) and longer needles (4 cm and 3.3 cm, respectively).

Scions from Hamlin had the lowest survival rate (80.5%), scions from Austin had the longest average candle lengths (4 cm), and scions from Raft River had the longest needle lengths (5.4 cm). These results, excluding the average candle lengths, were statistically significant, indicating that side-wedge grafts are more successful than side-veneer grafts for grafting pinyon pine seedlings. Considering that these scions were collected from wild grown trees rather than cultivated trees, the results are very promising for the establishment of stock blocks.

Combined Proceedings IPPS 68:234-236. 2018.

Traditional and Contemporary Propagation of Hawaiian Crops

Noa Lincoln
University of Hawai'i at Mānoa, 3190 Maile Way, St John 102, Honolulu, Hawai'i 96822, USA

nlincoln@hawaii.edu

Keywords: *Colocasia esculenta*, clonal propagation, *Saccharum officinarum*, sugarcane.

Abstract

Native Hawaiian horticulturalists introduced a suite of crops to Hawai'i that were cultivated in the diverse climates represented by the archipelago. The crops represent an endpoint of a unique selection process resulting from repeated sub-selection and diversification that occurred during the island hopping across the Pacific. Among the many results was decreased fertility and resulting asexual propagation of the vast majority of crops. Despite this, Native Hawaiians created a highly diverse collection of cultivars. Furthermore, they developed a high level of skill in vegetative propagation techniques. Today, revived interest in traditional crops, along with new challenges, has resulted in multiple efforts to improve vegetative propagation techniques.

INTRODUCTION

Island drift of crop cultivars

As Pacific voyagers colonized the vast area of Polynesia over the last 5,000 years, they carried with them important plants for crops and resources. Each departing voyage could only carry a small subset of the genetic diversity of each species, which was subsequently re-diversified upon arrival at a new island home. This process was repeated multiple times, resulting in a unique and rapid genetic drift. Among the many traits selected for was seedlessless, which for many of the crops increases predictably as a distance from Papua New Guinea, where many of the crop wild relative still occur. As seedlessness (and by extension, vegetative propagation) increased miotic defects and resulted in increased infertile propagules, asexual propagation techniques became necessary for many of the crops. Of the 21 edible crops introduced to Hawai'i, 12 are entirely sterile. Even the few crops that remained fertile to some degree were often propagated vegetatively in order to maintain elite cultivars that were true-to-form, although some evidence exists that eastern Polynesian farmers deliberately set and scattered seed in order to produce new varieties that, if favored, would be maintained through vegetative propagation. Of the 21 crops mentioned above, only 5 were commonly propagated by seed.

VEGETATIVE PROPAGATION: TECHINQUES, CONSEQUENSES, AND EXPERTISE

A range of vegetative techniques were used for vegetative propagation of Hawaiian crops, including shoot cuttings, rhizomes, tubers, and root runners. For the primary staple, kalo (*Colocasia esculenta*), a hybrid method was used in which a small portion of the tuber was used along with the defoliated stalk of the plant. At the joint of the stalk and the tuber a brightly colored seam is found, known in Hawaiian as the kohina; if this seam is intact, a new plant will grow.

There are multiple advantages and disadvantages to clonal propagation. Advantages include the assurance of favorable genotypes, avoidance of inbreeding depression, preservation of highly specific chemical compounds, ability to quickly propagate favorable mutations, reduction of gene-flow from wild populations, ease of propagation, and, in some cases, higher yields produced when compared to plants grown from seed. Disadvantages include the loss of certain components of diversity, deleterious mutations, competition between propagules and use, and accumulation of pathogens that are passed down through the vegetative propagule.

As stated by Dr. Craighill Handy, one of the first ethnographers to examine traditional Hawaiian agriculture, "in the matter of shrewd observation of varieties and careful, conscious selection of mutants in the creation of subvarieties of their plants, the Hawaiians were truly experimental horticulturalists (Handy et al., 1972)." From a handful of introductions, the Hawaiians developed over 650 names, likely representing over 200 cultivars of kalo (Winter, 2012), with similar diversification of other crops. This exemplifies a poorly studied phenomenon – that under strict clonality evolution is more dynamic than is usually thought, and that somatic mutations are so frequent that genetic identity of clones becomes virtually impossible. This can be seen in Pacific breadfruit, where exclusively clonally propagated plants in the eastern Pacific exhibit nearly 50% unique genotypes from their clonal ancestors in the western Pacific (Zerega et al., 2015).

The accumulated diversity of cell lineages drives significant expression of phenotypic diversity, even with minimal genotypic diversity. Somatic mutations could be an important source of genetic variation in clonally propagated plants, with diplontic selection accelerating the accumulation of favorable alleles in clonal lineages. However, empirical study of these aspects has been uncommon.

One reasonably well documented example is leaf variegation, where cells bear both chlorophyllous and achlorophyllous cell lineages, with specific manifestations empowered by human selection of propagules, environment, or chance. Over time, the achlorophyllous lineages are typically eliminated by diplontic selection when human selection in not maintained. This is again seen in the Hawaiian examples, with the term ʻāweu describing kalo that has escaped plantations and reverted back to its "wild form," loosing many of the desirable traits maintained during its cultivation.

In addition to the cultivation of cell lineage expression, epigenetic inheritance in vegetatively propagated plants plays a significant role in the expression of clones. In propagating plants, Native cultivators considered the selection of vigorous material to be of the upmost importance, with specific growth for each crop used for propagules. For kō (sugarcane, *Saccharum officinarum*), Hawaiians has a specific term, ʻēlau, referring to the upper portion of the stalk from which cuttings were taken (Lincoln, in press). A simple student experiment using the ʻēlau compared to other propagules for sugarcane shows that the usage of this propagule produced stalk volumes twice as great as lesser propagule material in a single generation. Presumably, this effect would be amplified with selections from multiple generations. This poorly explored aspect of the biology of clonal crops has been demonstrated well in forestry, where cuttings from vertical or horizontal growth manifest growth traits from those axes in the next clonal generation.

CONTEMPORARY SETTING OF HAWAIIAN CROP PROPAGATION

Hawaiʻi is ripe for study in the realm of clonal propagation. Cultivation of traditional crops and methods have revived rapidly and taken hold in a new generation of farmers that blend indigenous observation with new tools. Facebook groups support highly detailed discussions of minute morphological and phenotypical differences of crop varieties, driving intense debate on what drives those manifestations.

Bringing intensive genetic and quantitative study to this arena rich with ethnographic documentation could provide a wealth of knowledge into the poorly studied realm of evolution in clonally propagated plants.

The challenges of today are different, and in particular a need for clean propagation material, access to traditional varieties, and the ability to create larger quantities of propagules is present. Some solutions have been emerging. Tissue culture is a powerful tool for creating clean material and large quantities; however, there have been some pushback as some traditional farmers feel that they have observed changes in the somatic mutations associated with tissue cultured plants. More growers favor the higher volumes, but more stable propagation through single-node (vs. traditionally multiple-node) cuttings and adventitious shoot cuttings. However, gaining sufficient volume with these techniques remains challenging without dedicated facilities. Growing private and public germplasms have helped with access to varieties, but more impactful have been farmer-to-farmer connections fueled largely by social media platforms.

Literature Cited

Handy, E. C., Handy, E. G., and Pukui, M. K. (1972). Native Planters in Old Hawaii. Bernice P. Bishop Museum Bulletin 233.

Lincoln, N. (in press). Ko: An Ethnobotanical Guide to Native Hawaiian Sugarcane Varieties. Honolulu; University of Hawaii Press.

Winter, K. B. (2012). Kalo [Hawaiian Taro, *Colocasia esculenta* (L.) Schott] Varieties: An assessment of nomenclatural synonymy and biodiversity. Ethnobot. Res. App. *10*:423-447.

Zerega, N., Wiesner-Hanks, T., Ragone, D., Irish, B., Scheffler, B., Simpson, S., & Zee, F. (2015). Diversity in the breadfruit complex (*Artocarpus*, Moraceae): Genetic characterization of critical germplasm. Tree Genet. Genomes, *11*:4.

Combined Proceedings IPPS 68:237. 2018.

Propagation Techniques for Blue Elderberry (*Sambucus nigra* ssp. *cerulea*)

Tiffany Maughan, Brent Black, Larry Rupp, and Matt Yost
Plants, Soils, and Climate Department, Utah State University, Logan, Utah 84322, USA

brent.black@usu.edu

Keywords: Cold-callusing, *Sambucus cerulea*, specialty food products.

Abstract

Native fruiting species provide an interesting opportunity for sustainable diversification by fruit growers. However, efforts to develop commercial fruit production using the blue elderberry (*Sambucus nigra* L. ssp. *cerulea* (Raf.) R. Bolli (syn. *Sambucus cerulea* Raf.) have been hampered by difficulty in propagating this Western North American species.

We previously reported on a series of experiments to develop viable propagation protocols. Semi-hardwood cuttings collected from wild-grown plants at full bloom and treated with a commercial naphthaleneacetic acid (NAA) formulation had rooting success greater than 60%. Hardwood cuttings taken from greenhouse-grown stock plants and then cold-callused for 70 to 120 days had success rates ranging from 50 to 100%. In the latest experiment, cuttings were taken from clonal plant material grown in three different environments (greenhouse, field, wild) in December, January or February, and cold-callused from the collection date until March.

Survival was then evaluated after 6 weeks in the greenhouse. Survival from wild-grown cuttings was very low (< 10%) for all collection dates/callusing times. For cuttings from field-grown plants, survival was higher using earlier collection dates (longer callusing period) but did not reach commercially viable levels. For greenhouse-grown stock plants, survival ranged from 75% to 100%, with the highest survival rate for plants collected in February and cold-callused for 60 days.

Additional work is needed to optimize both mother plant environment and cold-callusing conditions. These methods will be useful for domesticating blue elderberry for both the landscape and the specialty food products industry.

Combined Proceedings IPPS 68:238-241. 2018.

A Decrepit Old Propagator Found New Life by Going Back to Basics

Greg McPhee
Eco Petals, Lismore, New South Wales 2480, Australia

greg.mcphee@hort.com.au

Keywords: Edible flowers, grow boxes, horticultural therapy, salad crops.

INTRODUCTION

Life Changes Forever

Friday, 13 August 2009, was not a good day for me. It started out okay as I was working on finishing our new house and had prepared a great dinner. About 6:30 PM I felt ill and took some advice from my daughter-in-law. For those that know me, this is unusual, but she is a nurse and knows her stuff. I went to the hospital to get checked out. I cleared the waiting room in a few seconds when I became violently ill and then collapsed. I got sicker, was intubated, placed on life support, and had emergency surgery.

I was sent home on New Year's Eve – some five months after admission. My life was changed forever. I was still very unwell. I had lost most of my memory and cognitive thinking skills. I was weak and unable to walk 100 m. I was back to basics!

A New Beginning

After months of rest (with unfortunately more operations) and plenty of thinking, my wife Lyndy and I (being the eternal optimists) started a new garden. The new garden had to be basic as I could not handle anything else. My friend and fellow IPPS member Des Boorman built some garden beds based on a design concept by Jacob Mittleder. He called it "grow box gardening".

Our beds are 2.4 m by 1.2 m (8 feet by 4 feet. The beds are 400 mm high and constructed from local untreated hardwood timber. We started at 200 mm high, but quickly realized 400 mm was easier on the back and reduced weeds enormously. Our soil mix contained one-part coarse river sand, two parts sandy loam, and two parts compost. To this we added 200 L of composted chicken litter per bed (about 10 tons per acre). It takes about 1.2 m^3 of this soil to fill a bed. The garden started with 12 beds, which was enough to get us into production.

We deviated from Jacob Mittleider's method and grew organically. I was and still am immunocompromised. Spraying is not an option. Pests are an ongoing issue, but not an onerous problem with high garden beds and regular maintenance. We started by growing salads for our own consumption.

Things went well. The beds grew all our salad needs, plus we had surplus and plenty of it. I was asked to sell at a local farmers market and we 'launched' our microgreens business in July 2012. You learn a lot when you speak with the people who use your product in their everyday life. We learned that being local, seasonal, fresh, and spray-free were important. Most important was trust. We found that many of our customers were buying our product for health reasons –they really needed to trust their suppliers.

We decided to keep things basic and only sell what we grow ourselves. We would pick fresh, would not apply insecticides or fungicides (spray-free), and used compost and manures for nutrients. We are still holding to this. Sadly, other growers were not. The lesson here is: BE HONEST (or DON'T SCREW THE LOCALS as they live close and have long memories).

Our salad business grew bigger. This created problems. To meet the increased demand, we needed more beds. Also, our bed arrangement was fine for a start, but it became apparent that 'ride on' mowing between the beds was needed. We decided to reconfigure the beds to make our life easier.

OUR PRODUCTION SYSTEM

Bed Layout

We have about 100 garden beds and the beds are still the same size as the original beds. The space between the beds is wide enough for a ute (our Australian term for a pick-up truck) to get to each bed and allow room for running a ride-on mower between the beds. We do not have bed covers or hot houses. We had an excavator expand our growing space slightly, but we have almost reached our limit.

Between the beds is a vegetation strip, mainly containing grass. We throw all old crop residues and weeds on this area, chop this up with a mower, and then use a catcher to pick up the clippings for mulch. There is a drainage line between the beds to remove excess water when we have rainy weather.

Water

We have about 800 m^2 of available roof and with an annual rainfall average of 1.5 m that results in potentially 1.2 Ml (megalitres) per year of fresh water. I have calculated that we need about 1 Ml of water per year. So, we capture as much roof water as we can. Most of our rainfall occurs in the summer (sometimes 200 mm per day) and our spring is typically dry. This lack of regularity can leave us short on irrigation water. After drilling a few bores (one went below sea level!), we found a small supply. It provides enough water for a solar pump that delivers 3,000 L per day and currently trickle-fills a 250,000-L tank.

We mainly water by hand. There are spray systems for each bed for use in hot weather. I use my time while watering by monitoring and texting on my mobile phone.

Labour

Our desire is to have garden that we can work in without the need to regularly employ people. We were able to access support for our garden from the WorldWide Opportunities on Organic Farms (WWOOF) program (http://wwoof.net/). Simply stated, we invited able-bodied people to help and exchanged their labour for our food, company, and accommodation. We received more than we initially bargained for and welcomed the enthusiasm and technical expertise given. Lifelong friendships have been made. I recommend you get involved if you grow organically.

Fertilizer

We are constantly adding compost and manure to the beds. As we run with very high amounts of organic matter (35%+), the soil level shrinks with decomposition. We add more compost a every crop change. For more long-lasting crops, we cut back and fertilize regularly. This keeps nutrient availability high and we are getting increased crop outputs as a result.

CROP CHOICES

From Greens to Color

Our first choice for selling was salad greens. We were growing for ourselves and salad greens give a high return per kilogram. We looked closely at the harvest from each bed to see which crops we should expand. Cut-and-regrow lines, such as rocket, basil, chard, and spinach were winners. Sadly, shallots, carrots, and beetroot were not. Kale was a real winner, but next year we were swamped by competitors who were copying our crop choices and undercutting our prices. Our garden bed system is relatively expensive to establish and is a high labour-use system. Things needed to change.

Lyndy said we should have more colour on our salad stall. My comment was that salads should be green, since that is what people expect! So, after discussion we added flowers to the mixed salad bags and had some bunches of garden blooms on the bench. I had argued that, since Lismore is not a rich town, no one would be interested. Let me tell you about Lismore. Lismore is a university town with about 50,000 people and no major industry and a high level of unemployment. Flower sales grew as salad sales declined. Lyndy was right and I was… well, not right. Let the market decide. We knew that our advantage would be short-lived if we grew the easy-to-grow flowers. We needed to seek out the more exclusive varieties and grow them better than others who copied our new crops.

Edible Flowers

One of our salad customers (who owned a restaurant) wanted to add flowers to his dishes. Edible flowers were quite novel in Lismore at the time. They have been a good addition to our operation as we now have a use for smaller cut flower stems. We even sell flowers to a vegan cake maker!

Our customers for edible flowers all require that the flowers be perfect, non-sprayed, colourful, and do not collapse when chilled. Some customers require that the flowers look good when frozen. It can be very time-consuming to pick and package flowers, which is not always worth the effort.

We dealt with a middleman who procured produce for high-end resorts. The prices paid were higher, but this was negated by the extra demands from super-picky and temperamental chefs. My suggestion is that you stay away from Rockstar chefs. The lesson we learned is that chasing chefs' demands can break your heart. It is good for your image, but hard on the pocket.

Chefs are not gardeners and (despite what they say) have no idea of the difficulty of growing plants. Chefs demand quality and consistency. Also, they want unique products and always something new. Edible does not mean they taste nice. It probably means non-toxic. I could not find figures on the number of people poisoned or killed from eating flowers. I assume it's not very significant. No one seems to eat flowers placed on their food anyway.

OUR CURRENT PRODUCTION SYSTEM IS STILL BASIC

We are still using and enjoying our grow boxes. We have found that 2.4 m by 1.2 m boxes to be ideal. Our planting and flower support grids fit perfectly. It makes our calculations easier because all beds are the same size.

My original calculations were based on 20 Ml of water per hectare. As our garden has grown, so too has our need for water. We now hold a 3-month water supply and have increased our bore-pump size. Most of this is fresh rainwater harvested from our roof. I sleep better at night if I do not need to contemplate running out of water.

Salads are almost gone from our garden as they cannot match the returns from cut flowers. Likewise, easy-to-grow flowers are gone as they are not 'special" enough. We are moving more towards flower crops that are 'protected', i.e., require written contracts before we can purchase plugs. Most are tissue-cultured and expensive. The real benefit of this is that competitors cannot chase us as easily and we can offer better flower choices.

We have become much more seasonal. It's an advantage to have 'something new' each season or month. Teaching customers to wait is difficult, but the offset is that they buy in season as they learn that some flowers won't be around for long. Seasonality (in my opinion) is the next big thing in fresh produce and flowers.

Our garden soils are improving. We add compost and manure (100 L per bed) at every crop turn. We also mulch with lawn clippings. It's common to have four bed turns per year, so we really have rich soils. Our production per bed is increasing and thc weeds are pretty much gone.

We no longer make our own compost. Our local council has invested in a large-scale, organically certified composting 'factory' and we buy it from them. I have converted my ute to a tip tray, so getting the compost is very easy. We hold compost and manures for a few months before using it. This allows for composting to settle and for beneficial microflora to establish. For more information on this, I suggest you read some of the work done by Harry Hoitink.

Our garden waste and the few weeds go under our mowers when we mow (which is done regularly to limit flowering) and the clippings are used as mulch.

I am also experimenting with 'grass alternatives.' Since it is easy to harvest lawn clippings, I have aimed at growing plants that will enrich the nutrient value of our mulch. So far, I have planted clovers, medic, pinto peanut, and Wynn (round-leaf) cassia. All have their positive and negative points.

Insects and diseases are minimal. We do regular monitoring and have many beneficial insects in the garden. If a crop is weak or infested, we will quickly remove it. We rotate crops. I am amazed at how well we do. We are the only spray-free flower growers we know of.

We monitored crop volumes are and are increasing bunch output per bed per annum (our measure of success). We are now also selecting crops to ensure year-round production and easier bunching labour.

Demand is changing. Posy tins are current a hot trend, as are wedding-u-pick. We have converted to compostable or recyclable packaging.

It was a good idea to start a garden. Not only has my health improved, but it provides for luxuries and fills a local demand for quality, just-picked flowers. We do have some pretty amazing, regular customers.

So, I suggest you consider going back to basics. Don't let a near-death experience be the reason you change your lifestyle.

Combined Proceedings IPPS 68:242-247. 2018.

New Zealand: Native Plants, Headford Propagators, Waimate, and a Muscle Car

Megan Robinson
Headford Propagators, 24 Horsnells Road, Morven 7980, New Zealand

meganrobinson901@gmail.com

Keywords: *Astelia chathamica*, medicinal plants, *Leptospermum scoparium*, *Myosotidium hortensia*.

Abstract

IPPS-New Zealand Region member Megan Robinson was the recipient of the 2018 New Zealand Region/Western Region exchange fellowship. Megan traveled from her home in Waimate, New Zealand, to the IPPS-Western Region Annual Meeting in Hawaii to participate in the Pacific Rim Conference. Megan works for Headford Propagators. Megan shares some information about Headford Propagators, New Zealand native plants, her hometown, and her interests outside of horticulture.

INTRODUCTION

My name is Megan (or Megs to my friends) and I'm a kiwi girl from New Zealand. I grew up in my little hometown of Waimate and am currently working at Headford Propagators as a nursery worker. I would like to tell you a bit about my nursery back home, my home country of New Zealand, and some of the native plants we grow at our nursery. I will also include some information about my hometown and my other interests outside of horticulture.

I started at Headford Propagators Ltd. in September 2015. My bosses are Grant and Robynne Hayman, who some of you may have met through IPPS. We are a production nursery growing and supplying liner plants of both New Zealand natives and ornamentals for plantings. Our plants are mostly sold in 5-cm or 7-cm pots or are grown under contract to suit customers' requirements.

Headford Propagators is located in South Canterbury, New Zealand. Our location provides an excellent climate for growing strong, resilient plants. We are at a latitude of 45° South. We have long days, especially during the spring and summer months, intermittent breezes, a good water supply, proximity to glacial rivers, and a central location that makes us competitive in the nursery industry.

New Zealand has the ninth longest coastline in the world (15,134 km), and no part of the country is more than 128 km (79 miles) from the seaside. About one-third of the country is protected national park land.

HEADFORD PROPAGATORS

Before I tell you more about our present nursery, I would like to share some background information, history, and how Headford Propagators has come to be the outstanding nursery it is today. In 1988, Grant, being the worker, and Robynne, a stay-at-home mum, owned a small trucking business with three trucks. Everything was going well until Robynne got a phone call informing her that Grant had been in an accident. He had just finished unloading logs from the truck when the bolster fell onto his foot and crushed it. This led to a removal of one of his toes and six weeks of bed rest with his leg up in the air in the hospital. After six weeks, Grant got the news that he could no longer drive a car, let alone a truck. With this, they had to sell up, that is, they sold their three trucks and whatever necessary to make ends meet for their family. Grant was on a New Zealand ACC policy which meant the government would pay to get an accident victim back to work. He had the option of taking a horticulture course, which he was very interested in doing, and with this Robynne could go back to nursing. Grant received a National Diploma in Horticulture certificate and, most importantly, this made the idea of building a nursery from scratch a reality. Headford Propagators started from a farming block in 1994, which was originally a sheep and crop farm. Grant bought the house, shed, yards with about 50 acres of flat land close to the coast. This was an opportunity for Grant and Robynne to build a business with very open minds. This is typical of the opportunities available in New Zealand for motivated people.

The Headford Propagators team is made up of about 30 people. Some of the nursery crew are married couples and some have been working at Headford Propagators for over 15 years. We all have designated areas and jobs but help out each other wherever necessary. In our nursery, we have two double-skinned houses and a large shade house. One house has heat beds inside for cuttings and is run by the propagation team. The other house is for newly potted plants and is run and looked after by the potting team. The double-skinned houses protect plants from the cold, especially on nights in winter which can get to -7°C (44.6°F). We also have a single-skinned house, outside standing areas, and an extensive range of stock plants. We grow a wide range of species, including New Zealand native shrubs and grasses, wetland plants, hedging shrubs and *Buxus*, a small selection of ornamental exotic shrubs, *Agapanthus*, and mondo grass.

The nursery also has a seed propagation area. One of the off-site jobs some of us do through the warmer months is seed collecting. We go to various places and collect seeds needed for the nursery. This seed is then processed and put in the seed area or plastic houses, depending on the variety. I have worked in all areas of production, doing jobs from seed processing to propagation, potting, and even dispatch. My main role at Headford Propagators is being second in charge of the potting department, where I work alongside my wee team. Nicky is mostly in charge and Denise also assists. During our winter season, five to six members of our team work daily during poplar season, removing plants from the ground. The rest of us process the different varieties of poplars. This includes trimming the roots, cutting to look sleek and to the measurement required, and then bundling into groups of 25. The bundles are either placed into cages which are delivered on to various customers or they get healed into sawdust to stop root growth and keep them alive until removed as required. The processing crew has daily targets that must be reached.

HEADFORD PROPAGATORS FREIGHT BUSINESS

Along with the nursery at Headford Propagators, there is also a freight business running alongside the nursery. We ship plants and do general freight with our regular nationwide service. Only a small part of what we freight is grown in our nursery. We freight all plants, from cell size to monstrous bagged or boxed trees. We have 11 trucks. Most have specialised curtain sides. Our freight services are used by many growers to move green goods throughout New Zealand.

I remind you that New Zealand has two main islands, so the trucks have to take a 3.5-hour ferry ride to cross one of the roughest seas in the world. The freight service does have its challenges.

Headford Propagators has evolved quite a bit since 1994 from a piece of farmland to an outstanding, very busy, and well-respected business. We are well-known nationwide, not only for plants, but also our specialised freight business as well.

NATIVE TREES AND PLANTS OF AOTEAROA

The native trees and plants of Aotearoa (the Māori name for New Zealand) make up much of what we grow. It is said that there are 200 native, edible plants in New Zealand. Eighty percent of our trees, ferns, and flowering plants are endemic, that is, found only in New Zealand. About 10% to 15% of the total land area of New Zealand is covered by native flora, from *Agathis australis* and *Podocarpus totara* to rainforest dominated by rimu (*Dacrydium cupressinum*), beech (*Fuscospora* spp., and *Lophozonia menziesii*), tawa (*Beilschmiedia tawa*), matai (*Prumnopitys taxifolia*), and rata (*Metrosideros* spp.), and ferns and flax (*Phormium* spp.). There are dunelands with their spinifex (*Spinifex sericeus*) and pingao (*Ficinia spiralis*), alpine and subalpine herb fields, and scrub and tussock. Although New Zealand looks like a small country, this is only because it is dwarfed on the world map by Australia.

Chatham Islands

A very special place in New Zealand is the Chatham Islands, which are off the east coast of the country. The Chatham Islands have been isolated for more than 80 million years, long enough to develop many plants found nowhere else. There are said to be 388 indigenous terrestrial plant species on the Chatham Islands. Forty-seven (about one-eighth) are endemic, meaning they cannot be found anywhere else in the world. Endemic species include forest trees and several giant herbs and seaweeds. Even the local flax is unique.

Most well-known among the endemics are the Chatham Island forget-me-not (*Myosotidium hortensia*), Chatham Islands kakaha (*Astelia chathamica*), and soft spear-grass (*Aciphylla dieffenbachii*). These are all plants we grow at the nursery because they are quite ornamental.

What makes Chatham Islands plants unique is that the plants show a much higher proportion of coloured flowers than those from mainland New Zealand. The leaves of Chatham Islands species are also often fleshier, and the trees are bigger than their New Zealand counterparts. The plants do not tend to show juvenile forms.

The environmental influences on the Chatham Islands include its oceanic setting, which has a profound influence on plant life. Winds sweep the islands, bringing gales, salt spray, cloudy skies, common showers, and occasional blasts of cold air, but temperature extremes, droughts, frosts and snow are very rare. Sunshine hours are about half of those of the sunny parts of the New Zealand mainland. Plants are adapted to these conditions. Protective leaf and twig furriness is a feature of the tree and shrub daisies, and the indigenous trees have the ability to layer themselves after having been blown over. Other species have developed giant leaves. Chatham Islands karamu (*Coprosma chathamica*), which grows into a forest tree, is by far the largest species in this genus of shrubs. Akeake (*Olearia traversii*), which we also propagate by cuttings, is one of the largest tree daisies on earth. This may be due to the climate, long isolation, and high soil fertility.

Propagation of Chatham Islands Forget-Me-Not

We produce *Myosotidium hortensia* from seed. When the seed is sprouted, we delicately pull the seedlings from the seed trays with a butter knife and, trying not to disturb the other small seedlings, we place them on a tray which is layered with paper. They are then potted into 7-cm plastic pots or 7-cm peat pots and placed into our plastic house where the plants will be grown. These plants have glossy green leaves and produce blue or white flowers as they get older. Plants grow to about 40 cm and require some shade or a location that is not too dry.

Propagation of Chatham Islands Kakaha

We grow *Astelia chathamica* from seed and, when ready to pot, we pot them into 7-cm pots. We plant with soil just up to the crown of the plant and leave about a 1-cm gap and sprinkle shingle on top of the soil in the pot. Plants have a striking form with silvery-green, flax-like leaves. They get grown in a shade house.

MEDICINAL USES OF PLANTS WE PROPAGATE AND GROW AT HEADFORD PROPAGATORS

Something I have always found interesting about native trees and wild plants, even before I started working at Headford Propagators, is the medicinal use of plants and how some plants can have physical healing and also specific uses. I became interested in this when I was a kid and learned what stinging needle was the first time. A way to reduce the skin irritation of stinging needle is to use a dock leaf. I discovered more about medicinal use of plants from a book I read recently titled "Native Plants and Wild Plants of Aotearoa". These medicinal uses were recognized by the Maori people long before the Europeans arrived.

***Coprosma robusta* (Karamu)**

Karamu nourishes the urinary tract and provides healing for the sexual organs of both women and men. Karamu is nourishing for the kidneys as well.

***Sophora microphylla* (Kowhai)**

Kowhai is beneficial for the kidneys as a homeopathic and has anti-cancer properties. It may be boiled up as a tea and is useful for skin issues, itching, and other sensitivities. My mum has this in her garden back home.

***Podocarpus totara* (Totara)**

Totara nourishes the womb, uterus, and ovaries. It may be used as a circulation tonic and for uplifting and grounding at the same time. It may also be used to prepare a digestive and kidney tonic.

Phormium tenax (Harakeke, New Zealand Common Flax)

Common flax is found throughout New Zealand, especially in wet areas. Flax is unique to New Zealand and is one of our most ancient plant species. Common flax grows up to 3 m high and its flower stalks can reach up to 4 m. It has seed pods that stand upright from the stems. Flax is used in soaps, hand creams, shampoos, and a range of other cosmetics. There have even been experiments conducted to make flax into wine! Flax was the most important fibre plant for the Māori in people of New Zealand. Floats or rafts were made out of bundles of dried flower stalks. Flax also had many medicinal uses. The sticky sap or gum that flax produces was applied to boils and wounds and used for toothache. Flax leaves were used in binding broken bones and matted leaves were used as dressings. Flax root juice was routinely applied to wounds as a disinfectant.

Leptospermum scoparium (Manuka, Tea Tree)

Tea tree is more universally known for its health properties than many of our other native plants. It may be used as a kidney tonic and blood cleanser (by clearing toxins from the system). It nourishes the urinary tract and can help with motion sickness. Tea made from the leaves is also a useful mouthwash and is said to reduce fevers. Skin disorders can be treated by rubbing the ashes of burnt manuka on the skin. An infusion made from the bark is used as a sedative and to treat burns. The inner bark boiled in water is used as mouthwash. Inhalation of leaves placed in boiling water is used for colds.

In recent years, manuka has gained attention for a tea made from its nectar. Specifically, honey that honey bees produce using nectar from the manuka tree has shown remarkable antibiotic properties. It can be used to treat minor wounds and cuts and to fight infections. Manuka honey is produced by European honey bees (*Apis mellifera*) foraging on the manuka which grows uncultivated throughout New Zealand and south-eastern Australia.

PLANTING ONE BILLION TREES

You might think THAT New Zealand has plenty of trees, but our government (under the current Prime Minister Jacinda Ardern) has set a goal of planting one billion trees over 10 years (between 2018 and 2027). Forestry has a range of benefits, such as helping to diversify income, invest in the future, improve land productivity, reduce the effects of climate change, improve water quality, moderate river flows, provide important habitats for a range of native species, and enhance natural landscapes. Grant, my boss, says that this is a great opportunity for our industry, so Headford Propagators is gearing up to meet the increased demand.

INFORMATION ABOUT WAIMATE AND ME

Now I would like to to tell you a little about my hometown and my other interests.

Waimate

I have grown up in the district of Waimate in a rural area called Willowbridge, which is about 10 minutes from the small town of Waimate. Waimate is a town in South Canterbury, New Zealand. It is situated just inland from the eastern coast of the South Island and not far from the Waitaki River. The population is roughly about 4,000 people. Waimate is well known for its population of Bennett's wallabies. These marsupials were introduced from Australia and now live wild in the countryside surrounding the town. It is, in fact, a pest now, but has remained fairly local as it does not cross rivers. Waimate is also recognized for the White Horse Monument, which was built in 1968. Retired farmer Norman Hayman, who is one of my boss's ancestors, returned from a European holiday inspired by a Friesian cow statue in Holland. It is a silhouette of a white horse that can be seen on the hills behind the town. It commemorates the Clydesdale horses that helped to break in the land in earlier days. The monument's lookout has panoramic views of the town and the district's green plains out towards the Pacific Ocean. The district is a very productive agricultural area with a mix of pastoral, cropping, dairy farming, and fruit and vegetable growing areas.

In March 2018, our engineer Bill Scott painted four beautiful murals on the Waimate silos. The silos were built in 1934, are 35 m high (including the little house on top). The four murals were painted to become a tourist attraction for Waimate. The four paintings represent: Margaret Cruickshank joining World War 2 hero Eric Bachelor, Chief Te Huruhuru, Michael Studholme, and Norman Kirk. Margaret Cruickshank was the first woman to be registered as a doctor in New Zealand. She sacrificed her life while fighting the 1918 influenza pandemic. She served Waimate from 1897 until her death in 1918 at age 45. The meeting of Chief Te Huruhuru with Michael Studholme occurred in 1854. Michael Studholme was the first European settler in Waimate. Norman Kirk was a former Prime Minister in the 1970s. All these people are buried in Waimate in the old Waimate Cemetery.

My Other Interests

I grew up in a car yard with my dad, mum, and brothers. My dad is a mechanic and so was my grandad. I have always loved the mechanic trade and ever since I can remember I was sitting up with my dad under the hood of a car passing him tools of all sorts. I grew up with my two brothers and what we always loved doing was building new things, from go-carts to paddock bashers to double-ended cars. All this has led up to my dream of buying and owning a muscle car. I've had my car for about two years, with about a year and a half on the road. It is a 1982 Ford Fairlane with a Cleveland v8 engine. Burt Munro (who you may know as the World's Fastest Indian if you have seen the movie) was cousin of my dad's grandma Isabelle Munro.

FINAL THOUGHTS

It has been an absolute pleasure to be a part of this IPPS conference, meeting new IPPS members from around the world, and seeing the familiar faces of those I already know. I just want to say a big thank you to the IPPS for giving me this opportunity, I never thought in my lifetime I would travel, not only overseas, but to a beautiful, exotic, and dreamy place like Hawaii.

And bring on the rest of the exchange program! I can hardly wait!

Combined Proceedings IPPS 68:248. 2018.

The Influence of Varying Concentrations of Boron on the Growth of Excised Shoots of Walnut and Pistachio Clones in vitro

Dharam P. Sharma
Dry Creek Laboratories, Duarte Nursery, Hughson, CA 95326, USA

dharam@duartenursery.com

Keywords: *Juglans*, micropropagation, *Pistacia vera*.

Abstract

Boron nutrition of walnut (*Juglans* L. spp.) and pistachio (*Pistacia vera* L.) is challenging because, although a required element, it is needed in only small quantities and has a narrow optimum range that varies by variety. Unlike with many other species, boron is immobile in pistachios and walnuts, accumulating in the older foliage.

The present study was undertaken to see if excised, in vitro shoots from clones of walnut and pistachio differ in tolerance to varying levels of boron in the growing medium. A total of nine clones of walnut and nine clones of UCB-1 pistachio rootstock were selected and the shoots placed in standard tissue culture media supplemented with 0, 10, 50, 100, 200, 300, 400, 500, or 1000 mg l^{-1} of boron and placed in a controlled, aseptic environment (16/8 hours day/night at 30 μmol m^{-2} s^{-1} light, 25°C).

Growth was measured in terms of shoot length and leaf count as a percentage of control. Boron toxicity symptoms appeared as necrosis on tips on older leaves, extending to the margins and followed by leaf drop. Boron concentrations above 50 mg l^{-1} stopped growth of most of the nine pistachio clones under study and induced leaf drop on all of them. Among the walnut clones, the rootstocks 'Vlach B' and 'VX211' and the scion 'Tulare' could grow slowly even at 300 mg l^{-1} boron in the growth medium after four weeks of culture. 'D154W' pistachio and 'VX211 1C' walnut were the most boron-tolerant rootstocks, whereas 'Chandler' was the most boron-tolerant scion.

The differences in growth among clones in response to boron concentrations will be tested further in field trials.

IPPS Vol. 68 – 2018

Combined Proceedings IPPS 68:249-251. 2018.

Nursery Certification Programs and How They Aid in Shipping Clean Plant Material: A Canadian Example

Valerie Sikkema
Van Belle Nursery, 34825 Hallert Road, Abbotsford, British Columbia V3G 1R3, Canada

valerie@vanbelle.com

Keywords: Canadian Nursery Certification Program, crop traceability, phytosanitary certificates, phytosanitary management, plant quarantine.

Abstract

Van Belle Nursery started the process towards certification in 1995, allowing the nursery to conduct their own inspections. The nursery began to write their own phytosanitary certificates in 2007. The nursery's certification manual was completed and accepted by the Canadian Food Inspection Agency in 2008 and undergoes regular revisions and adjustments as system and staff changes are made and new products are added. The system includes full traceability of plant material and prevents non-certified crops from crossing the border. The main benefit of the certification program is that it allows for efficient shipping, but it also promotes excellent record keeping, encourages training for all growers and IPM personnel, and promotes professionalism in the nursery.

INTRODUCTION

Van Belle Nursery was established in 1973 by Bill and Grace Van Belle on Matsqui Prairie in Abbotsford, British Columbia. The nursery rapidly grew from a landscape supply company (with the majority of crops grown in the field) to supplying liners to other wholesale nurseries and retail-ready material chiefly to independent garden centres. In the 1990s, the nursery was increasingly shipping container-grown plant to the United States and had a strong desire to do this efficiently and with the greatest flexibility to adjust orders at the last minute. There was talk in the industry about starting a certification program to allow nurseries to do their own inspections and write their own phytosanitary certificates, so Van Belle Nursery decided to participate right from the beginning. Since this was long before I was hired, I asked Bill Van Belle why they started this process and he said: "We have always been a progressive company and this [was] part of the path."

IPPS Vol. 68 – 2018

Van Belle Nursery started the process towards certification in 1995 and had the first audit in 1996. The benefits started shortly thereafter as the nursery went from inspections of every load to field inspections every few weeks (or as needed). The Canadian Food Inspection Agency (CFIA) would write the phytosanitary certificates ("phytos" for short) based on those inspections. By 2006, the nursery had the certification manual almost completed and decided to hire me as the Crop Protection Manager to take over responsibility for the program from Grace Van Belle. In 2007, we were allowed to begin writing our own phytos and in 2008 the manual was complete and accepted by CFIA. We continue to revise and adjust based on changes in the company, but the Canadian Nursery Certification Program (CNCP) continues to serve us well.

THE CERTIFICATION PROGRAM: A SYSTEMS APPROACH

The CFIA directive D-04-01 is the regulatory document that governs the certification program and outlines the requirements for phytosanitary and pest management systems, certifying plant material, facilities and staff, audits, manual, documentation, administration, and corrective actions. It can be found at:
http://www.inspection.gc.ca/plants/plant-pests-invasive-species/directives/date/d-04-01/eng/1323820371646/1323820675188.

The manual is a constantly changing document as we make system and staff changes and as we grow in area and product offerings. All the changes need to be made as soon as possible, and there is a complete review of the manual at least once each year. There are four internal and three external surveillance audits per year, plus one internal and external systems audit. Each surveillance audit does include some aspects of systems, just to make sure there are no systemic errors occurring that could jeopardize the program.

Since this is a systems-based approach to phytosanitary management, it requires all aspects of planning and production to work properly. Before any new plants are ordered they need to be compared to the restricted list to ensure that they will be certifiable, and if not, that they be prevented by holds in the inventory system from being shipped to the United States. All incoming plants are inspected carefully before being allowed into the nursery. We do regular, scheduled pest monitoring of all growing locations, carefully recording all findings, treatments, and treatment results. Finally, all outgoing plants are carefully inspected, especially if they will be getting a phyto or an Interfacility Stamp (IFS) to allow them to go seamlessly to another CNCP nursery.

In case of issues (regulated or newly introduced pest, or other problems), we need to have full traceability of our plant material. Where did the product come from? Where has it been located on the nursery? Where has it gone? Each audit includes several trace-forward (from purchase) and trace-backward (from sales) exercises. Thankfully, now this is all tracked in our computer inventory and we no longer have to dig through boxes of paper records.

SHIPPING RESTRICTIONS

Probably the most important part of this process is preventing crops that are not certified for one reason or another from crossing the border. There are three categories of restrictions that are established by the USDA that we must follow, namely: A) Any plant imported into Canada with growing media except if from the US or grown under Canadian or US Growing Media Programs; B) A list of specific genera, no matter what the source, and C) A list of specific genera if sourced anywhere except Canada or the US. We therefore need to ask all Canadian suppliers about the source of their plants so we can ensure that if they do not accidentally get shipped as certified material if they fall on any of these lists.

The second thing that can prevent a crop from being certified and allowed to cross with a CNCP phyto is the length of time that it has been at the nursery. Generally, it is one growing cycle (4 months), but it can be as little as 28 days for annuals or tropicals that have only been grown in a greenhouse. *Phytophthora ramorum* hosts require a full 12 months if they are sourced from outside North America. If they come from another certified nursery with an Interfacility Stamp or from the US on a US phyto, they can be shipped on a CNCP phyto right away; in the case of US origin, this must be stated. If there is an item that we would like to include, but it has not completed the residency requirement, we can request a CFIA inspection and get a 1337. Essentially all these requirements come from the USDA PPQ document which can be found at: https://www.aphis.usda.gov/import_export/plants/manuals/ports/downloads/plants_for_planting.pdf.

We added a module for Post Entry Quarantine (PEQ) for hydrangeas when we had an opportunity to bring in new varieties from Europe that would be distributed all over North America. The quarantine lasts 9 months; during that time, we are watching closely for any development of rust, which is the disease of concern in these plants. Right now, the origin needs to be either the Netherlands or Germany and the plant material needs to be inspected by CFIA upon arrival. For the following 9 months, the plants must remain a minimum of 3 meters from any other hydrangea, they need to be uniquely identified by a special label which includes all their entry information, and all debris must be collected and inspected by CFIA before disposal. We have permission to take cuttings and increase the stock during this time, but all offspring are considered to have come from the same crop as the mother plants and must continue to be identified accordingly (with the code in the inventory system and proper label). No PEQ plant can be included in CNCP and will always require a CFIA phyto to cross the border. To solve this problem, we ship plants to the US and then buy them back as stock plants or cuttings after an appropriate amount of time; after this, they can go into CNCP.

My role with the certification program also includes separate Canadian and other US and international regulations. This means that I apply for import permits when necessary, watch for other CFIA movement restrictions (for example, *Berberis* and *Rhamnus*). Some states have their own additional restrictions outlined in all the summaries listed here: http://nationalplantboard.org/laws-and-regulations/.

In addition, there is CITES (Convention on International Trade in Endangered Species), other border issues (rice hulls), and constant vigilance for changes in regulations and new pest finds.

BENEFITS OF THE CERTIFICATION PROGRAM

This seems like a lot of investment in people, time, and money. Is it worth it? Yes, we still think so. The main benefit is still the same one that got us on this path in the first place: it allows for efficient shipping. But I think the side benefits are the ones that keep us going. A systems-based approach works with the product from the time it enters the nursery until it leaves, helping us maintain an efficient Integrated Pest Management (IPM) program. It promotes excellent record keeping, encourages training for all growers and IPM personnel, and promotes professionalism in our nursery. It gives us the advantage of being allowed to complete post-entry quarantine for US-bound hydrangeas on our property, and our customers have one more reason to trust the nursery to provide them with clean, healthy plants.

Combined Proceedings IPPS 68:252-253. 2018.

Propagating Ornamental Plant Cuttings with Foliar-Applied Auxin

Benjamin D. Taylor and Benjamin K. Hoover
Department of Horticulture and Crop Science, California Polytechnic State University, San Luis Obispo, CA 93407, USA

bkhoover@calpoly.edu

Keywords: *Abelia* ×*grandiflora*, *Anisodontea capensis*, *Cistus ladanifer*, *Coprosma* ×*kirkii*, *Euryops pectinatus*, *Loropetalum chinense*, *Salvia mellifera*, *Teucrium chamaedrys*.

Abstract

In the propagation of many plant species, the basal end of cuttings is dipped in auxin powder diluted with talc to promote rooting. While the talc dip method has provided satisfactory results for many years, it is labor intensive and requires significant employee training and personal protective equipment.

Foliar application of auxin in aqueous solution could reduce employee exposure to auxin, since the treatment may be performed by a few workers for large numbers of cuttings after they are stuck. However, the foliar spray method has not been tested for most plant species. In this study, we tested the effects of auxin applied as a talc dip or a foliar spray at 0, 500, or 1000 ppm indole-3-butyric acid (IBA) to cuttings of cape mallow [*Anisodontea capensis* (L.) D.M. Bates], Chinese fringe flower (*Loropetalum chinense* R. Br.), and glossy abelia [*Abelia* ×*grandiflora* (André) Rehd.], and at 0, 1000, or 3000 ppm IBA to cuttings of bush daisy [*Euryops pectinatus* L. (Cass)], crimson-spot rockrose (*Cistus ladanifer* L.), mirror plant (*Coprosma* ×*kirkii* Cheeseman), 'Terra Seca' black sage (*Salvia mellifera* E. Greene 'Terra Seca'), and wall germander (*Teucrium chamaedrys* L.).

Root growth was assessed by two-dimensional root area, root dry weight, manually outlined root area, and rooting index. Chinese fringe flower and rockrose cuttings exhibited no difference in root growth in response to auxin treatment regardless of application method. 'Terra Seca' black sage root areas were generally not affected by auxin application, but root numbers were greater with auxin treatment (either 1000 or 3000 ppm) and with a talc dip versus a foliar spray.

IPPS Vol. 68 – 2018

Bush daisy and mirror plant exhibited increased rooting in response to auxin treatment, with generally similar root response whether applied by foliar spray or talc dip. Glossy abelia, cape mallow, and wall germander exhibited increased rooting in response to auxin treatment, with superior root response to foliar spray versus talc dip application. Our results suggest that auxin may not aid propagation of Chinese fringe flower and rockrose cuttings taken in late spring and early fall, respectively, while the talc dip method may be superior to the foliar spray method for propagating cuttings of 'Terra Seca' black sage. The foliar spray method resulted in equal or superior root growth compared to the talc dip method for bush daisy, cape mallow, glossy abelia, mirror plant, and wall germander cuttings.

Due to the potential for cost savings, improved employee comfort, and simplification of workplace safety compliance, we recommend that nurseries adopt the foliar spray method for propagating bush daisy, cape mallow, glossy abelia, mirror plant, and wall germander cuttings and conduct trials with additional species.

Combined Proceedings IPPS 68:254-266. 2018.

Plant Propagation and Commercial Cultivation in the Micronesian Region: Challenges and Measures for Sustainable Black Pepper Production

Virendra M. Verma
Micronesia Plant Propagation Research Center, Kosrae Agricultural Experiment Station (USDA Land Grant Program), College of Micronesia-FSM, Kosrae, Micronesia

vmv_vmv@hotmail.com

Keywords: Micropropagation, peppercorns, *Piper nigrum*, tropical horticulture.

Abstract

Black pepper (*Piper nigrum* L.), a flowering vine of the Piperaceae family, is valued for its dried berries called peppercorns, which are known for their health benefits and are used as a spice and seasoning. Native to the humid jungles of the Malabar Coast of Southwestern India, the plant is cultivated in the tropics worldwide. In the Micronesian region, it is gaining commercial importance as an important cash crop because of the premium price of peppercorns. However, the limited availability of disease-free black pepper seedlings and the trunks of the native tree fern (*Cyathea nigricans*) that are used as supports for black pepper vines are becoming limitations for sustainable commercial black pepper cultivation in the region. Therefore, to ensure the year-round availability of uniform, disease-free and high-quality planting material in Micronesia, an efficient micropropagation and acclimatization protocol was developed for a local, commercially important black pepper cultivar (*Piper nigrum* cv. Srilanka). Shoot apical meristems were used as explants for culture establishment. The best culture initiation was observed on Murashige and Skoog medium augmented with 5 µM 6-benzylaminopurine (BA). For further growth and subsequent multiplication, the established cultures were transferred onto 7.5 µM BA and 5 µM indole-3-acetic acid (IAA). The number of multiple shoots produced from each explant after two subcultures varied from 8 to 20. The best rooting was observed on 2 µM indole-3-butyric acid (IBA). Plantlets were acclimatized with 68% survival rate in 10 weeks. Research trials for sustainable commercial black pepper cultivation were designed and implemented, and vigorous vegetative growth was observed. To overcome the limitations of live tree-fern supports, non-living supports such as reinforced cement-concrete standards were specifically designed and used to support the vines for commercial cultivation. First harvesting was done after 12 months of planting

IPPS Vol. 68 – 2018

and data collection and analysis is underway. The outcome of this analysis will be used to provide assistance to the regional farming communities to promote sustainable commercial cultivation of black pepper in the region.

INTRODUCTION

Black pepper (*Piper nigrum* L.) is a perennial woody climbing liana. The plant is native to India, Indonesia, Malaysia, South America, and the West Indies, but is also widely cultivated in tropical regions worldwide. Black pepper is a universal table condiment used to flavor all types of cuisines worldwide. It is considered as the 'King of Spices' (Nair, 2004; Srinivasan, 2007). The spicy taste is mainly due to the presence of the compound piperine. Piperine is a pungent alkaloid (Tripathi et al., 1996) that enhances the bioavailability of various structurally and therapeutically diverse drugs (Khajuria et al., 2002). Increasingly popular modern-day uses of piperine, the active principal of black pepper, are to stimulate metabolism, aid absorption of nutrients, and boost the efficacy of drugs (Szallasi, 2005).

In most pepper-producing countries, black pepper is a smallholder's crop and many farmers depend on it for their livelihood. Cultivation varies from intensive monoculture to extensive homestead gardens. The use of reliable standards (supports) for the successful establishment of black pepper plantations is a common practice in producing countries. Standards are of two types: living and non-living. The use of non-living (dead) standards (reinforced cement concrete posts, granite pillars, and teak poles), though often resulting in higher black pepper yields, is less widely practiced by smallholders, mainly due to the high capital investment required (Dinesh et al., 2005). Non-living standards have been used in Malaysia, Vietnam, Brazil, Thailand, and Indonesia, facilitating closer spacing and higher yields (Kurien et al., 1985; Menon et al., 1982; Reddy et al., 1992).

Black pepper is considered an important cash crop in the Pacific, specifically in Micronesia. The Micronesian islands lying just above the Equator enjoy a tropical climate with relatively even, warm temperatures throughout the year. The Federated States of Micronesia (FSM) is made up of 607 small islands spread over a million square miles of the Western Pacific Ocean with a total land area of only about 271 square miles. Rainfall is generally plentiful, and Pohnpei, the capital state, is reputedly one of the wettest places on Earth, with up to 330 inches of rain per year. Nevertheless, drought conditions do occur periodically throughout FSM, especially when the El Niño condition moves into the Western Pacific. During these times, groundwater supplies have dwindled to emergency proportions. Tropical typhoons constitute an annual threat, particularly to the low-lying atolls of FSM (Government of the Federated States of Micronesia, 2014).

Agriculture is an important industry and can greatly help with economic development and growth, and in bringing food self-sufficiency to Micronesia. However, current agricultural programs in the country are mostly on a subsistence level and economic development is largely dependent on the outside world. Serious damage caused by natural calamities, such as wave surges, saltwater flooding, and drought, continually pose challenges for the local farmers. Moreover, a lack of technical know-how and changing lifestyle and food habits of the islanders are causing an increase in the consumption of imported foodstuff, leading to an overall decline in local agricultural production.

One recent example of such decline in agricultural production is the ceasing of local production of black pepper in Pohnpei, Micronesia. Black pepper from Pohnpei, Micronesia, is regarded as a relatively rare commodity of exceptionally high quality. The "Pohnpei Pepper", a pepper product that was unique to Pohnpei was marketed successfully from Pohnpei for a short time. The product was admittedly a high-end, niche product without a large volume of sales, but the potential of the product was barely tapped before its production ceased (Cheshire, 2003). Therefore, to promote sustainable black pepper cultivation practices in Micronesia, a project was developed to support local farmers and enhance agricultural productivity of black pepper in the region.

Based on the inputs of the stakeholders, farmers, agricultural professionals, and direct observations during black pepper farm visits, the project team identified the following issues that have caused serious decline in black pepper production in the Micronesian region:

- Non-availability of elite and disease-free seedlings;
- Limited traditional tree fern supports (standards).
- Poor soil fertility management and fertilizer applications.
- Occurrence of pests, diseases, and nutrient deficiencies.
- Shortage of trained agricultural professionals.
- Inadequate knowledge for fast propagation of cash crops.
- Limited skills for commercial production and basic crop management.
- Insufficient storage and processing facilities.
- Stiff competition between local and imported products.

The non-availability of disease-free and elite seedlings is a major bottleneck in quality black pepper production in the region. Micronesia is a small islands state; therefore, quarantine measures are very strict, and the entry of any planting material is strictly prohibited. Considering the difficulty in maintaining disease-free parental stocks in the tropics, meristem culture is increasingly being appreciated as a potential means of germplasm preservation and for the production of elite and disease-free planting materials on a mass scale.

Plant biotechnology is a powerful tool. Appropriate and skillful use of biotechnological approaches, such as in vitro multiplication, could help in successful multiplication of the elite and disease-free seedlings of cash crops and provide a means for germplasm conservation in an inexpensive way. In vitro multiplication is the best way to multiply the planting material of vegetatively propagated crops in bulk and produce disease-free, elite seedlings throughout the year.

The project aims to address some of the prime agricultural issues affecting the commercial black pepper production in Micronesia. It is specifically designed to develop micropropagation and nursery management systems in the region. The objectives of the project include: produce elite black pepper seedlings in bulk quantities to ensure the year-round availability of identical, disease-free and high-quality planting material; find alternatives for tree fern supports; determine appropriate fertilizer type and rates; and publish a commercial black pepper cultivation guide appropriate for Micronesia. Adoption of new practices, such as micropropagation of black pepper for improvement and enhanced productivity, will ultimately help in reviving the local pepper industry. The following sections describe the health benefits, in vitro multiplication, and commercial cultivation of black pepper.

IN VITRO MULTIPLICATION OF BLACK PEPPER

Plant Material

To establish cultures for in vitro multiplication of black pepper (*Piper nigrum* cv. Srilanka), 10- to 15-cm-long vines with shoot apex were collected in the morning from visually healthy, one-year-old plants that were growing in farmers' fields. The collected vines were kept in a greenhouse in which 100% relative humidity was maintained. Explants from these vines were obtained within 2 to 4 hours of their collection.

To obtain explants, vines were further trimmed to 4-cm-long tips containing the shoot apex. These trimmed vine tips were thoroughly washed with running tap water and were surface sterilized by immersion in 50% (v/v) ethanol for 10 minutes, which was followed by a treatment with 2% (v/v) sodium hypochlorite solution with 5 drops of Tween 20 for 5 minutes. Sterilized vine tips were then rinsed 5 times with sterile distilled water and were kept immersed in the solution until shoot apical meristem explants (including surrounding base tissue of 0.5 × 0.5 cm size) were excised for in vitro culture establishment.

Culture Medium

Murashige and Skoog (1962) medium (MS) was used in this study as a basal medium. All media contained 0.8% agar and 3% sucrose. The pH was adjusted to 5.8 prior to autoclaving. Different concentrations and combinations of cytokinins and/or auxins were used to augment the media for culture establishment, multiplication, and rooting.

Micropropagation

Shoot apical meristem explants were inoculated on MS medium without any growth regulator and supplemented with 50 µM cupric sulfate pentahydrate (CSP). Excellent establishment of black pepper cultures was observed in 6 weeks. Adding CSP in the media reduced the rate of fungal and bacterial contamination rate to less than 1%. Micropropagation of black pepper occurred in three distinct phases: 1) shoot initiation phase (SIP); 2) shoot multiplication phase (SMP); and 3) shoot elongation phase (SEP) (Figure 1a, 1b, 1c). To induce shoot initiation, the established cultures were transferred onto MS medium augmented with 5 µM 6-benzylaminopurine (BA) on which they were kept for 6 weeks and a passage (additional transfer) was given for another 6 weeks. After 12 weeks of incubation on SIP media, 100% shoot initiation was observed in all cultures. SIP cultures were then transferred onto MS medium augmented with 7.5 µM BA and 5 µM indole-3-acetic acid (IAA) for 6 weeks to induce shoot multiplication, and two passages at 6 weeks each were given for subsequent multiplication during SMP. Multiple shoots induced during SMP were then transferred onto MS medium without any growth regulators for 6 weeks, and two passages were given at 6 weeks each for further growth and shoot elongation during SEP. The total incubation time for culture establishment and all three phases was 56 weeks, with the duration of culture establishment being 6 weeks long, SIP being 12 weeks long, SMP being 18 weeks long, and SEP being 18 weeks long. The number of elongated shoots of 7 to 9 cm size produced from each explant after two subcultures varied from 8 to 20. Every 6 weeks, data were recorded, and each experiment was replicated three times with 25 explants per replication. A photoperiod of 16 hours with 40 µmol m^{-2} s^{-1} light intensity along with 24° C day and 22° C night temperatures were maintained during all phases of micropropagation.

Rooting

MS medium augmented with 2µM indole-3-butyric acid (IBA) proved best for inducing rooting of multiple shoots obtained through SEP (Figure 1c). Fully elongated shoots of 7 to 9 cm in height were transferred onto the rooting medium in groups of 20 to 25 shoots per culture (Figure 1c). After 4 weeks on rooting medium (Figure 1d), the percentage of rooting, number of roots per shoot, and root length were recorded. Each experiment was replicated three times with 100 cultures per replication.

Figure 1. Black pepper culture establishment (a), black pepper multiplication (b), rooting of black pepper multiple shoots (c), and complete black pepper plantlets (d).

Acclimation

Complete plantlets with 8- to 10-cm-long roots were transferred into 10-cm pots containing a sterilized soil:vermiculate (1:1, v/v) mixture. These potted plantlets were kept in the greenhouse for the first 6 weeks and then transferred into the screen house for the next 12 to 18 weeks. The screen house was covered with 60% green, knitted shade cloth and the temperature was maintained between 26°C and 28°C with 50% to 55% humidity. During these phases of acclimatization, the plants were irrigated once every 2 days with tap water and once per week with a one-fourth strength solution of MS basal salts. The survival rate of the plants was recorded after 10 weeks and a 68% survival rate was achieved. Each experiment was replicated three times with 100 plantlets per replication. After 12 to 18 weeks of ex vitro growth in the screen house, completely acclimatized plants were transferred to the nursery where they were kept until transfer to the field.

Figure 2. Acclimatization of black pepper plantlets (a), disease-free acclimatized black pepper seedlings in nursery (b), raised beds for proper water drainage and durable reinforced cement-concrete standards to support black pepper vines (c), and planting holes 1.5 feet across and 1.5 feet deep (d).

Statistical Analysis

Each experiment was replicated three times. A one-way analysis of variance was used to determine the level of significance between experimental treatments. Statistical significance of the results was determined using the least significant difference (LSD) test at α=5%.

BLACK PEPPER: COMMERCIAL CULTIVATION

Black pepper is valued for its dried berries, called peppercorns, which are used as a spice and for medicinal purposes. Native to the humid jungles of the Malabar Coast of Southwestern India, the plant is cultivated in the tropics worldwide. In Micronesia, it is gaining commercial importance as an important cash crop because of premium price. Traditionally, the trunks of two cultivars of large native tree fern (*Cyathea nigricans*) are used as supports for black pepper cultivation. However, short lifespan of these tree ferns, along with the rapid decline in their population due to increasing use of tree trunks for construction, is becoming a limitation for commercial black pepper cultivation in the region.

An in vitro multiplication protocol for locally preferred and commercially important black pepper cultivar *Piper nigrum* cv. Srilanka was developed and utilized for the multiplication and production of elite, uniform, and diseases-free black pepper plantlets in Micronesia. An efficient nursery management system was also standardized for the acclimatization of hundreds of plantlets into uniform and disease-free seedlings for sustainable commercial cultivation.

In Micronesia, the trunks of tree fern (*Cyathea nigricans*) are traditionally used as living supports for commercial black pepper vines. These large, native tree ferns are important sources of wood and are used for traditional house construction and as supports for commercial black pepper cultivation. Of the two cultivars of tree ferns that are traditionally recognized, one produces a red staining juice and is preferred over the other cultivar, which produces a greyish juice. The increasing construction in Pohnpei, along with the short life span of the desired tree ferns, has resulted in a drastic reduction in their lowland population. With newer roads now providing access to several inland locations, the upland populations of tree ferns are also threatened (University of Hawaii, 2014).

Considering the increasing demand for commercial black pepper cultivation and the extremely limited availability of traditional tree fern supports, non-living supports such as reinforced cement-concrete standards have been specifically designed and constructed at the pilot site to support the vines of fully acclimatized black pepper plants in the field. Standards of reinforced cement-concrete (6-inch length and 6-inch width in an octagonal shape and 13- to 15-foot height) were constructed and used as a support for each plant. In addition, raised beds, which ensure perfect water drainage, were used for the establishment of black pepper commercial plantations. To provide perfect nutrition and maintain these plantations, organic fertilizers, along with organic mulching and automatic fertilizer injectors, were used for soil amendment (Figure 3).

Figure 3. Soil amended with organic fertilizers for black pepper planting (a); liquid fertilizer application to each plant at regular intervals by automatic fertilizer injector (b, c), and healthy and vigorous black pepper vine vegetative growth (d).

This project is integrating and employing the latest tools and technologies, such as plant biotechnology, horticulture, microbiology, plant physiology, and plant pathology, for the sustainable, climate-smart and organic commercial cultivation of black pepper in Micronesia. The project team is utilizing plant biotechnological techniques, such as in vitro multiplication, for uniform black pepper plantlet production and greenhouse acclimatization of multiplied black pepper plantlets for elite, disease-free seedling production. The team is also using automatic fertilizer injectors for uniform fertilizer application, organic fertilizers to provide essential nutrients and maintain beneficial soil microorganisms, and appropriate site-specific and climate-smart horticultural, plant physiological, and integrated pest and disease management practices.

Climatic Conditions

Black pepper originates from tropical, warm, humid latitudes, where temperatures of 77°F and 80 to 120 inches annual rainfall predominate. Evenly distributed rainfall is ideal. Supplemental irrigation is necessary in dry, low-rainfall areas. Due to its tropical climate and adequate rainfall, pepper can be grown throughout the year in Micronesia.

Soil Characteristics

Black pepper can be grown on a wide range of soil types, but best results are obtained on deep, well-drained soils with good water-holding capacity. The best soil characteristics are sandy clay loam to clay loam with adequate essential plant nutrients and high organic content. Suitable soil pH is between 5.0 to 6.5. A slope not exceeding 10 ° to 15° is recommended for better soil conservation, easier harvesting, and farm management.

Field Preparation

Soil preparation for black pepper is similar to that for most dryland crops, such as corn. Existing vegetation is turned under with a moldboard or disc plow, or by spading. Most soils benefit from adding compost at this stage. During cultivation, phosphate fertilizer can also be added if required. After turning, leave the soil for a few days to allow for decomposition, and then break soil clods by harrowing or rotovating (or with a hoe or rake in small gardens). After the soil has been pulverized, the surface should be smoothed in preparation for black pepper planting. Black pepper can be planted on ridges, in furrows, or on flat ground.

Preparation of Planting Materials

Traditionally, black pepper has been propagated through cuttings that are prepared from stock plants. The cuttings consist of the upper 5 to 7 nodal segments. Selected planting materials should come from varieties that are disease- and pest-resistant, vigorous, and high yielding, with good productivity with respect to the final product. In recent years, owing to the advantages of disease-free planting material and uniformity in growth and higher yields, the use of tissue cultured plantlets as planting material for black pepper has become increasingly popular among farmers.

Supports and Planting

In Micronesia, the trunks of the tree fern (*Cyathea nigricans*) are traditionally used as living supports for commercial black pepper vines. Considering the extremely limited availability of tree fern supports and their very short lifespan, non-living supports such as reinforced cement-concrete supports are a good alternative. Supports should be planted well before planting black pepper at a depth of 2 to 3 feet. The planting holes should have a depth of 1.5 feet and a radius of at least 1.5 feet from the support. Prior to planting, the soil should be amended adequately with organic fertilizers, such as compost. Disease-free seedlings should be planted in prepared holes at the onset of a rainy day or in the evening. Young vines should be tied loosely to the support and shaded with suitable plant material.

Considering the frequent and heavy rains and poor drainage in the Micronesian region, planting black pepper seedlings in rows on raised beds is recommended (Figure 4a). The plants should be spaced in the rows at 8 feet apart with a 10-feet-wide alley maintained between rows.

Figure 4. Healthy and vigorous black pepper vines: One year old after planting (a); black pepper vine pruning (b); black pepper flowering (c); and ready-to -harvest drupes of black pepper (d).

Pruning

A couple rounds of pruning should be carried out during the vegetative phase of vine growth. Initial pruning of terminal shoots is done 4 to 6 months after planting. The next pruning is done when the vines are about 1 year old (Figure 4b), and the last pruning is done when the terminal shoots have reached the top of the standards.

Irrigation

Often grown in areas with high rainfall, black pepper is generally a rain-fed crop. Black pepper plantations do not require irrigation under normal conditions, except perhaps during the initial establishment period or in drought-prone areas. The plantations should not be allowed to become waterlogged for any extended length of time. For best results, maintain soil moisture at or near field capacity (moist but fully drained) throughout the growing period.

Fertilizer Application

Soils should be analyzed for nutrition status to determine nutrient requirements for growth and productivity of black pepper vines. In a tropical climate, it is better to apply small quantities of fertilizer often, rather than to add a large quantity in one treatment. This makes the fertilizer application more profitable and prevents too much rapid growth. Black pepper requires good soil fertility. In the first year, organic fertilizers such as compost may be applied at the rate of 4 to 6 pounds, along with 0.25 pounds of inorganic fertilizer, such as 12:2:14 NPK plus micronutrients every 3 months. In the second year, organic fertilizers may be applied at a rate of 8 to 10 pounds, along with 0.50 pounds of inorganic fertilizer, such as 16:16:16 NPK, plus micro-elements every 3 months. In the third year and onwards, organic fertilizers may be applied at a rate of 10 to 12 pounds, along with 1.0 to 1.5 pounds of inorganic fertilizer, such as 12:12:17 NPK plus micro-elements every 3 months.

To apply compost or organic fertilizers, scrape the soil surface around the circumference of the canopy. Apply the fertilizer along with the organic fertilizers at the recommended rate and then cover it with soil taken from the inter-spaces. Ensure sufficient moisture availability during fertilizer application.

Weed Control

Black pepper is susceptible to weed competition, especially during the first 8 to 12 months after planting when the leaf canopy is being formed. During this time, control weeds by hand-pulling or cultivating with a hoe. After the crop has attained the maximum vegetative stage, the lush foliage will shade out weed growth and cultivation for weed control should be minimized to avoid injuring the roots. When necessary, limited weeding by hand may be done in the inter-spaces and around the base of the vine.

Insect Pests and Diseases

Nematode infestation by *Meloidogyne* spp. causes the main problem in conventional pepper cultivation. Soil-borne fungi are the most significant cause of disease to black pepper. They possess a wide spectrum of hosts and can affect practically all of the crop types. Therefore, carry out constant and frequent scrutiny to identify any incidence of disease or pest at an early stage and take immediate action to control them. Integrated pest and disease management principles need to be applied at all stages to maximize productivity and minimize crop loss. Phytosanitary measures, such as physical removal of pests, affected plant parts, infected plants (virus-infected plants, severely disease-infected or pest-infested plants, including plants affected by *Phytophthora* spp. or slow decline or yellow wilt) are important to control these incidents.

Organic plant products and biocontrol agents, such as neem oil, neem cake, hot chili pepper solutions, and recommended predators for insect pest control may be used. Agrochemicals for control of pests and diseases may be used only when all other measures have been exhausted. Chemicals used should comply with all state regulations. Application of chemicals should follow recommended practices and chemicals should be applied only under the supervision of qualified professionals.

Harvesting

Each flower on the pendulous inflorescence of black pepper can develop into a single-seeded drupe (Figure 4c). Drupes that are almost mature with all green berries can be picked to process as green pepper. Drupes on an inflorescence with one or two berries beginning to turn yellow can be picked to process into black pepper (Figure 4d). To process into white pepper, drupes should be fully mature, with one or two ripe yellow-orange berries on each drupe. Drupes should be picked selectively and harvesting rounds should be carried out frequently throughout the year. Harvested drupes of pepper should be handled hygienically, collected, and transported in clean and closed baskets for processing into peppercorns (International Pepper Community, 2008).

Processing, Drying, and Storage

To ensure high quality, threshing of green pepper berries from the drupes is done manually in Micronesia. Separated green pepper should be washed in clean water to remove field dirt, insects, or other contaminants that may be present. Washed, cleaned pepper should be soaked for 1 to 2 minutes in water at 194°F to eliminate contaminants. Soaking in hot water would also facilitate drying and improve the appearance of the dried peppercorns. In Micronesia, solar dryers and electric dehydrators are used because of frequent rain and extremely high relative humidity. Black peppercorns should be dried to a moisture level of 10% for long storage. To avoid the loss of volatiles in peppercorns, drying must not be done at temperatures above 131°F.

Texture and Color

Different harvesting times and processing techniques can result in various colors and textures of peppercorns (Naturland, 2001). For green pepper, green peppercorns are produced when almost-mature green berries are harvested, processed, and conserved in brine (salt water, vinegar, and citric acid). For black pepper, black peppercorns are produced when mature green berries are harvested, processed, and dried (Figure 5a, 5b, 5d). For white pepper, white peppercorns are produced when ripe yellow-orange berries are harvested, processed and dried (Figure 5c).

Figure 5. Harvested drupes of black pepper (a), processing of black pepper (b), processed white peppercorns (c), and processed black peppercorns (d).

ACKNOWLEDGEMENTS

To promote sustainable black pepper cultivation in the region, an integrated research, outreach, and education project entitled, 'Black pepper micropropagation for elite seedling production: Comparison of local practices and commercial cultivation methods' was financially supported by the United States Department of Agriculture-National Institute of Food and Agriculture (USDA-NIFA). This project is of great significance, as it is specifically designed to develop black pepper micropropagation and nursery management systems to produce and ensure the year-round availability of identical, disease-free, and high-quality planting material. The objectives of the project include: finding alternative supports to overcome the limitations caused due to shortage of tree ferns, determining appropriate fertilizer type and rates, and development and publication of a commercial black pepper cultivation guide appropriate for Micronesia. The author would like to thank administrators, staff, and colleagues for their support.

Literature Cited

Cheshire, C.L. (2003). An alternative strategy for developing a Micronesian export industry. Micronesian Seminar, Micronesian Counselor #47. http://www.micsem.org/pubs/counselor/frames/altstratfr.htm

Dinesh, R., Kandiannan, K., Srinivasan, V., Hamza, S., and Parthasarathy, V.A. (2005). Tree species used as supports for black pepper (*Piper nigrum* L.) cultivation. Focus on Pepper *2*(1):39-47.

Government of the Federated States of Micronesia (2014). Geography. http://www.fsmgov.org/info/geog.html

International Pepper Community. (2008). Report of the Meeting of Experts' Group on Good Agricultural Practices for Pepper (*Piper nigrum* L.). Institute of Agricultural Sciences for Southern Vietnam, HCM City, Vietnam.

Khajuria, A., Thusu, N., & Zutshi, U. (2002). Piperine modulates permeability characteristics of intestine by inducing alterations in membrane dynamics: Influence on brush border membrane fluidity, ultra-structure and enzyme kinetics. Phytomed. *9*:224-231.

Kurien, S.A., Babu, N.M., Cheeran, A., and Nair, B.P. (1985). A note on crop standard interaction in pepper. Indian Cocoa, Arecanut, Spices J. *9*:35-36.

Menon, K.S., Nair, M.K., and Sharma, O.P. (1982). Preliminary report on the performance of black pepper on non-living standards. Indian Spices *19*:3-5.

Murashige, T., and Skoog, F. (1962). A revised medium for rapid growth and bioassays with tobacco tissue culture. Physiol. Plant. *15*:473-497.

Nair, K.P.P. (2004). The agronomy and economy of black pepper (*Piper nigrum* L.) - the 'King of Spices.' In: D.E. Sparks (ed.). Adv. Agron. *82*:271-389.

Naturland. (2001). Organic Farming in the Tropics and Subtropics - Pepper. Naturland e.V, Grafelfing, Germany.

Reddy, B.N., Sivaraman, K., and Sadanandan, A.K. (1992). High plant density approach to boost black pepper production. Indian Cocoa, Arecanut, Spices J. *15*:35-36.

Srinivasan, K. (2007). Black pepper and its pungent principle - piperine: A review of diverse physiological effects. Crit. Rev. Food Sci. Nutr. *47*:735-748.

Szallasi, A. (2005) Piperine: Researchers discover new flavor in an ancient spice. Trends Pharmacol Sci. *26*:437-439.

Tripathi, A.K., Jain, D.C., and Kumar, S. (1996). Secondary metabolites and their biological and medical activities of *Piper* species plants. J. Med. Aromat. Plant Sci. *18*:302-321.

University of Hawaii (2014). People and Plants of Micronesia: *Cyathea nigricans* and *Cyathea ponapensis*. http://manoa.hawaii.edu/botany/plants_of_micronesia/index.php/full-database/344-cyathea-nigricans .

Combined Proceedings IPPS 68:267-269. 2018.

The Horticulture Industry in China: Situation and Trends

Julienne Zhu
Nursery Stock Branch, China Flower Association, Hangzhou, Zhejiang, China

juliennezhu0419@aliyun.com

Keywords: Chinese nurseries, international horticulture, nursery exports, nursery imports, *Photinia*.

Abstract

The Nursery Stock Branch of the China Flower Association was founded in 2009 and is one of 14 branches of the CFA. The Nursery Stock Branch focuses on its 300 corporate, nursery stock-producing members and covers almost 62% of the horticulture business in China. In China, the government and society play quite different roles. The government focus on the policy making, which is a powerful section and takes active position. Society helps the government to standardize industry development without any authority and simply links business partners together by activities, conferences, exhibitions, etc., which is passive behavior. Since 2015, China has changed the approach from a government-led position to a market-leading position and provided policies to meet supply-side reform. The horticulture industry has much room for further development. The Chinese government pays much attention to environment protection, including water, soil, and even air improvement. A greening industry will be more prosperous.

TYPES OF NURSERIES IN CHINA

Most nurseries in China are privately owned, except for forestry nurseries, which belong to the state government. China has different kinds markets than do other countries, but only the horticulture production nurseries provide products to the final customers; these are the retail and wholesale garden centers, which may also market plants via an e-commerce platform. The other types of nurseries are all wholesale businesses, but with different types of marketing distribution. The storage nurseries provide plants to their own landscape construction projects. The business nurseries focus on production for cash-and-carry business. The investment nurseries pursue high quality and high value products to maximize their property, such as large bonsai plants of *Podocarpus* introduced from Japan and container-grown plants of special varieties from

IPPS Vol. 68 – 2018

the United States. Under a changing approach, China has started public-private partnership programs upon which many characteristic towns are being built and includes some specialized nurseries around these areas. These nurseries attract some tourists from the small town and enhance the landscapes.

PLANT PRODUCTION IN CHINA: AN EXAMPLE

I would like to use an example to explain plant production in China and help you to understand how incredible the market is in China. *Photinia* 'Red Robin' was introduced from the United States and Japan in 2002 by the Senhe Company. At that time, I was working in the international trade business. The president of the Senhe Company was Mr. Zheng Yong Ping, (his English name is Paul), and he had a chance to visit the American state of Florida in 2001. During his trip, he noticed some red-leaved plants growing in gardens and along the streets, some grown as specimen shrubs and some grown as hedges. "This plant is fantastic!" Paul said to himself. "I want to buy these plants!" he told his friend Dr. Alan Lang, who worked for Speedling in Florida.

Something magical then happened in China in the winter of 2002. The Senhe Company imported one 40-foot shipping container of *Photinia* seedlings. One hundred workers were waiting at the company's nursery warehouse and made cuttings as soon as the container arrived at midnight at the nursery. The next year, millions of young plants were produced, although the survival rate was not good because Senhe was not very experienced with this variety. However, a new variety was successfully introduced and promoted in the China market in 2004. The price was 5 RMB per plant (about one Australian dollar). Madam Yang, one of the growers in Xiaoshan of Hangzhou, who bought some of these plants, stated excitedly, "I have earned the first pot of gold in my life."

In 2005, photinia started to be used in the landscape and the market gradually began to accept the new plant. A local grower wanted to grow this plant, but it was expensive, so the grower had some workers cut branches for propagation, since the variety was not covered by intellectual property protection and the Senhe Company could not protect this variety.

Over the next few years, Senhe began to promote other plants with red and yellow colors, but none achieved as much success as the photinia. Today, photinia is planted in every landscape project in China from north to south and east to west. Huge numbers of photinia have been sold over the past ten years.

In China, if you can produce and provide a good variety with large-scale production, you can cover the market quickly. But the variety must be marketed with a strong presentation and have widespread application.

CURRENT CHARACTERISTICS OF HORTICULTURE IN CHINA

China's horticulture industry is expanding. The industrial structure has been formed and the cluster effect has appeared. Research systems have gradually brought improvements and innovations. The market system has undergone its initial establishment. The business model has been enriched. The brands of special events have been set up. Flower culture has become increasingly prosperous.

Today in China, the business of flower festivals is on the increase. Production of potted plants is on the rise. Cut flowers are available with better quality at a good price. Flowers are being sold through both online and offline business. Landscape programs are creating demand and there is a big demand for nursery stock.

Over the past 30 years, China have been developing quickly and changing fast. By the end of 2017, China's horticulture planting area was about 140 million hectares, with nursery stock plants area at about 78,000 hectares (about 58%) of this total area. Total sales have been reported by China's Agriculture Ministry at about $23.59 billion US dollars. According to the statistics from China Customs, imports and exports together have reached $560 million US dollars: $270 million USD dollar in imports and $290 million in exports. Comparing with 2016, these numbers have increased by 26.5% and 0.67%, respectively.

China has become one of the largest flower and plant production bases in the world. It will be one of the most important consumption and trade countries. China's horticultural products are exported to Japan, Holland, Germany, the United States, Canada, Australia, South Korea, Thailand, Vietnam, Singapore, and another 105 countries and regions.

In China there are now about 3,220 fair markets, and 84,978 enterprises. Online business continues to expand, and the turnover of cut flowers reached $3.63 billion US dollars in 2017. According to the data released by alibaba.com, the flowers and plants category showed the highest growth rate (200%).

CHINA NURSERYMAN MAGAZINE

China Nurseryman magazine is one of the important information media for the Nursery Stock Branch of CFA, working to spread information about propagation technology and nursery industry developments and trends. It has four editions per year and is sent to all the members for free. But we still believe the magazine has room for improvement and can be even more professional.

CONCLUSION: TRENDS AND OPPORTUNITIES

New trends and opportunities in horticulture continue to develop in China. There is top-level policy support, as seen with the Beautiful Chine Green Life program. There is support for ecological and environmental protection for creating safe living environments. A large population in China brings a huge consumer market. Standardization and mechanization are improving efficiency. Branding is becoming important in marketing. Rules are regulations are developing for intellectual property protection. Landscape projects are enhancing construction of towns, parks, and resorts.

PROCEEDING'S PAPERS

SOUTHERN REGION OF

NORTH AMERICA

Dr. Fred T. Davies, Jr., Regional Editor

Forty-third Annual Meeting - 2018

Chattanooga, Tennessee USA

Combined Proceedings IPPS 68:270-271. 2018.

Southern Region – IPPS Opening Presentation

David Creech
Stephen F. Austin State University Arboretum, P.O. Box 13000, Nacogdoches, Texas 75962, USA

Email: dcreech@sfasu.edu

Keywords: Annual Meeting, Southern Region of North America (SRNA)

INTRODUCTION

President Creech welcomed everyone to Chattanooga, Tennessee for the 43rd Annual Meeting of the International Plant Propagators' Society-Southern Region of North America (SRNA). He thanked Local Site Committee Chair, Amy Fulcher and her committee and volunteers for the long hours in arranging the excellent tours, hotel, other planning activities and all their attention to detail.

He welcomed students, first time attendees and new members, asking them to stand and be recognized. Creech thanked the Executive Committee, and Brie Arthur's Sponsorship Committee, which raised $41,500 in cash and $8,000 in-kind sponsorships - which was outstanding. Creech encouraged the membership to visit and show their support of our sponsors during the meeting. He encouraged all members to make new members and first-time attendees feel welcome — share with them and seek from them. He called for good questions and enthusiastic participation at the Tuesday night question box.

Creech announced that the SRNA is in its second year of the Southern Region Educational Endowment, which will be discussed in greater detail later in the meeting. With a base donation of $20,000 from an anonymous donor, the Education Endowment balance is now $40,512 – and growing. It will greatly enhance our region's ability to support students and early career professionals – and ensure continued quality of the outstanding educational programs our region is known for.

Creech announced that this is the sixth year the SRNA has participated with the European Region in the *Early-Career Propagator Exchange* program between the two regions. He recognized Jasper Schermer from the European Region, who was hosted by Judson LeCompte and J. Berry Nursery of the SRNA. Lindsay Day of the SRNA was our designee to the European Region. Both of these early-career professionals had an incredible exchange experience in our respective regions.

IPPS Vol. 68 – 2018

This is the seventh year we are doing the Vivian Munday Young Horticultural Professional Scholarship Work Program (Vivian Munday Scholarship).

We currently have four young professionals: George Grant (University of Florida), Kaitlin Barrios (University of Georgia), Nastaran Basiri Johromi (University of Tennessee) and John Nix (North Carolina State University) - who are making a strong contribution to this year's program. Creech thanked Program Chair and 1st Vice-President, Elliott Hallum, for the excellent program and slate of speakers he assembled (Fig. 1).

Figure 1. President David Creech (left) with Elliott Hallum (right), program chair of the 2018 Chattanooga, Tennessee 43rd annual conference.

Combined Proceedings IPPS 68:272-280. 2018.

A Survey of the Volatile Profiles of Daylily Species and Hybrids

S.A. Keene, T.S. Johnson, C.L. Sigler, T.N. Kalk, P. Genho, and T.A. Colquhoun
1523 Fifield Hall, PO Box 110670, Gainesville, Florida 32611, USA

keene284@ufl.edu

First Place - Parkerson Student Research Paper Competition

Keywords: *Hemerocallis*, fragrance, gas chromatography-mass spectrometry, hybridization.

Abstract

Daylilies (*Hemerocallis*) lost their fragrance as a result of many years of hybridization that singularly focused on flower color and form. Using a field collection system and gas chromatography-mass spectrometry, this study assessed the fragrance profiles of 147 daylilies. Major volatile constituents and their variations in the daylily study populations were determined and suggest that fragrance could be a trait pursued in a breeding program to enhance the sensory phenotypes of new daylily varieties.

INTRODUCTION

Daylilies belong to the genus *Hemerocallis* and are monocotyledonous herbaceous perennial plants. The genus and common name reflect the blooming habit of daylilies: their flowers only last for one day. Daylilies are native to sub-tropical and temperate Asia, arising mainly from China, Korea, and Japan (Rodriguez-Enriquez and Grant-Downton, 2013). Approximately 20 species are recognized, the colorations of which are limited to yellow, orange, and fulvous red (Gulia et al., 2009). Daylily hybridization began in earnest in the early 20th century, mostly by amateur breeders that focused on increasing the diversity of flower colors, shapes, and forms (Gulia et al., 2009).

Nowadays, daylilies come in a dazzling array of colors, patterns, shapes, and sizes, and many of these "hybrids" bear little to no resemblance to the modest species from which they were derived. The American

Hemerocallis Society (AHS) is the official daylily registrar, and currently maintains an online database of more than 87,000 registered cultivars. Over the years, however, such a singular focus on flower color, pattern, and form in daylily breeding resulted in an unfortunate unintended consequence: the loss of fragrance. While a number of the original daylily species possess noticeable, distinct fragrances, those fragrances are greatly reduced or largely absent in many modern hybrids (Jiao et al., 2016). As daylily breeders seek to create novel hybrids, they have turned their attention to the long-ignored trait of fragrance (P. Genho and J. Gossard, personal communication).

Floral fragrance is composed of mixtures of volatile organic compounds (VOCs or volatiles), mostly lipophilic liquids with high vapor pressures at ambient temperatures that typically fit in families of terpenes, phenylpropanoids, or benzenoids, as well as derivatives of amino acids and fatty acids (Dudareva et al., 2013). Plant volatiles serve a number of biological functions, including attracting pollinators or seed dispersers, acting as defense compounds, protecting the plant during certain abiotic stresses, and acting as signaling molecules (Dudareva et al., 2006). Floral volatiles also serve as a sensory attractant to people. A study of consumer preference for floral attributes found that a flower that does not make fragrance at all had the largest negative effect on consumer interest, indicating that consumers prefer fragrant flowers (Levin et al., 2012).

Two studies have been conducted on daylily aroma, both in China (Jiao et al., 2016; Lin et al., 2003). Lin et al. (2003) evaluated the essential oil of a single daylily species; however, depending on the extraction process, the aroma of an essential oil can differ from the aroma experienced by a person smelling the flower from which the oil came (Tholl et al., 2006). Jiao et al. (2016) evaluated 46 daylilies and identified 37 volatiles; however, the authors used authentic compound standards to verify only three of the compounds, thus casting some doubt on the veracity of the identity of the remaining compounds (Tholl et al., 2006). These studies provide a point of reference for further investigation of daylily scent, but ultimately only reflect the volatiles emitted by a tiny fraction of the expansive modern daylily germplasm.

The objective of this research was to evaluate the volatile profiles of a larger number of daylily hybrids and a small number of daylily species in three locations across the U.S., to identify volatile compounds in those profiles, assess the variation of volatile emissions among different daylily hybrids, and finally to determine which species or hybrids may be genetic resources for different volatiles. Daylily hybridizers could use this information to selectively breed for daylilies with enhanced fragrance.

MATERIALS AND METHODS

Volatile Collection. Volatiles were collected *in situ* from three privately owned populations of daylilies in Florida, Ohio, and Utah (Table 1).

Table 1. Study population and collection details.

	Florida	Ohio	Utah
Collection Dates	5/13/16 – 7/11/16	7/10/17 – 7/12/17	7/14/17 – 7/19/17
Avg. High Temperature (°F)	91	84	96
Avg. Relative Humidity (%)	70	87	42
Number Daylilies Sampled	64	33	50
Collection Owner	J. and E. Salter, Rollingwood Gardens	J. and D. Gossard, Heavenly Gardens	P. Genho, Private Collection

The presence of two fully open flowers was the main selection criterion for volatile sampling. Beyond that, daylilies of as many colors, color patterns, and flower forms as possible were selected at random from the populations. Volatiles were sampled via headspace sorption for 1-2 hours between 1000 and 1400 hours (Huber et al., 2005). Inflorescences were inserted into a nylon resin cooking bag and the bag was gathered around the scape beneath the flower and cinched with a twist tie (Stewart-Jones and Poppy, 2006). A glass column containing approximately 50 mg HaySep Q 80-100 porous polymer adsorbent (Hayes Separations Inc., Bandera, TX) was inserted into a slit at the top of the bag above the flower and secured with a twist tie. The glass column was fastened to a wooden stake to prevent the collection apparatus from collapsing on the flower (Fig.1).

Following volatile enrichment, a single-setting vacuum pump (Barnant Company, Barrington, IL, USA) was used to pull the air out of the bag through the adsorbent trap for three minutes. On each collection date, volatiles were sampled from empty nylon resin bags to account for background contaminants. Volatiles were collected in duplicate from each daylily. Volatiles were eluted from the adsorbent polymer within 12 hours with 150 μL of methylene chloride spiked with 2 μL of nonyl acetate as an elution standard. Samples were stored at -80° C until analysis by gas Chromatography-mass spectrometry (GC-MS).

Figure 1. Schematic diagram of the collection system. The column containing the adsorbent polymer is attached to the upper part of the stake, and an empty bag, from which volatiles were sampled to account for background contaminants, is attached to the lower portion of the stake.

Volatile Analysis. Volatile samples were analyzed on an Agilent 7890A gas chromatograph fitted with a DB-5 column (5% phenyl, 95% dimethylpolysiloxane, 29 m length x 0.25 mm internal diameter x 1 μm film thickness) and coupled to an Agilent 5973A mass spectrometer (Agilent Technologies, Santa Clara, CA). Compounds were tentatively identified by comparing their mass spectra to the National Institute of

Standards and Technology (NIST) mass spectral library.

Volatile identification was achieved by comparing the retention times and mass spectra of peaks in the samples to those of authentic standards. Analysis of volatile data was performed using MassHunter Qualitative and Quantitative software programs (Agilent Technologies; Santa Clara, CA, USA). Calculation of relative amount of volatile emission was based on individual peak area relative to the peak area of the elution standard within each sample. Calibration curves for authentic standards were run in duplicate on the GC-MS under the same conditions described above.

RESULTS

The volatile profiles of 147 daylilies, six species, 98 registered cultivars, and 43 unregistered "seedlings" were evaluated, and 18 volatile organic compounds were identified. Table 2 provides information about these compounds, including their prevalence and variation within the total study population, as well as the daylily cultivar or species that emitted the greatest amount of each compound.

Table 2. Summary statistics of volatile compounds emitted by daylily study population (N = 147).

Compound	n[1]	Freq.	Mean[2] ± SE	Median	Max	Daylily Emitting Max
Acetoin	98	66%	0.39 ± 0.03	0.31	2.39	'Bright Blaze of Magic'
2-Methyl-1-butanol	26	17%	0.26 ± 0.03	0.19	0.69	'Celtic Witch'
(*E*)-2-Methyl-2-butenal	111	75%	0.35 ± 0.04	0.21	2.31	'Bright Blaze of Magic'
3-Methyl-2-butenal	123	83%	1.88 ± 0.19	1.24	11.24	'Cheddar Explosion'
Hexanal	49	33%	0.22 ± 0.03	0.13	0.70	'Midnight Crossroads'
Benzaldehyde	34	23%	0.15 ± 0.02	0.11	0.46	'Cheddar Explosion'
6-Methyl-5-hepten-2-one	6	4%	0.03 ± 0.01	0.03	0.05	'Jalapeno Crunch'
β-Myrcene	136	92%	0.77 ± 0.04	0.64	3.00	'Blue Vibrations'
2-Ethyl-1-hexanol	139	94%	0.24 ± 0.01	0.21	0.67	'Wind Rider'
(*E*)-β-Ocimene	147	100%	9.22 ± 0.59	8.00	40.52	*Hemerocallis citrina*
(*Z*)-β-Ocimene	147	100%	33.67 ± 1.81	32.47	94.33	*Hemerocallis thunbergii*
β-Linalool	92	62%	0.30 ± 0.03	0.21	1.60	*Hemerocallis thunbergii*
Phenylethyl alcohol	102	69%	0.33 ± 0.01	0.28	0.66	'Winter Halo'
allo-Ocimene	115	78%	0.04 ± 0.003	0.03	0.18	*Hemerocallis citrina*
Indole	61	41%	0.49 ± 0.07	0.25	2.74	'Oh Great One'
β-Caryophyllene	20	13%	0.22 ± 0.03	0.17	0.50	'Micro Magic'
α-Farnesene	97	65%	4.27 ± 0.58	2.31	41.35	*Hemerocallis thunbergii*
(*E*)-Nerolidol	71	48%	0.34 ± 0.05	0.20	1.90	'Wind Rider'

[1]n denotes the number of daylilies that emitted a given compound, and Freq. denotes the frequency of that compound's occurrence in the total study population (n/147).
[2]Amount of volatiles emitted is given in μg/inflorescence.

Average emissions of each volatile compound are shown by study location in Figs. 2 and 3. Table 3 provides information about the total volatile emissions in each study location. Finally, Table 4 lists the top ten most fragrant daylilies, as determined by total volatile emissions, for each study location.

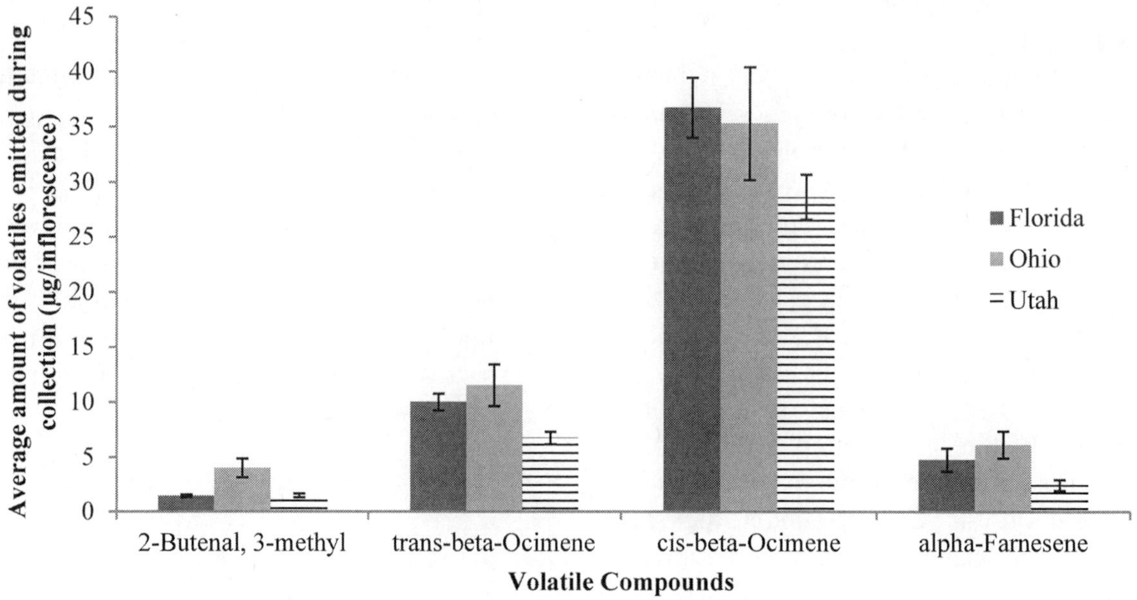

Figure 2. Average amount of high abundance volatile compounds emitted by daylilies in each study location. Standard error bars are shown for each volatile mean.

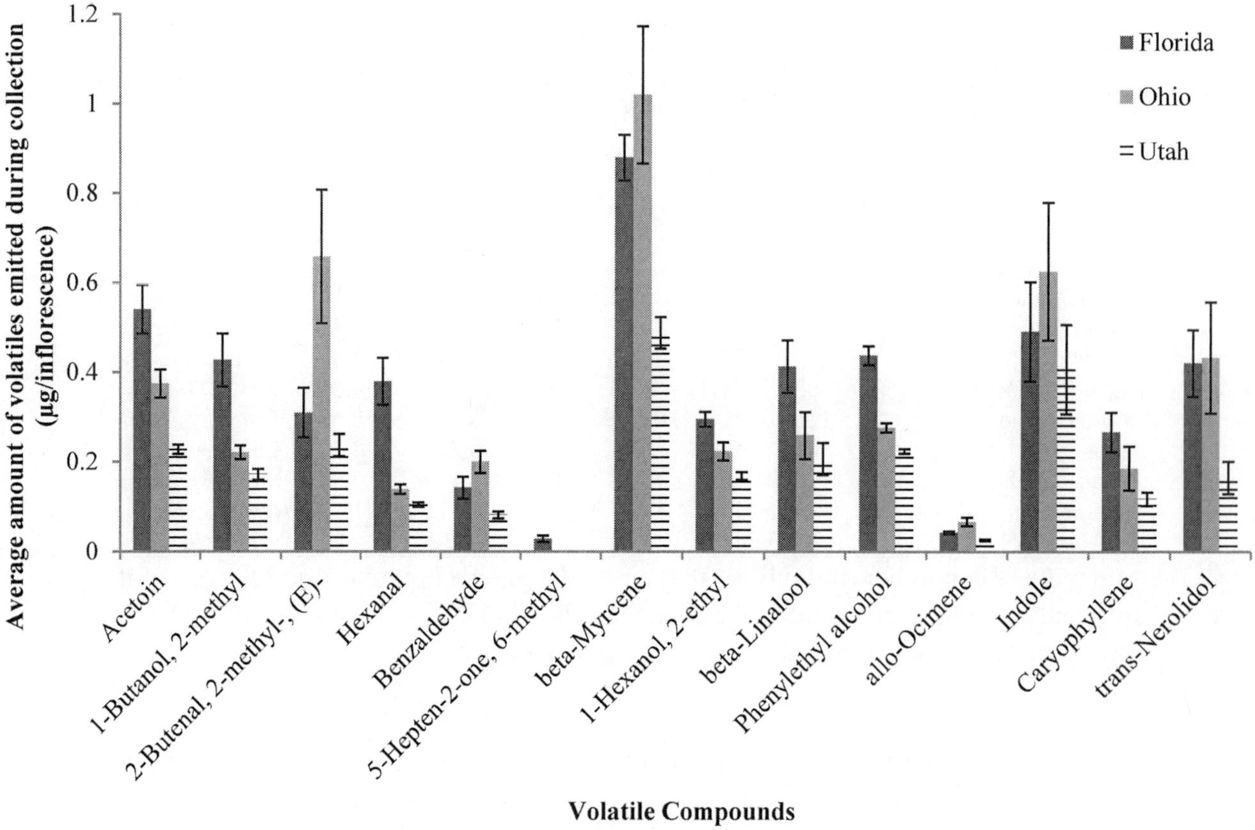

Figure 3. Average amount of low abundance volatile compounds emitted by daylilies in each study location. Standard error bars are shown for each volatile mean.

Table 3. Summary statistics for total relative volatile emissions, in μg/inflorescence, for each study location.

	Florida (n = 64)		Ohio (n = 33)		Utah (n = 50)	
	Emission	Daylily	Emission	Daylily	Emission	Daylily
Minimum	2.04	'Rim of Fire'	0.92	*H. fulva* 'Korean'	1.48	'Cheers for Now'
Mean	54.13		56.01		39.98	
Median	51.77		49.42		41.88	
Maximum	159.54	*H. thunbergii*	152.45	*H. citrina*	84.06	'Cranberry Daiquiri'

Table 4. The ten daylilies emitting the greatest total amount of volatiles, in μg/inflorescence, by location.

	Florida		Ohio		Utah	
Rank	Daylily	Total Emission	Daylily	Total Emission	Daylily	Total Emission
1st	*H. thunbergii*	159.54	*H. citrina*	152.45	'Cranberry Daiquiri'	84.06
2nd	'Spacecoast Blue Eyed Majesty'	129.59	'Out of Balance'	126.46	'Glistening Accent'	80.16
3rd	'Bridge of Dreams'	112.56	'Popcorn at the Movies'	116.30	'William Seaman'	70.63
4th	'Jalapeno Crunch'	101.29	'Blue Vibrations'	116.27	'Samurai Jack'	68.77
5th	'Spacecoast White Chocolate'	98.66	'Mystical Elf'	113.56	'Love and Marriage'	61.86
6th	'The Fantastic Barbara Watts'	82.56	'Spacecoast Devil's Eye'	113.48	'Viva Piñata'	60.65
7th	'Breakfast with Santa'	81.15	'Blackwater Captain Jack'	103.94	'Sailing'	58.71
8th	'Winter Halo'	78.90	'Double Yellow Thunder'	97.33	'Born to Run'	58.25
9th	'Heavenly Bengal Tiger'	64.58	'Cheddar Explosion'	87.20	'Ultimate Design'	58.16
10th	'Midnight Crossroads'	58.66	'Double Rays of Sunshine'	86.13	'Lover's Lemonade'	56.55

DISCUSSION

The majority of the 18 volatiles identified in this study were terpenoid compounds, including monoterpene hydrocarbons and alcohols, sesquiterpenes, and terpenoid derivatives. Similarly, Jiao et al. (2016) found that terpenoids represented over 80% of the total volatiles released by the daylilies they evaluated. While some geographic variation in the average emission amounts of the volatiles was observed, the differences between the three study locations in terms of climate, management practices like fertilization and irrigation regimens, soil type, and other factors, makes direct comparisons between the daylily populations unfeasible. Rather, each population was assessed individually.

Of the daylilies sampled in Ohio, four of the most fragrant cultivars exhibited "double" flower forms, which the AHS defines as a form with extra whorls of petals or petaloid tissue inside the normal petal whorl. As floral volatiles are emitted from petal tissue, these double daylilies may be more fragrant in part because they have more petal tissue. However, since the daylilies were part of an active breeding program, destructive sampling methods could not be employed so emissions per gram of fresh weight were not determined.

In Florida, *H. thunbergii* had a total volatile emission of almost 80 times that of the least fragrant daylily, while in Ohio, *H. citrina* had a total volatile emission of more than 150 times that of the least fragrant daylily. In comparison, over 80% of the total study population emitted less than half the amount of volatiles emitted by *H. citrina*. Jiao et al. (2016) obtained comparable results: all 38 daylily hybrids they evaluated were classified as having low or no floral aroma, with only five species exhibiting "intense or medium" floral aroma. The stark contrast between the fragrance of the species and hybrids illustrates the effect of hybridizers' traditional breeding objectives: the focus on flower form and coloration has indeed resulted in daylilies with drastically reduced fragrance. Nonetheless, out of only 147 daylilies, this study identified hybrids that do have heightened aromas, some of which are nearly as fragrant as the species already. Given the vast number of registered cultivars, many other fragrant hybrids certainly exist. As genetic resources of certain volatiles, these daylilies could be used by hybridizers in a breeding program to selectively breed for enhanced fragrance. In a practical sense, hybridizers do not need fancy analytical equipment to screen for fragrance. All they need is a decent sense of smell.

Because this research was conducted in the field on daylilies managed by different people, there were several uncontrolled variables, including soil type and fertilization regimens, among others. While the collection system was economical and practical for a field setting, it may not have been sensitive enough to detect volatiles at very low levels. Moreover, volatiles were collected at a single time point. Despite these limitations, this study yielded useful qualitative and relative quantitative data about the volatile profiles of almost 150 daylilies, highlighted the aromatic variation that exists in a slice of the germplasm, and identified daylilies that are potential genetic resources of volatiles. Hybridizers could use this information to potentially create "novel," highly fragrant daylilies that stand out from the 87,000+ existing cultivars. Daylily hybridizers are an especially avid community of plant breeders that have wrought incredible transformations in the visual characteristics of daylilies. If they increase their focus on aromatic characteristics, they will undoubtedly transform and enhance the fragrance of daylilies, too.

Figure 4. The two most fragrant daylilies, in terms of total volatile emissions, *sampled in Florida*. The species *Hemerocallis thunbergii* is shown on the left, and the hybrid 'Spacecoast Blue Eyed Majesty' is shown on the right.

Figure 5. The two most fragrant daylilies, in terms of total volatile emissions, *sampled in Ohio*. The species *Hemerocallis citrina* is shown on the left, and the hybrid 'Out of Balance' is shown on the right.

Figure 6. The two most fragrant daylilies, in terms of total volatile emissions, *sampled in Utah*. The hybrid ‘Cranberry Daiquiri’ is shown on the left and the hybrid ‘Glistening Accent’ is shown on the right.

Literature Cited

Dudareva, N., Klempien, A., Muhlemann, J.K., Kaplan, I. (2013). Biosynthesis, function, and metabolic engineering of plant volatile organic compounds. N. Phytolog. 198:*16*-32.

Dudareva, N., Negre, F., Nagegowda, D.A., and Orlova, I. (2006). Plant volatiles: recent advances and future perspectives. Critical Rev. Plant Sci. *25*:417-440.

Gulia, S.K., Singh, B.P., and Carter, J. (2009). Daylily: botany, propagation, breeding. Hort. Rev. *35*:193-220.

Huber, F.K., Kaiser, R., Sauter, W., and Schiest, F.P. (2005). Floral scent emissions and pollinator attraction in two species of *Gymnadenia* (Orchidaceae). Oecologia *142*:564-575.

Jiao, F., Liu, Q., Sun, G.F., Li, X.D., Zhang, J.Z. (2016). Floral fragrances of *Hemerocallis* L. (daylily) evaluated by headspace solid-phase microextraction with gas chromatography-mass spectrometry. J. Hort. Sci. Biotechnol. *91*:573-581.

Levin, L.A., Langer, K.M., Clark, D.G., Colquhoun, T.A., Callaway, J.L., and Moskowitz, H.R. (2012). Using Mind Genomics® to identify the essential elements of a flower product. HortScience *47*:1658-1665.

Lin, P., Cai, J., Li, W., Su, Q. (2003). Constituents of the essential oil of *Hemerocallis flava* day lily. Flavour Fragrance J. *18*:539-541.

Rodriguez-Enriquez, M.J., and Grant-Downton, R.T. (2013). A new day dawning: *Hemerocallis* (daylily) as a future model organism. AoB PLANTS *5*: pls055.

Stewart-Jones, A. and Poppy, G.M. (2006). Comparison of glass vessels and plastic bags for enclosing living plant parts for headspace analysis. J. Chem. Ecol. *32*:845-864.

Tholl, D., Boland, W. Hansel, A., Loreto, F., Rose, U.S.R, and J. Schnitzler. (2006). Practical approaches to plant volatile analysis. Plant J. *45*:540-560.

Combined Proceedings IPPS 68:281-289. 2018.

Seed Germination of *Crocanthemum arenicola* (Coastal Sand Frostweed)

Gabriel Campbell-Martínez, Anna Dicks, Mack Thetford and Debbie Miller
University of Florida, 5988 Highway 90, Building 4900, Milton, Florida 32583, USA

camp5595@ufl.edu

Second Place - Parkerson Student Research Paper Competition

Keywords: Physical dormancy, *Helianthemum*, beach dune, scarification, endemic.

Abstract

Crocanthemum arenicola (coastal sand frostweed) is a southeastern USA native plant with ornamental and restoration potential. Propagation information for this plant is lacking. However, other species of Cistaceae have seeds with physical dormancy that can be alleviated by scarification. The objective of the present study was to test the effects of scarification via sandpaper abrasion with an electric seed scarifier and photoperiod on germination of *C. arenicola*. Scarification for 50-200 seconds with an electric seed scarifier lined with sandpaper was sufficient to break physical dormancy with ≥90% germination for graded seed. Additionally, no difference in germination (63-64%) was detected between non-graded seeds exposed to a 12-hour photoperiod or left in the dark.

INTRODUCTION

Crocanthemum is a new world genus composed of 15 species within Cistaceae, the majority of which are native to the southeastern USA (Sorrie, 2011; USDA and NRCS, 2018). Plants are fire-tolerant perennial herbs to subshrubs that resprout readily from a woody caudex (eFloras, 2018). The species of interest in this study, *C. arenicola* (coastal sand frostweed), is an herbaceous, perennial groundcover species found on beach dunes, scrub, and sandhill plant communities (Wunderlin and Hansen, 2011; Figure 1). It is endemic to 12 coastal counties of the Florida panhandle, Mississippi, and Alabama (USDA and NRCS, 2018). This plant has 2 types of flowers, showy (chasmagamous) yellow flowers and non-showy (cleistogamous) flowers (eFloras, 2018; Figure 1).

Propagation information for *C. arenicola* is generally lacking. However, other *Crocanthemum* and the closely related

Helianthemum species have been reported to have physical dormancy that may be alleviated via scarification. It has been proposed to move *Crocanthemum* species from the *Helianthemum* genus (Sorrie, 2011). Mechanical scarification via sandpaper abrasion has been reported as an effective form of dormancy alleviation for several species of *Helianthemum* (Pérez-García et al., 1995; Pérez-García and González-Benito, 2006; Yeşilyurt et al., 2017).

In these studies, seeds germinated in both light and dark conditions with mixed results. The objective of the present study was to describe basic information about seeds and fruits, physical dormancy, and germination for *C. arenicola.* We hypothesized that seeds of this plant are physically dormant and that photoperiod may affect germination. Additionally, we informally discuss a success story of containerized production of this plant from stem cuttings.

Figure 1. (A) 1-gal containerized plants and (B) chasmagamous flower of *Crocanthemum arenicola*.

MATERIALS AND METHODS

Stock Plants. Stock plants were derived from spring-collected softwood-stem cuttings from wild plants growing in secondary dunes on Santa Rosa Island and Perdido Key, Florida. Rooted cuttings were grown in 1-gallon plastic containers with a 2:1 mix of pine bark:Sungro MetroMix® 830 and top-dressed with 0.5 tablespoon of Osmocote® 18-6-12. Plants were hand watered as needed and subjected to a natural photoperiod in a research greenhouse (University of Florida, Milton Campus).

Fruit and seed collection, characteristics, and storage. Fruits were collected from stock plants as fruits turned brown and fell off or as fruits fell off when the plant was shaken slightly throughout the fall and winter of 2017. Seeds were separated from fruit by hand and air dried under laboratory conditions (~25°C and 70% RH) for at least 2 weeks. A total of 30 fruit were used to determine average number of seeds per fruit. A total of 18 samples, each containing 20 seeds, were weighed (mg) to determine average seed weight. Seeds were placed in an airtight jar in the dark for storage.

Imbibition experiment. An imbibition experiment was conducted on scarified and non-scarified (control) seeds to determine if physical dormancy was present. Seeds were scarified using an electric seed scarifier (Seedburo Equipment, Des Plaines, Illinois) lined with sandpaper (80 grit aluminum oxide, GatorGrit™, USA) for 50 seconds. Control seeds received no sandpaper abrasion. Seeds were placed in Petri dishes on two sheets of blotter paper that were saturated with 5 mL of distilled water and maintained at room temperature (25°C) in a laboratory. Four replicate dishes, each with 90 seeds, were used per treatment. Seed mass was measured to the nearest 0.01 mg after 0, 0.25, 0.5, 1, 2, 3, 6, 12, and 24 hours. Water was added when the film of water around the seeds decreased to <1 mm.

Scarification experiment. An experiment was conducted to determine the time of abrasion with sandpaper in an electric seed scarifier needed to break physical dormancy and allow germination. Seeds for this experiment were graded and seeds that appeared empty, deformed, or discolored were discarded. All seeds were surface sterilized via a 1-minute 70% isopropyl bath (10 mL in a 50 mL glass beaker), followed by a 10-min bath in sodium chloride (5 mL of 8.25% sodium hypochlorite solution diluted with 45 mL distilled water), and finally triple rinsed with distilled water. Additional seeds were not surface sterilized to determine if the sterilization process alone would break physical dormancy. Four Petri dishes per treatment, each with 25 seeds, were kept in the dark (double wrapped in aluminum foil) and maintained at room temperature in a laboratory. Germination percentage was recorded after 14 days.

The experimental design was a single factor experiment with 5 (0, 50, 100, 150 and 200 seconds of scarification) treatment levels plus a non-scarified, non-sterilized control. Experimental units were Petri dishes containing seed, and germination percentages per dish was the response variable. Data were transformed prior to analysis via an arc-sine square-root transformation. Statistical analysis was conducted using RStudio 1.1.419. Data were subjected to a one-way analysis of variance to determine differences among treatments. Differences among means were determined using the least squares means package with a Bonferroni correction at an alpha level ≤5%.

Photoperiod Experiment. An experiment was conducted to test the effects of photoperiod on germination. Ungraded and non-sterilized seed were scarified using an electric seed scarifier lined with sandpaper for 50 seconds. Seeds were exposed to a 12-hour photoperiod of cool-white fluorescent light or kept in the dark. There were five replicate Petri dishes containing 25 seeds for each treatment. An additional 2 Petri dishes with non-scarified seeds were exposed to dark to confirm that seeds remained physically dormant throughout the experiment. Petri dishes were placed in growth chambers at 25°C and evaluated for germination (1 mm radical emergence) and disease (presence of contamination) tri-weekly for 28 days. Petri dishes wrapped in aluminum foil were unwrapped for no more than 5 minutes and exposed to ambient laboratory lighting.

Data for this experiment were analyzed using semi-parametric (Cox regression models) and non-parametric (Kaplan-Meir estimators) time-to-event analyses. These methods were used in lieu of an analysis of variance because we had several data collection dates and removed diseased seeds throughout the experiment. More information about time-to-event analysis and its utility in germination studies is provided by Allison (2010) and McNair et al. (2013), respectively. The experimental unit was a seed and germination was the response variable. Seeds that germinated during the course of the experiment were coded as 1. Diseased seeds were coded as 0 at infection day. Seeds which did not germinate by day 28 were coded as 0. Cox regression model construction (analyzed using PROC PHREG in SAS 9.4) and Kaplan-Meier estimates of the survivor functions (analyzed using survival package in RStudio 1.1.419) were implemented using the same methods as described in Campbell-Martínez et al. (2017).

RESULTS

Fruit and seed characteristics. Fruits are 3-valved capsules with stellate trichomes on the distal end. There were 10.9 ± 3.2 seeds per fruit. Average seed weight was 0.68 ± 0.05 mg. The number of seeds per fruit ranged from 3 to 19. Within fruits, seeds form a discrete unit which remains intact in a spherical structure that may be disassembled upon the application of pressure (Fig. 2). The vast majority of seeds were <2 mm wide (Figure 2).

Imbibition. Scarified seed imbibed 3 times the amount of water in 24 hours compared to non-scarified seed (Fig. 3). However, a small portion of non-scarified seeds did imbibe by hour 24. Imbibed seeds were easily identifiable by the naked eye as they were more swollen and lighter in color and seed coats were more translucent than non-imbibed seeds.

Scarification experiment. Regardless of time of abrasion in the seed scarifier (50 – 200 seconds), germination percentages for scarified seeds (≥90%) were significantly higher than non-scarified seeds (9-11%) (Figure 4; Table 1). There was no difference in germination between non-scarified seed that were surface sterilized or non-surface sterilized indicating the sterilization process did not break dormancy (Figure 4).

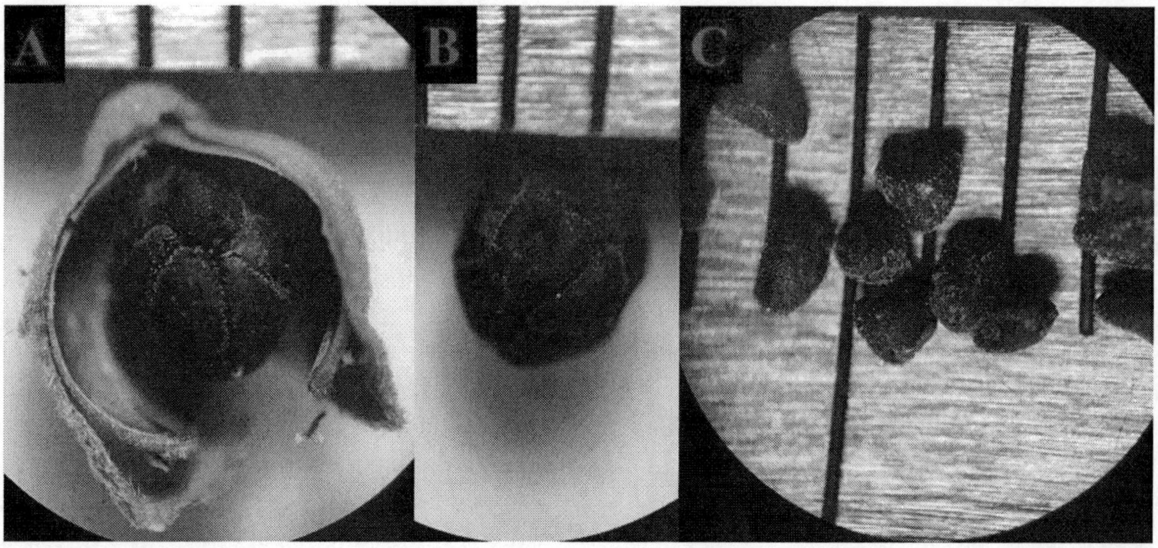

Figure 2. Fruit (A), intact cluster of seeds (B), and individual seeds (C) of *Crocanthemum arenicola* (coastal sand frostweed). Note 1-mm lines for reference.

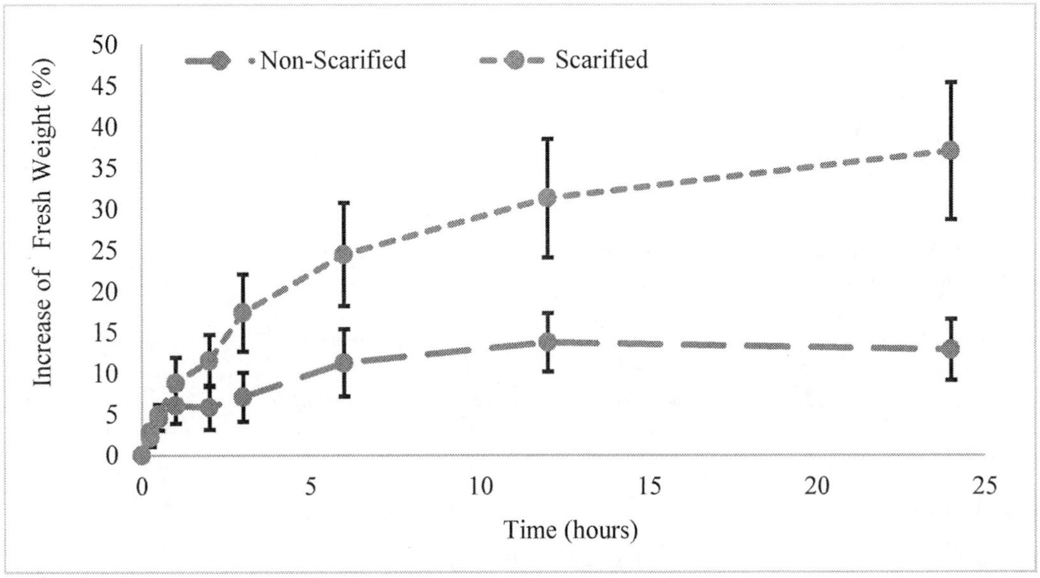

Figure 3. Seed imbibition curves for non-scarified (blue line) and scarified (orange line) *Crocanthemum arenicola* seeds. Scarified seeds were placed in an electric seed scarifier lined with sandpaper for 50 seconds while control seeds received no treatment.

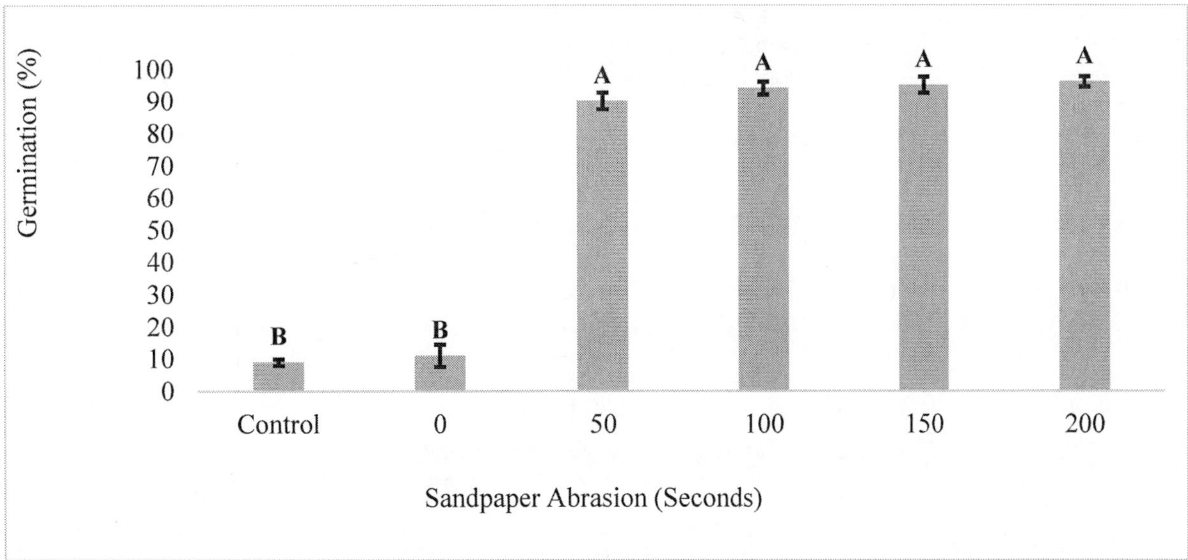

Figure 4. Effects of surface sterilization and exposure to abrasion by sandpaper in an electric seed scarifier on germination percentages of *Crocanthemum arenicola*. Control seeds were not surface sterilized or scarified. Seeds were classed as germinated following a 1 mm protrusion of the radicle. Seeds were placed in the dark at room temperature (~25°C) and observed for 14 days. Different letters above bar graph denote significant ($P \leq 0.05$) treatment differences using a pairwise comparison test with a Bonferroni correction (analyzed using lsmeans package in RStudio 1.1.419).

Photoperiod experiment. Photoperiod had no effect on germination probability across time (Figure 5; Table 2). Germination began at week 1 and continued through the end of the experiment (Figure 5). At the end of the experiment, 63% and 64% of scarified seed germinated, respectively, under light and dark conditions (Figure 6). Only a small portion (14%) of non-scarified seeds germinated by the end of the experiment. Disease ranged from 2-8% (Figure 6).

Table 1. Analysis of variance table (analyzed using RStudio 1.1.419) indicating significance of scarification treatments on germination of *Crocanthemum arenicola*. Seeds were germinated at 25°C for 28 days.

	df	Sum of squares	Mean Square	F-value	P-value
Scarification	5	5.677	1.1355	107.7	<0.0001
Error	18	0.19	0.0105		

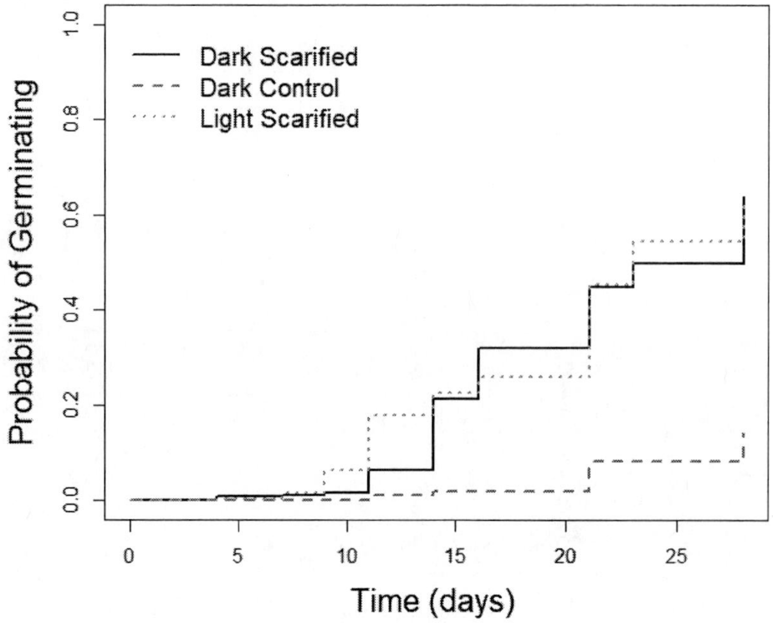

Figure 5. Inverted Kaplan-Meier curves (analyzed using survival package in RStudio 1.1.419) showing effects of photoperiod (seeds were either exposed to a 12-hour photoperiod or left in the dark) on germination probability of *Crocanthemum arenicola*. Ninety-five percent confidence intervals are excluded for clarity. Seeds were germinated within growth chambers at 25°C for 28 days. Control seeds were non-scarified.

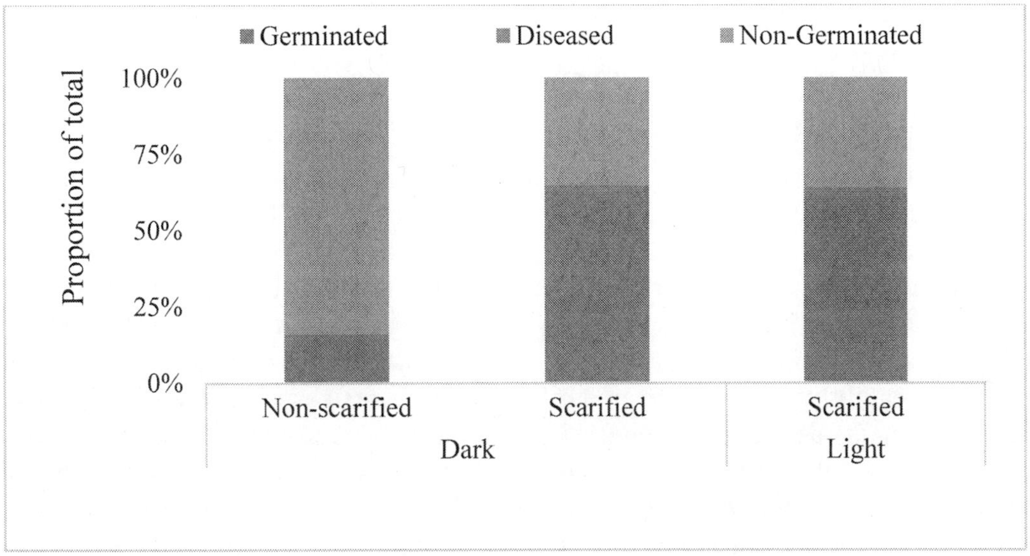

Figure 6. Effects of photoperiod (seeds were either exposed to a 12-hour photoperiod or left in the dark) and scarification on relative percentages (proportion of total) of germinated, diseased, non-germinated seeds of *Crocanthemum arenicola*. Seeds were classed as germinated following a 1 mm protrusion of the radicle and diseased when seeds showed visual pathogen infection. Seeds were placed in a growth chamber in the dark at 25°C for 28 days.

Table 2. Effects of photoperiod on germination probability of *Crocanthemum arenicola.* Seeds were germinated at 25°C for 28 days. Data were analyzed via cox regression models using PROC PHREG in SAS 9.4

Treatment	Coefficient (βi)	SE of (βi)	Wald χ^2	*P*-value	Hazard Ratio[1]
Photoperiod (Dark vs. Light)[2]	<0.01	0.11	<0.01	0.99	1.00
Experiment Run (Run 1 vs Run 2)[3]	0.06	0.11	0.27	0.60	1.06

[1]Hazard ratios > or < one with p value ≤0.05 indicate increased or decreased relative germination probability.
[2]The second row represents comparisons of seeds germinated in the dark or light (12-hour photoperiod).
[3]The third row represent comparisons between experiment runs.

DISCUSSION

Seeds of *C. arenicola* are large enough to be seen with the naked eye but may be too small to be sown by hand. Seeds are orthodox and may be dried and stored without complicated procedures. Like other seeds of Cistaceae, seeds of *C. arenicola* have physical dormancy due to a water-impermeable seed coat (Thanos et al., 1992). However, a small amount of *C. arenicola* seed germinate readily without pretreatment (i.e., are not dormant) as has been reported for *C. scoparium* (Keely, 1985). Similar to the closely related *Helianthemum,* seed dormancy of *C. arenicola* can be alleviated via abrasion by sandpaper (Pérez-García et al., 1995; Pérez-García and González-Benito 2006). Likewise, other *Helinathemum* have germinated readily with exposure to light (Escudero et al., 1997) or in the dark (Yeşilyurt et al., 2017) - as was the case for *C. arenicola*. We recommend at least 50 seconds of abrasion within an electric seed scarifier prior to sowing with or without light exposure for seeds of *C. arenicola.*

The authors have successfully grown containerized plants of *C. arenicola* (some of the stock plants used in this study). Rooted stem cuttings (no auxins were used on stem cuttings) have been successfully transplanted to multiple container types (3.5 in. square and tall, 4 in. -round, 4 in. wide, and 1-gal pots); extensive root growth occurred in pots within weeks to a few months.

Plants were grown using a mix of peat-based horticulture media (Sungro MetroMix® 830) and milled pine bark with fines at a 2:1 ratio. Plants were kept within a climate-controlled greenhouse during the winter months of 2018 and flowered prolifically for extended periods of time (Fig. 1). Plants responded well (vigorous growth and flower and seed production) to moderate amounts of slow-release fertilizer (1 teaspoon 18-6-12 Osmocote® per gallon) and fertigation (100 ppm N Peters Professional® 20-20-20). Plants tolerated frequent substrate saturations as well as frequent periods of substrate drying. Aphids and white flies were seen feeding on plant tissue but did not appear to diminish the appearance, flowering, or fruiting of stock plants. Insect populations were kept low by the application of horticultural oils and soaps as needed. The use of the southeastern USA native *Crocanthemum* in restoration and as an ornamental plant is currently limited. This limited use is possibly due to a lack of propagation, production and out-planting information. Here, we present information that informs the grower on how to collect, store, and germinate, *C. arenicola* seeds as well as how to produce containerized plants of *C. arenicola*.

Literature Cited

Allison, P. (2010). Survival Analysis Using SAS®: A Practical Guide, 2nd ed. SAS Institute Inc., Cary.

Campbell-Martinez, G., Thetford, M., Miller, D., and Pérez, H.E. (2017). Follicle maturity, seed size, temperature and photoperiodic effects on seed germination of *Asclepias humistrata* (sandhill milkweed). Seed Sci. Tech. *45*:1-17.

eFloras. (2018). http://www.efloras.org [accessed 15 September 22018] Missouri Botanical Garden, St. Louis, MO and Harvard University Herbaria, Cambridge, MA.

Escudero, A., Carnes, L.F., and Pérez-García, F. (1997). Seed germination of gypsophytes and gypsovags in semi-arid central Spain. Journal of Arid Environments *36*:487-497.

Keeley, J.E., Morton, B.A. Pedrosa,A. Trotter, P. (1985). Role of allelopathy, heat and charred wood in the germination of chaparral herbs and suffrutescents. J. Ecol. *73*:445-458.

McNair, J.N., Sunkara, A., and Frobish, D. (2013). How to analyze seed germination data using statistical time-to-event analysis: non-parametric and semi-parametric methods. Seed Sci. Res. *22*:77-95

Pérez-García, F. and González-Benito, M.E. (2006). Seed germination of five *Helianthemum* species: Effects of temperature and presowing treatments. J. Arid Environ. *65*:688-693.

Perez-Garcia, F., Iriondo, J.M., Gonzalez-Benito, M.E., Carnes, L.F., Tapia, J., Prieto, C., Plaza, R., and Perez, C. (1995). Germination studies in endemic plant species of the Iberian Peninsula. Israel J. Plant Sci. *43*:239-247.

Sorrie, B.A. (2011). Transfer of North American *Helianthemum* to *Crocanthemum* (Cistaceae): new combinations. Phytologia 93:270-271.

Thanos, C.A., Georghiou, K., and Kadis, C. (1992). Cistaceae: A plant family with hard seeds. Israel J. Plant Sci. *41*:251-263.

USDA and NRCS. (2018). The PLANTS Database, http://plants.usda.gov [accessed 15 April 2018]. National Plant Data Team, Greensboro, NC 27401-4901 USA.

Wunderlin, R.P. and Hansen, B.F. (2011). Guide to the vascular plants of Florida. 3rd ed. Gainesville (FL): University Press of Florida.

Yeşilyurt, E.B., Erik, S., and Tavşanoğlu, Ç. (2017). Inter-population variability in seed dormancy, seed mass and germination in *Helianthemum salicifolium* (Cistaceae), a hard-seeded annual herb. Folia Geobotanica *52*:253-263.

Combined Proceedings IPPS 68:290-295. 2018.

Improving Nursery Crop Weed Control by Assessing Herbicide Movement Through Organic Mulch

Debalina Saha[1], Chris Marble[1], Brian J. Pearson[1], Hector E. Perez[2], Gregory E. Macdonald[3], and Dennis Odero[4]

[1]Environmental Horticulture Department, Mid-Florida Research and Education Center, University of Florida, 2725 S. Binion Road, Apopka, Florida 32703, USA. [2]Environmental Horticulture Department, PO Box 110670, University of Florida, Gainesville, Florida 32611, USA. [3]Agronomy Department, 2089 McCarty Hall D, University of Florida, Gainesville, Florida 32611, USA. [4]Agronomy Department, Everglades Research and Education Center, University of Florida, 3200 E Palm Beach Road, Florida 33430, USA

debalina@ufl.edu

Third Place - Parkerson Student Research Paper Competition

Keywords: Pinebark, pinestraw, hardwood, bioassay, chemical assay, preemergence.

Abstract

Experiments were conducted in 2018 to assess herbicide movement through organic mulch materials including pinebark, pinestraw, and hardwood. Weed species evaluated were crabgrass (*Digitaria sanguinalis*), garden spurge (*Euphorbia hirta*), and eclipta (*Eclipta prostrata*). Liquid formulations of prodiamine, dimethenamid-P + pendimethalin, and indaziflam were evaluated in combination with mulch materials applied at a depth of 5.1 cm (2 in.). Quantification of these herbicides was performed using biological and chemical assays from the soil samples collected from below the mulch layers. Results showed that only 67% eclipta control was observed in pots originally mulched with hardwood, which indicates that indaziflam was more tightly bound to this mulch. Crabgrass data showed that pinebark (65% control) was the only mulch type that caused a significant reduction in prodiamine efficacy. Dimethenamid-P + pendimethalin efficacy on garden spurge was reduced in pots originally mulched with hardwood or pinebark, but all treatments provided ≥94% control. Chemical assays showed that approximately 20% of pendimethalin, prodiamine, and indaziflam that was applied reached the soil surface when mulch was present during the application. More dimethenamid-P reached the soil surface than any other herbicide, with 69% being retained by the pinebark mulch.

INTRODUCTION

Weed management is a costly endeavor in container nursery production as hand weeding can be laborious, time-consuming and expensive. Growers most often rely on preemergence herbicides and supplemental hand weeding for control. In some cases, mulch may also be used for weed control in containers, either as a sole means of weed suppression or in addition with preemergence herbicides (Mathers, 2003). When mulch and herbicides are both used for weed control, long-term weed control can be achieved, especially when lower mulch depths are applied (Bartley et al., 2017). Use of these herbicide + mulch combinations can also provide environmental benefits, such as reducing off-target movement of herbicides after application (Chalker-Scott, 2007; Fawcett et al., 1994). However, it is unknown if the efficacy of preemergence herbicides is reduced when they are applied to mulched containers.

The objective of this research was to assess herbicide movement through organic mulch to determine the most appropriate mulch type for use with preemergence herbicides.

MATERIALS AND METHODS

Bioassay. Research was conducted at the Mid-Florida Research and Education Center, Apopka, FL in summer 2018. Nursery containers [946 ml (1 qt.)] were filled with a pinebark: peat substrate and amended with Osmocote® Plus 15-9-12 at the rate of 4.7 kg m^{-3} (0.03 lbs. /gal.). After filling containers, twenty seeds of either crabgrass (*Digitaria sanguinalis*), garden spurge (*Euphorbia hirta*) or eclipta (*Eclipta prostrata*) were sown to the surface of each container. Pinebark, pinestraw or hardwood mulch were then applied at a depth of 5.1 cm (2 in.) on top of each container. Liquid formulations of indaziflam (Marengo® 0.622 SC, Bayer Crop Science, Research Triangle Park, NC), prodiamine (Barricade® 4 FL, Syngenta Crop Protection, Greensboro, NC), and dimethenamid-P + pendimethalin (Tower® 6 EC + Pendulum® 3.3 EC, BASF Corp., Research Triangle Park, NC) were applied with a CO_2 backpack sprayer calibrated to deliver 561 liters per hectare (60 gal. per acre) using a 8004 flat fan nozzle (TeeJet Technologies, Wheaton, IL) at a pressure of 30 psi on June 06, 2018 (round 1) and July 12, 2018 (round 2) - at their labeled rates to pots seeded with eclipta, crabgrass, and garden spurge, respectively. A separate group of nontreated pots were maintained for each herbicide and mulch combination.

All containers were placed on a full sun container nursery pad and received 3.5 cm (1.4 in.) of irrigation via two irrigation cycles through overhead sprinklers. Following irrigation, mulch was carefully removed from each pot so that only the herbicide reaching the soil surface was available for weed control and the presence of mulch did not confound results. The experiment consisted of a completely randomize design with six replicates per treatment. Data collection included weed counts at 2 and 4 weeks after treatment (WAT). At 4 WAT, all weed species were cut at the soil line and shoot fresh weights were determined for each weed species. Shoot fresh weights were converted to percent control by using the formula [(Nontreated control – treated) / nontreated control) × 100].

All percent control data were subjected to analysis of variance using the PROC GLM procedure in SAS® (SAS 9.4, SAS Institute, Inc., Cary, NC). Fisher's Least Significance Difference Test was used to compare between individual means of experimental variables. All differences were considered significant at $p \leq 0.05$ and each weed species was analyzed separately.

Significant differences observed in biweekly weed counts were reflected in fresh weight data; therefore, for the sake of brevity only percent control of shoot fresh weight data will be discussed.

Chemical assay. In addition to the bioassay described previously, chemical assays were performed to quantify herbicide movement through pinebark mulch. Nursery containers (946 ml or 1 qt.) were filled with substrate and amendments as previously described. Pinebark mulch was then applied at a depth of 5.1 cm (or 2 in.) on top of each container.

Liquid formulations of indaziflam, prodiamine, and dimethenamid-P + pendimethalin were applied, and pots were irrigated as described above. Following treatment and 3.5 cm (1.4 in.) of irrigation, mulch was removed carefully without disturbing the underlying substrate. The substrate was then sampled to a depth of 2.5 cm (or 1 in.) out from each container.

Quantification of each herbicide in the soil layer underneath mulch was determined using previously described methods (EPA 1996; EPA 2007; EPA 2018). All herbicide quantification data were converted to percent retention by pinebark mulch by using the formula [(No mulched control – herbicide treated) / no mulch control) × 100] to determine which herbicide was more tightly bound by the pinebark mulch. Data were analyzed as described previously.

RESULTS

Bioassay. When indaziflam was applied to pots initially mulched with pinebark (89% control) and pinestraw (99% control), eclipta control was similar to that of nonmulched pots (100% control), indicating that these two mulch types had no detrimental effect on indaziflam efficacy of eclipta (Table 1). Only 67% eclipta control was observed in pots originally mulched with hardwood, which indicates that indaziflam was more tightly bound to this mulch. Crabgrass data showed that pinebark (65% control) was the only mulch type that caused a significant reduction in prodiamine efficacy. Prodiamine provided similar crabgrass control when pots were originally mulched with hardwood (80% control), pinestraw (91%), and when no mulch was present. The combination of dimethenamid-P + pendimethalin provided similar control of garden spurge when it was applied to bare soil and pots originally mulched with pinestraw, which both treatments resulting in 100% control. Pots that were originally mulched with either hardwood or pinebark (94 to 95% control) provided commercially acceptable control, but to a less degree than was observed in non-mulched pots or pots originally mulched with pinestraw.

Chemical analysis. Over all three herbicide treatments (four active ingredients), pinebark reduced the amount of herbicide reaching the soil surface by 85% (Table 2). Approximately 20% of pendimethalin, prodiamine, and indaziflam that was applied reached the soil surface and was detected using chemical assay (Table 3). This indicates that 10 to 20% of the herbicide that was applied was available for weed control. More dimethenamid-P reached the soil surface than any other herbicide, with only 69% being retained by the pinebark mulch.

Table 1. Percent control of three herbicides applied to different mulch materials.

Weed Species	Herbicide	Mulch types[z]	Percent control[y]
Eclipta	Indaziflam	Hardwood	67 b[x]
		Pinebark	89 a
		Pinestraw	99 a
		No-mulch	100 a
Crabgrass	Prodiamine	Hardwood	80 ab
		Pinebark	65 b
		Pinestraw	91 ab
		No-mulch	100 a
Garden spurge	Dimethanamid-P + pendimethalin	Hardwood	94 b
		Pinebark	95 b
		Pinestraw	100 a
		No-mulch	100 a

[z]Liquid formulations of each herbicide (or combination) were to applied to pots mulched with hardwood, pinebark, pinestraw, or contained no mulch. Two days after application, mulch was removed prior to bioassay.

[y]Percent control was calculated as a percent decrease in shoot fresh weights in pots receiving no herbicide or mulch treatment by using the formula:

[((nontreated control - treated)/ nontreated control) x100)]

[x]Means within each weed species followed the same letter are not significantly different in based upon Fisher's Protected LSD test ($P \leq 0.05$).

Table 2. Average amount of herbicide[y] detected in soil samples following application to pots mulched with pinebark and those containing no mulch.

Mulch type	Herbicide detected (mg/Kg)
Pinebark	49.4 b[z]
No-mulch	321.7 a

[z]Means followed the same letter are not significantly different in a column ($P \leq 0.05$).

[y]Herbicides applied were indaziflam, prodiamine, and dimethenamid-P + pendimethalin. The amount of herbicide presented in this table is the average of all four herbicides.

Table 3. Percent of preemergence herbicides retained by pinebark.

Mulch type	Herbicides	% Retained in mulch[z]
Pinebark	Pendimethalin	88 a[y]
	Prodiamine	84 a
	Indaziflam	80 a
	Dimethenamid-P	69 b

[z]Percent retention in mulch was calculated as a percent decrease in herbicide amount in soil samples receiving no mulch treatment by using the formula: [((no mulch control - treated)/ no mulch control) x100)]

[y]Means followed the same letter are not significantly different based upon Fisher's Protected LSD test ($P \leq 0.05$).

DISCUSSION

Similar to previous findings, herbicides evaluated in these experiments provided a high level of control of each bioassay weed species when applied to the soil surface when no mulch was present at the time of application (Johnson, 1997; Marble, 2011). When mulch was present during the application, results differed by herbicide as some herbicides will bind tighter to organic matter than others. Pinestraw was the only mulch material in which control of the target weed species was similar to control achieved when no mulch was present across all herbicide treatments and weed species evaluated. This indicates that the herbicides evaluated in this study may move more effectively through pinestraw compared with hardwood or pinebark mulch.

Results from this study show that many of the most commonly used preemergence herbicides are bound in organic mulch materials. Only 10 to 30% of the herbicide that was applied reached the soil surface, at least following only approximately 3.8 cm (1.5 in.) of irrigation over a short time period. As most herbicide labels indicate that 0.6 to 1.3 cm (0.25 to 0.5 in.) of irrigation is needed to water in herbicides following application, more irrigation may be needed with preemergence herbicides applied to mulched nursery containers or landscape beds. While only a small portion of the total herbicide applied reached the soil surface. In most instances, commercially acceptable weed control resulted over a 4-week evaluation period. It is unknown, however, how efficacy would be affected over a longer period and/or if weed seed were sown on multiple dates.

While this data shows a high degree of herbicide binding to mulch, use of preemergence herbicides to mulched containers or landscape beds would still offer significant advantages. In many cases, weed germination and growth significantly increase when seeds are placed on top of mulch compared to seeds below mulch (Richardson et al., 2008). Therefore, herbicide that is retained in the mulch layer would be available to prevent growth from weed seeds introduced on top of the mulch layer that may germinate/grow within that mulch layer. Weed control from these different herbicide + mulch combinations (when mulch was left on the soil surface) was not determined in the current study - but these combinations have been shown to provide season long weed control in previous studies (Bartley et al., 2017; Somireddy, 2012). While this data shows that pinestraw may be the most compatible mulch for use with the preemergence herbicides evaluated here, more data is needed to evaluate long-term control with this combination in a variety of environments. Cost, aesthetics, availability, and consumer acceptance should also be considered when evaluating mulch either with or without the use of preemergence herbicides.

Literature Cited

Bartley, III., P.C., Wehtje, G.R., Murphy, A.M., Foshee, W.G. III., and Gilliam, C.H. (2017). Mulch type and depth influences control of three major weed species in nursery container production. HortTech. *27*: 465-471.

Chalker-Scott, L. (2007). Impact of mulches on landscape plants and the environment- a review. J Environ Hort. 25:239-249.

EPA. (1996). Method 3540C. Soxhlet extraction. https://www.epa.gov/hw-sw846/sw-846-test-method-3540c-soxhlet-extraction Accessed August 22, 2018.

EPA. (2007). Method 8321B. Solvent-extractable nonvolatile compounds by high-performance liquid chromatography/ thermo-spray /mass spectrometry (HPLC/TC/MS) or ultraviolet (UV) detection. https://www.epa.gov/sites/production/files/2015-12/documents/8321b.pdf. Accessed August 22, 2018.

EPA. (2018). Method 8270E. Semivolatile organic compounds by gas chromatography/ mass spectrometry. https://www.epa.gov/sites/production/files/20191/documents/8270e_revised_6_june_2018.pdf Accessed August 22, 2018.

Fawcett, R.S., Christensen, B.R., and Tierney, D.P. (1994). The impact of conservation tillage on pesticide runoff into surface water: a review and analysis. J. Soil Water Conserv. *49*:126-135.

Johnson, B.J. (1997). Sequential applications of preemergence and postemergence herbicides for large crabgrass (*Digitaria sanguinalis*) control in tall fescue (*Festuca arundinacea*) turf. Weed Technol. 11: 693-697.

Marble, S.C., Gilliam, C.H., Wehtje, G.R., Van Hoogmoed, A.J., and Palmer, C. (2011). Early postemergence control of spotted spurge in container production. J. Environ. Hort. *29*:29-34.

Mathers, H.M. (2003). Novel methods of weed control in containers. HortTech. *13*:28-34.

Richardson, B., Gilliam, C.H., Fain, G.B., Wehtje, G.R. (2008). Container nursery weed control with pinebark mini-nuggets. J. Environ. Hort. *26*:144-148.

Somireddy, U. (2012). Effect of herbicide-organic mulch combinations on weed control and herbicide persistence. Ph.D. dissertation. Columbus, OH: Ohio State University.

Combined Proceedings IPPS 68:296-298. 2018.

Storage and Germination of *Quercus Virginiana* Seeds

Bill Barr
Tree Town USA, Glen Flora, Texas 77443, USA

wcbarr146@gmail.com

Keywords: Live oak, seed propagation, seed storage, cold storage.

INTRODUCTION

Quercus virginiana (Live Oak) seeds are commonly planted soon after collection. We wanted to know how long we could store seeds and still obtain a reasonable germination rate. Our production goal was to have two planting dates for young seedlings.

METHODS AND MATERIALS

For this experiment, two groups of seeds were used. One group was collected by nursery personnel directly from the trees and placed in cold storage immediately after collection. The second group was bought from a respected seed supplier. Our best estimate was that the purchased seeds were received about 30 days after the vendor collected them. Both groups were floated, put into two-quart bags, and stored in a cooler until planting. During storage, the average temperature was 2°C (35.5°F) with an average low of -2°C (28.5°F) and average high of 5.7°C (42.3 °F).

Seeds for each group were taken from the cooler and sown each month over an 8-month period. Seeds were sown into 32-cell Rootmakers using a medium common in southeastern U.S. consisting of mostly pine bark with some peat and incorporated fertilizer. For each group and sowing month, two Rootmaker trays were sown.

RESULTS

Seeds stored immediately after seed collection germinated well over the entire 8-month germination period (Figure 1.). Seeds held for 30 days germinated well initially but thereafter, germination declined with time (Figure 2).

SUMMARY

Results suggest that Live Oak seeds can be stored for up to 8 months and still have good germination if they are put into cold storage soon after collection (Figure 3). Seeds held 30 days, before cold storage, did not store well.

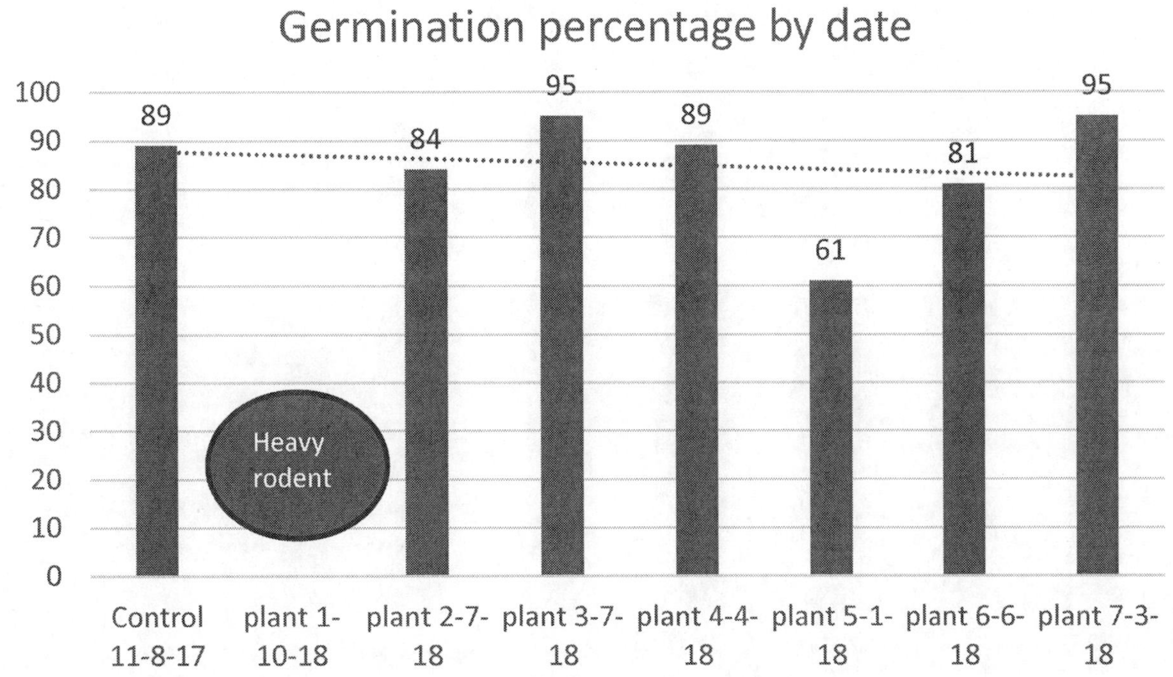

Figure 1. *Quercus virginiana* germination. Seeds were placed into cold storage immediately after seed collection. Oak seed were harvested and stored in cooler on 8 November 2017. Heavy rodent damage occurred on 10 January 2018.

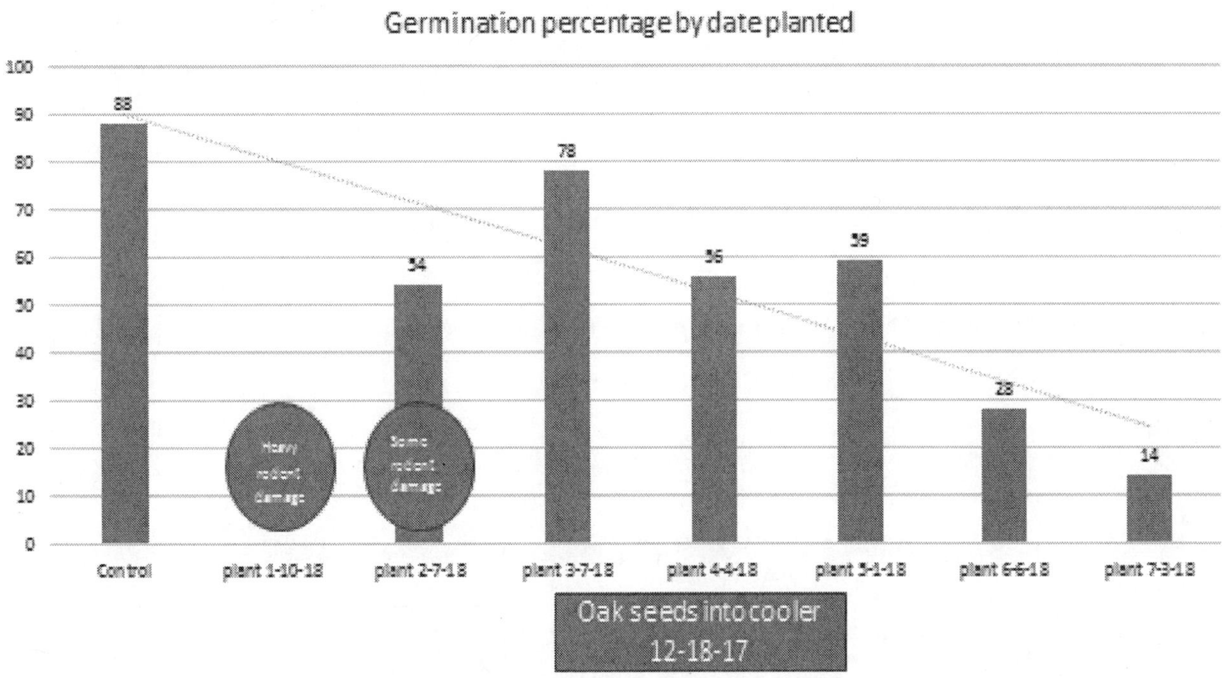

Figure 2. *Quercus virginiana* germination in seed held for 30 days after collection under ambient conditions before placing into cold storage. Oak seed was placed in cooler 30-days after collection on 18 December 2017. Heavy rodent damage occurred on 10 January 2018, and some damage on 7 February 2018.

Figure 3. Live oak seedlings with extensive root system, ready for shifting-up to larger containers at 3-months after sowing.

Combined Proceedings IPPS 68:299-303. 2018.

Home Grown Innovation at Lancaster Farms Inc.

Christopher J. Brown, Jr.
Lancaster Farms, Inc., 5800 Knotts Neck Road, Suffolk, Virginia 23435, USA

chrisbjr@lancasterfarms.com

Keywords: Mechanization, labor efficiency, pruning, machine design, technology adaption.

INTRODUCTION

In the modern-day nursery business when we hear the word innovation the first thing that comes to mind is technology. Whether it is sensors helping us make better irrigation decisions. Smart sprayers allowing us to quickly and efficiently apply chemicals. Pumping stations that have turned into water treatment facilities. Potting machines that increase output while decreasing labor or drones that give us a bird's eye view. They are all great examples of innovations that allow us to be more efficient while producing the same if not higher quality product. While all of these aforementioned innovations allow us to produce higher quality products more efficiently while reducing our impact on our environment and bottom line; they are big investments for our companies - and all too often we must update our facilities to accommodate them.

At Lancaster Farms we love to innovate and strive to find better, more cost-efficient ways to produce our products. While we have implemented several of these innovations - we embrace home grown ideas and I hope that some of ours will spark you to come up with some of your own.

Traditionally, every year when we plant our trees, we have a process - like everyone else. For us, the process has always started with prepping our beds drip lines and cleaning them up. Lastly, we will spray them with a pre-emergent herbicide, using our roadway sprayer, which typically gives us 10 to 14 months weed control. This process works great for our faster crops. The problem comes with our slower crops. Historically we would apply granular herbicide to older beds 4 times a year. The problem with this was that in our pot-n- pot production we do not have overhead irrigation. This left us trying to apply our herbicides when rain was in the forecast and we all know how that goes. It never rains when you need it to, and we never know how much we are going to get when it does!

So, we decided to start spraying the beds with backpacks, because moving the trees out of the bed was unrealistic. This was painstakingly slow and inefficient; 'painting' the beds - 3.5 gallons at a time - took forever and was poster child for waste. We knew that we had to find a better way. We looked everywhere and at everything, it was either too expensive and needed modifications, or too cheap and did not work. So, we went back to the drawing board. Honestly, our first attempt was a huge bust!

We took an old riding lawn mower modified the tires and added a pull-behind 40-gallon tank. The idea actually worked but driving in the beds with this piece of equipment was deadly to the trees. Even with a trusted employee using the equipment - we still would run over or scare trees by the dozens. It was just too hard to control. So, we went back to the drawing board and I stumbled across the Jacto PJB16 battery powered backpack (Figure 1).

Figure 1. Jacto PJB16 battery powered back-pack. Used for watering in granular herbicide applied to pot-n-pot containers.

It had everything we were looking for. It had preprogrammed setting for speed and pressure, and a battery that could run all day. All it was lacking was capacity. With a 3- gal capacity, it would require countless refills. So, we took off the tank, put some wheels on a frame and added a bigger tank. The backpack comes with everything you need, and we have killing weeds ever since!

Weeding on a nursery has always been one of the most hated jobs. It is painstakingly slow and hard on your back. Luckily, we have herbicides which allow us to prevent weeds and the awful job of weeding. They work great when used properly and can really save us some money. There are several issues with these chemicals though. I am not going to go into all the problems - just one. Not all crops are labeled for herbicide use. Our perennial and annual herbicide program needed help. We were not using herbicides either because we were not comfortable using them or they were not labeled for use. So, we stumbled upon rice hulls as a weed barrier. They worked great. The problem was applying weeds to the surface of containers was even slower than weeding to put them out. So again, we went to the drawing board. I do not remember who came up with this idea- but it has worked great for us! We simply use an insulation blower to apply rice hulls as a weed barrier (Figure 2). While it is surely not perfect, it gets the job done in a timely manner!

In today's nursery industry so many things are changing! Even how we load and ship our plants! Over the years rolling racks have become the standard - and using racks adds new variables to loading that were not previously a problem. When we started using racks years ago, they were solely for our annuals and perennials. The racks we purchased were perfect for them. However, when we started loading all our plants on racks, we started to realize that these were not necessarily the best "fit" for all our different plant sizes.

Figure 2. Using an insulation blower to apply rice hulls as a weed barrier to the tops of containers.

So, we started looking for different sized racks - looking at different dimensions to find the rack that worked the best for us! Once we found a rack that was to our liking - we realized that we would have to wrap our racks because they had no sides - and plants were falling off during every step of the process. After a spring of hand-wrapping our racks - we knew there had to be a better way! Our loading dock manager found out how other industries were wrapping pallets. The problem was the high cost of the wrapping machines! So as always true to our roots - we started building our own machine - and now have our own for a fraction of the cost (Figure 3)!

Figure 3. On-site, nursery-built machine for wrapping pallets with plastic to prevent plants from moving and falling during shipping.

Our retail customers are demanding more of us now than ever - and it is just not the quality of our plants that they require. Every year it seems that something else gets added to the list. One of our customers thought that our pots were to dirty and did not like the way they looked sitting on their retail nursery lot. I know it seems stupid; we do all work in dirt, soil and mud - and as soon as the plant is sold and planted - the pot is not going to matter. But if it matters to our customers - we should at least look into how we can better serve them. So yet again we went back to the drawing board! Our new solution allowed us to clean our pots without adding another step in our process. We simply modified our loading belt and added a spray system to clean the pots as they moved down the conveyor system for loading (Figure 4). We were off to the races! The best part about it was that almost all of the parts needed for the spray system were already on the farm. So it worked ideally with the conveyor system - and was relatively inexpensive to construct!

Figure 4. To make sure container walls are clean and free of dirt, the spray system in the shipping area was built which cleans the container as they move along the conveyor belt system during the loading of trucks.

These are just a few examples of how with a little thought and ingenuity - we can solve problems on our own. Not every problem we have can be solved in house, but when we can we like to for many reasons. I hope the some of our ideas will get your creative juices flowing and allow you to come up with a better version of what we have built or something else to help your company become more efficient.

Here are some other ideas we also use that are inexpensive and very simple! Scales can be used to "count" everything. For instance, we weigh several bundles of 50 unrooted cuttings held with a rubber band - to get an idea of the "average weight" per 50 unrooted cuttings. With the "average" weight, we can weigh cuttings to reach the approximate cutting numbers needed- without having to individually count cuttings for making bundles of 50 (Figure 5).

Figure 5. Scales are great for "counting" anything. Rather than counting 50 unrooted cuttings per bundle, propagators can weigh cutting bunches to determine the number of cuttings needed by weight.

It is much faster. Rather than manually picking up blown over shrubs and trees by hand - inserting containers into used tires is a great way to prevent plants from blowing over (Figure 6).

Figure 6. Inserting containers into used tires is a great way to prevent plants from blowing over.

Likewise, for pruning larger shrubs and trees, having the worker walk around on drywall stilts is a great way to quickly and efficiently prune (Figure 7).

Figure 7. Drywall stilts are a great way to quickly and efficiently prune trees.

Combined Proceedings IPPS 68:304-307. 2018.

Overused and Underutilized Landscape Plants

Ben Cecil
Ingleside Plantation Nursery, 5870 Leedstown Road, Colonial Beach, VA 22443

ben.t.cecil@gmail.com

Keywords: Plant diversity, monoculture, overplanting.

INTRODUCTION

Our landscapes have lost diversity. When considering the limited variety of species and selections that are currently being used in the landscape, the development of relative monocultures becomes apparent. This is disheartening considering the sheer number of viable landscape plants available that could be utilized.

It is an easy scenario to fall into. Take the introduction of Knockout® Roses, which is an extraordinary plant. Their disease resistance, ease of propagation and long bloom period make them an ideal candidate for the landscape. The problem arises when they are used in almost every landscape as a monoculture. This leads to the explosion of pests or diseases, such as Rose Rosette; or increases the severity of effects from introduced maladies through a greater loss of mature specimens. Examples of the latter are the overplanting of Ash trees with their susceptibility to the Emerald Ash Borer, and the devastation of Elms to Dutch Elm Disease.

So why do we not diversify our plant palette in production? The answer is, more often than not - financial. It is difficult to allocate space and dollars to a plant species that customers are not actively requesting. I am not certain how to overcome this obstacle. It is necessary to address, but perhaps not in this particular presentation.

All the above is not to imply we grow poor selections now. That could not be further from the truth. *Hydrangea paniculata* is a great species of plant for production. Given the plant's ease of propagation, ability to be maintained and lack of significant diseases and pests, it has all the characteristics we need to create a profitable plant. This is but one example. The next step is to not over-produce the item and flood landscapes with them.

Following are some plants that I feel are landscape worthy, but seem to be missing from production in sufficient numbers:

Platanus x acerifolia 'Suttneri' has variegated leaves, which really does not matter since the foliage is out of eyesight. The branching of this hybrid is typical but has a very striking white bark which really stands out in the landscape.

***Platanus x acerifolia* Exclamation!®** has excellent branching and a high resistance to leaf anthracnose (Figure1).

Figure 1. *Platanus* x *acerifolia* Exclamation!™

Figure 2. *Platanus occidentalis* Silverwood™.

***Platanus occidentalis* Silverwood™** was selected by John Nickel, owner of Greenleaf Nursery, out of a sycamore allee in Tulsa, OK. While it maintains the typical branching and size habits of the species, Silverwood reveals bright white bark after it exfoliates during the growing season (Figure 2).

Sophora japonica is a species of tree that deserves more attention. I have seen issues with cankers developing after being transplanted bareroot. It has been my observation that growing the liners in a container then planting into a field production system will alleviate canker issues to a significant degree.

***Sophora japonica* 'Gardiner's Weeping'** is a pendulous form found at the entrance of Bernheim Arboretum in Clermont, KY. I placed the name of Gardiner's Weeping on this plant to note its more prolific flowering tendencies when compared to others. Propagation seems to be best with either bench grafting or spring budding.

***Sophora japonica* 'Winter Gold'** sports chartreuse foliage during the growing season and bright yellow bark during the winter (Figure 3). This form is somewhat "twiggy" when young, so some correctional pruning for a solid central leader is necessary during the first year. Propagation is the same as described above.

Figure 3. *Sophora japonica* 'Winter Gold'.

Figure 4. Metasequoia 'Amber Glow'.

Metasequoia glyptostroboides **'Amber Glow'** is an improvement over 'Ogon'. It has yellow foliage that holds up better to bleaching when exposed to bright sunlight (Figure 4). Unlike many yellow-foliaged plants, 'Amber Glow' still has the same level as vigor as the straight species - making it a quick-to-finish plant in production.

Hydrangea arborescens **Invincibelle® Wee White** is a dramatic improvement over Annabelle. Wee White has a much smaller stature, struggling to reach 46cm (18-in.) tall x 61 cm (24-in.) wide in the container. The foliage stays clean with little leaf spotting, and without the sporadic leaf dying, characteristic of Annabelle. This selection also has heavier flowering tendencies than other available *arborescens* selections.

Hydrangea arborescens **'Haas Halo'** is a lacecap hydrangea with some serious flower power! While a lacecap flower style is not what the public typically prefers, this selection would definitely change their mind. A regular sized plant, 'Haas Halo' produces a good number of flowers with each flower being extraordinarily large.

Adina rubella is an unknown shrub that has some very admirable characteristics. The foliage on this species is fine in texture and glossy in appearance. Sporadically throughout the growing seasons, buttonbush-like flowers appear. As fall approaches, the foliage turns an excellent maroon color which persists for an extended period. There is an opportunity for breeding a more compact, heavier flowered plant if one could find viable seed.

***Viburnum carlesii* Sugar N' Spice™** displays a significant improvement over the typical species (Figure 5). While the flower clusters are about 1/3 smaller than normal, the sheer quantity make up for the size in effect. The high flower count leads to another characteristic that makes this plant a production gem. The branching on 'Sugar N' Spice' make the cull rate rather low on a species that is typically very high.

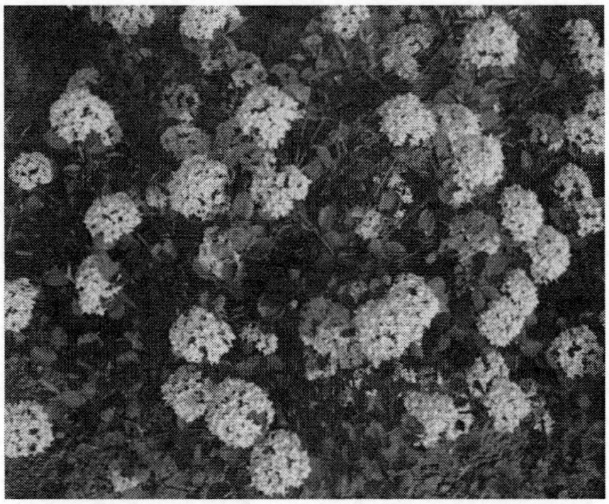

Figure 5. *Viburnum carlesii* 'Sugar N' Spice™'

***Microbiota decussata* 'Fuzzball'** is a subtle winner. Separating itself from the crowd, Fuzzball exhibits significantly less loss during the growing season when compared to the species along with better winter color that does not bronze. Liners are hard to find right now.

Indigofera heterantha should be used more. This plant has excellent drought tolerance and few pest or disease issues. The fine-textured blue-green foliage is quite attractive. Purple flower spikes begin around the first of June and continue until about Mid-August. It is best used as a cutback shrub in the landscape.

***Echinacea* 'Southern Belle'** is a standout (Figure 6). When visiting Mt. Cuba Center in August of this year, this selection was still in full bloom - while every other coneflower was brown and well past its peak. A double flower with deep, bright magenta petals. Unfortunately, I have yet to find liners of this plant in the trade.

Figure 6. Echinacea 'Southern Belle'.

Amelanchier sp. (weeping form) is located at Simpson Nursery in Vincennes, Indiana. I am not aware of a cultivar name for this plant, but it does truly have a pendulous habit. This particular selection is quite susceptible to the various diseases that attack this species; however, it does offer some good breeding material with some of the more resistant cultivars.

ACKNOWLEDGEMENT
Thanks to Kris Stone, Star Roses and Plants, Mike Hayman and Eric Clark for sharing their photographs used in this presentation.

Combined Proceedings IPPS 68:308-313. 2018.

Safety and Efficacy of Postemergence Herbicides for Container-Grown Landscape Groundcovers

Crystal J. Conner, Chris Marble, and Annette Chandler
Mid-Florida Research and Education Center, University of Florida, 2725 S. Binion Road, Apopka, Florida 32703, USA

c.conner@ufl.edu

Keywords: Landscape nursery, weed management, Asiatic jasmine, *Trachelospermum asiaticum* 'Minima', perennial peanut, *Arachis glabrata* 'Ecoturf'.

Abstract

Research was conducted on crop tolerance of Asiatic jasmine (*Trachelospermum asiaticum* 'Minima') and perennial peanut (*Arachis glabrata* 'Ecoturf') to postemergence herbicides including bentazon, sulfentrazone, iron HEDTA, indaziflam (a preemergence herbicide), sulfosulfuron, and clopyralid. Efficacy of these herbicides was evaluated on flowering eclipta (*Eclipta prostrata*) and hairy beggarticks (*Bidens pilosa*). All herbicides with the exception of bentazon caused no significant damage to Asiatic jasmine; injury resulting from bentazon was minimal. In perennial peanut, the highest injury was noted in plants treated with indaziflam, sulfosulfuron, or clopyralid, but injury was less than 30% and considered acceptable. All herbicides evaluated provided poor control of either weed species with the exception of clopyralid, which provided over 90% control of hairy beggarticks. Results indicate that several postemergence herbicides labeled for use in either nurseries or landscapes could be used to manage weeds in Asiatic jasmine or perennial peanut groundcovers, but further testing is needed.

INTRODUCTION

Turfgrass is the most widely planted irrigated crop in the United States and occupies the vast majority of most residential and commercial landscapes in Florida (NTRI, 2003). However, the common mantra of landscape design is "right plant right place".

In many neighborhoods, parks, and other areas containing significant tree canopy, turfgrass is not suitable due to limited sunlight. Additionally, many homeowners may opt for more low maintenance groundcovers if they do not want to install irrigation, mow, or make regular fertilizer applications. Topography, drainage, soil health, and other factors may also make it difficult for some to have healthy and sustainable lawns in all or part of their landscape.

In cases where turf is not suitable or desirable, many different groundcovers are available as turf alternatives. In Florida, the most common turf alternatives are perennial peanut (*Arachis glabrata*) and asiatic jasmine (*Trachelospermum asiaticum*). These groundcovers offer several advantages in that they are both very drought tolerant once established, require little to no fertilization, and in the case of Asiatic jasmine, can be planted in heavy shade or full sun conditions. Both groundcovers also offer significant advantages in terms of resistance to arthropods or fungal pathogens that infect almost all of the warm season turfgrass species planted in Florida. While these groundcovers offer many advantages over turfgrass in certain scenarios, the biggest disadvantage is that there are few postemergence options for weed control in non-turf groundcovers - while either in production or after landscape establishment.

Several herbicides can be used in perennial peanut (Sellers and Ferrell, 2018), but most are not labeled for use in nurseries or landscapes with the exception of certain graminicides, such as clethodim. In both production and landscapes, glyphosate is often applied for weed control in asiatic jasmine which may be safe in some instances but can cause significant injury after two applications or when applications are made at certain times of the year (Van Hoogmoed, 2012). The objective of this research was to identify potential postemergence herbicides that could be applied over-the-top of these groundcover species while in production (container-grown) and determine efficacy of these herbicides on common broadleaf weed species.

MATERIALS AND METHODS

These trials were conducted at the Mid-Florida Research and Education Center in Apopka, FL in 2018. Nursery containers (946 ml) were filled with a 70:30 pine-bark:peat substrate and topdressed with Osmocote® (ICL Specialty Fertilizers) 17-5-11 fertilizer (8-9 mo.) at a rate of 4 grams per pot. After filling, approximately 30 seeds of hairy beggarticks (*Bidens pilosa*) or eclipta (*Eclipta prostrata*) were surface sown by hand. Pots were placed on a full sun nursery pad and received 1.3 cm of overhead irrigation per day. These two weed species were allowed to grow for approximately 5 weeks until flower development. When the two weed species began to reach the reproductive growth stage, trade gallon (3.0L) containers of asiatic jasmine (*Trachelospermum asiaticum* 'Minima') and perennial peanut (*Arachis glabrata* 'Ecoturf') were purchased from a local nursery and placed on the nursery pad described previously.

On 30 July [29°C (85°F), 72% relative humidity, winds calm, partly cloudy] herbicides were applied over-the-top of both weed and groundcover species using a CO_2 backpack sprayer calibrated to deliver 468 L ha^{-1} (50 gal $acre^{-1}$) (Table 1). Following treatment, plants were grouped by species in a completely randomized design with 10 replications per herbicide treatment for each weed species and 6 replications per herbicide treatment for each groundcover species. Data collected included visual injury ratings at 7, 14, and 28 days after treatment (DAT) for the groundcover species. For weed species, visual ratings (control ratings for weeds, injury ratings for groundcovers) were recorded at 7, 14, and 21 DAT. In both cases, ratings were based on a 0 to 100 scale, 0 = no injury (or control) and 100 = complete plant death. At

21 DAT, shoot weights were collected for both weed species by cutting plants at the soil line. Fresh weights were converted to percent control using the formula:
[(fresh weight non-treated – fresh weight treated)/fresh weight non-treated) x 100].

Data were analyzed using a mixed model analysis in JMP with herbicide treatment as a fixed effect and replication as a random effect. Means were separated using Fisher's LSD at a 0.05 significance level.

RESULTS

At 7 DAT, only very minor injury was noted in asiatic jasmine treated with bentazon at either rate (Table 1). However, injury ratings were ≤5% and considered commercially acceptable. All other treatments resulted in injury ratings similar to the non-treated plants. Similar trends were noted at both 14 and 28DAT with minor injury observed in jasmine treated with bentazon and no injury following treatment with any other herbicide. Injury observed included chlorosis and some minor burning on new growth.

In perennial peanut, at least some injury was noted in all herbicide treatments at 7DAT but applications of sulfentrazone, iron HEDTA (50 fl. oz.), indaziflam, and sulfosulfuron resulted in injury ratings similar to the nontreated control. Injury ratings increased in most treatments at 14DAT. At this time, the highest injury was noted in plants treated with indaziflam, although peanut treated with bentazon (either rate), sulfosulfuron, and clopyralid displayed similar injury. At 28DAT, recovery was noted in most treatments with the exception of peanut treated with indaziflam, sulfosulfuron, and clopyralid, which had injury ratings of 23 to 27%. All other treatments had injury ratings of <8% and were similar to nontreated plants.

Herbicides evaluated in these studies generally provided poor control of eclipta. Clopyralid (47%) provided the greatest reduction in shoot fresh weights along with sulfentrazone (12 fl. oz.) (35%) and indaziflam (37%) which provided similar control (Table 2). All other treatments provided approximately 30% reduction in shoot fresh weights or less. While clopyralid provided approximately 50% reduction in shoot fresh weights, this level of control was considered unacceptable. Clopyralid provided a high level of hairy beggarticks control (91%) and outperformed all other herbicide treatments. The next most efficacious treatment was indaziflam providing 56% control. All other herbicides provided 15% control or less and were similar. In a nursery or landscape setting, the only treatment that would have been considered acceptable was clopyralid.

DISCUSSION

Results from this research indicate that several postemergence herbicides could be applied over the top of asiatic jasmine or perennial peanut with minimal to no crop damage. Currently the only herbicide labeled for use in either of these groundcovers is sulfosulfuron, which is labeled for use in asiatic jasmine growing in landscapes (Anonymous, 2016). Bentazon is labeled for use in peanuts (as Basagran®) in cropping systems, but currently Basagran® T/O is not labeled for use in perennial peanut in landscape or nursery situations. Of the herbicides tested, all are currently labeled for use in landscapes. Bentazon, sulfentrazone, iron HEDTA, and indaziflam are also labeled for use in nurseries, but not as over-the-top applications in container-grown ornamentals. It should be noted that while indaziflam was evaluated in this study as a postemergence herbicide, it is a preemergence herbicide but has shown some postemergence activity (Anonymous, 2013; Brosnan et al., 2012).

Table 1. Tolerance of minima jasmine (*Trachelospermum asiaticum* 'Minima') and perennial peanut (*Arachis glabrata* 'Ecoturf') to selected postemergence herbicides.

			Injury ratings (0 to 100%)[z]					
			Asiatic jasmine			Perennial peanut		
Trade name	Herbicide	Rate[y]	7DAT[x]	14DAT	28DAT	7DAT	14DAT	28 DAT
Basagran T/O	bentazon	24 fl. oz.	5 a	7 b	5 b	10 ab	12 a-e	8 b
Basagran T/O	bentazon	32 fl. oz.	3 ab	16 a	18 a	17 a	17 abcd	7 b
Dismiss	sulfentrazone	8 fl. oz.	0 c	0 c	0 c	7 bc	10 bcde	8 b
Dismiss	sulfentrazone	12 fl. oz.	0 c	0 c	0 c	3 bc	7 cde	7 b
Fiesta	iron HEDTA	25 fl. oz.	2 bc	0 c	0 c	12 ab	7 cde	3 b
Fiesta	iron HEDTA	50 fl. oz.	0 c	0 c	0 c	7 bc	5 de	7 b
Marengo SC	indaziflam	9 fl. oz.	0 c	0 c	0 c	3 bc	23 a	25 a
Certainty	sulfosulfuron	1.25 oz.	0 c	0 c	0 c	7 bc	22 ab	23 a
Lontrel	clopyralid	1 pint	0 c	0 c	0 c	12 ab	18 abc	27 a
Control	NA	---	0 c	0 c	0 c	0 c	0 e	0 b

[z]Injury ratings were recorded on a scale of 0 to 100, 0 = no injury, 100 = dead plant.

[y]Rate shows amount of formulated product applied on a per acre basis. Rates were generally low and high recommended labeled rates.

[x]DAT = days after treatment.

[w]Means within a column followed by the same letter are not significantly different according to Fisher's Least Significant Differences test ($p = 0.05$).

Table 2. Efficacy of selected postemergence herbicides for control of hairy beggarticks (*Bidens pilosa*) and eclipta (*Eclipta prostrata*).

			Eclipta	Hairy beggarticks
Trade name	Herbicide	Rate[z]	Percent control (0 to 100%)[y]	
Basagran T/O	bentazon	24 fl. oz.	17 cd	12 c
Basagran T/O	bentazon	32 fl. oz.	17 cd	12 c
Dismiss	sulfentrazone	8 fl. oz.	31 bc	6 c
Dismiss	sulfentrazone	12 fl. oz.	35 ab	12 c
Fiesta	iron HEDTA	25 fl. oz.	6 d	8 c
Fiesta	iron HEDTA	50 fl. oz.	6 d	4 c
Marengo SC	indaziflam	9 fl. oz.	37 ab	56 b
Certainty	sulfosulfuron	1.25 oz.	6 d	15 c
Lontrel	clopyralid	1 pint	47 a	91 a

[z]Rate shows amount of formulated product applied on a per acre basis. Rates were generally low and high recommended labeled rates.

[y]Percent control was calculated based on percent reduction in shoot fresh weights in relation to the non-treated control group.

[x]Means within a column followed by the same letter are not significantly different according to Fisher's Least Significant Differences test ($p = 0.05$).

The only herbicide that showed a high level of control of either weed species tested was clopyralid, which provided over 90% control of hairy beggarticks. Weeds tested in this trial were mature and had reached the flowering stage. Treating weeds at this mature stage of growth most often results in reduced efficacy, especially for contact action herbicides (Senseman, 2007). While the herbicides evaluated in this trial generally provided less than ideal control, all have shown a high degree of efficacy on numerous broadleaf and sedge weeds (Neal et al., 2017).

Some injury was noted in these trials but at all evaluations, injury was within acceptable levels (<30%). In some landscape (or production) scenarios, even minor injury may not be acceptable and as these herbicides are not labeled for over-the-top application to the groundcovers evaluated (with the exception of sulfosulfuron), the applicator would assume liability. Additionally, greater or less injury could potentially be observed under different environmental conditions. A major deterrent in the use of these groundcovers is the lack of postemergence herbicides that can be used, and the fact that weeds often become highly problematic once these species are transplanted, especially during the establishment phase.

Further testing is needed to determine safety of these productions under different scenarios and to determine if these products could be candidates for label expansion in the future.

However, this preliminary work indicates that several currently available herbicides can be used in these species for postemergence weed control both while in production and following landscape transplanting.

Literature Cited

Anonymous. (2016). Certainty herbicide label. Valent U.S.A. Corp. Walnut Creek, CA. 9 p.

Brosnan J.T., Breedon, G.K., McCullough, P.E., and Henry, G.M. (2012). Pre and post control of annual bluegrass (*Poa annua*) with indaziflam. Weed Tech. *26*:48-53

National Turfgrass Research Initiative (NTRI). (2013). National turfgrass research initiative. Accessed 28 Aug. 2018. http://www.ntep.org/ntep/pdf/turfinitiative.pdf

Neal, J., Chong, J.C., and Williams-Woodward, J. (eds.). (2017). Southeastern US pest control guide for nursery crops and landscape plantings. Accessed 25 Feb. 2018. https://content.ces.ncsu.edu/southeastern-us-pest-control-guide-for-nursery-crops-and-landscape-plantings

Sellers, B.A. and Ferrell, J. (2018). Weed control in perennial peanut. University of Florida EDIS pub. AGR-261. http://edis.ifas.ufl.edu/wg216. Accessed 12 Aug. 2018.

Senseman, S.A. (2007). Herbicide Handbook, 9th edition. Weed Science Society of America. Lawrence, KS. 458 p.

Van Hoogmoed, A., Gilliam, C.H., Wehtje, G.R., Knight, P.R., Foshee, W.G., Olive, J.W., and Murphy, A.M. (2012). Roundup Pro over the top of nursery crops: rates and timing. J. Environ. Hort. *30*:93-102.

Combined Proceedings IPPS 68:314-318. 2018.

Slow-Release and Controlled-Release Fertilizers: An Overview of the Market Today

Kyle Creamer
Harrell's LLC, 720 Kraft Road, Lakeland, Florida 33815-3244, USA

kcreamer@harrells.com

Keywords: Organic, synthetic, resin coated, polymer coated, fertilizer release characteristics, cost analysis.

INTRODUCTION

Commercially available slow-release fertilizers (SRFs) are different than controlled-release fertilizers (CRFs). They differ in coating materials, characteristics of fertilizer release, longevity as influenced by temperature, hydrolysis and/or microbial activity - and unit cost compared to actual usage/longevity cost in a commercial, production setting (Figure 1). Unit cost is not the only deciding economic factor.

Slow Release Fertilizers (SRF) are Not Controlled Release Fertilizers (CRF)

- One difference is in the coating characteristics.

Slow Release	Controlled Release
Natural Organics	Resin Coated
Synthetic Organics	Polymer Coated
Polymer Coated Sulfur Coated Urea - PCSCU	

Figure 1. Slow release fertilizers (SRF) differ from controlled release fertilizers (CRF) in coating characteristics.

SLOW-RELEASE FERTILIZERS (SRFS)

The different SRFs on the market today are (1) natural organic products, such as Nature Safe and other true organic compounds; (2) synthetic organic products, such as Nitroform Blue-Chip and IBDU-Isobutylidenediurea; these break down through hydrolysis and mineralization (Figures 2 and 3).

Slow Release Fertilizers (SRF)

Slow Release	Product example	Released via
Natural Organics	Milorganite, Nature Safe	Microbial Action
Synthetic Organics	IBDU	Hydrolysis
Synthetic Organics	Nitroform, Nutralene	Mineralization
Polymer Coated Sulfur Coated Urea - PCSCU	Poly-S, XCU®	Catastrophic, rupture

Figure 2. Examples or slow release fertilizer (SRF) products and their release mechanism.

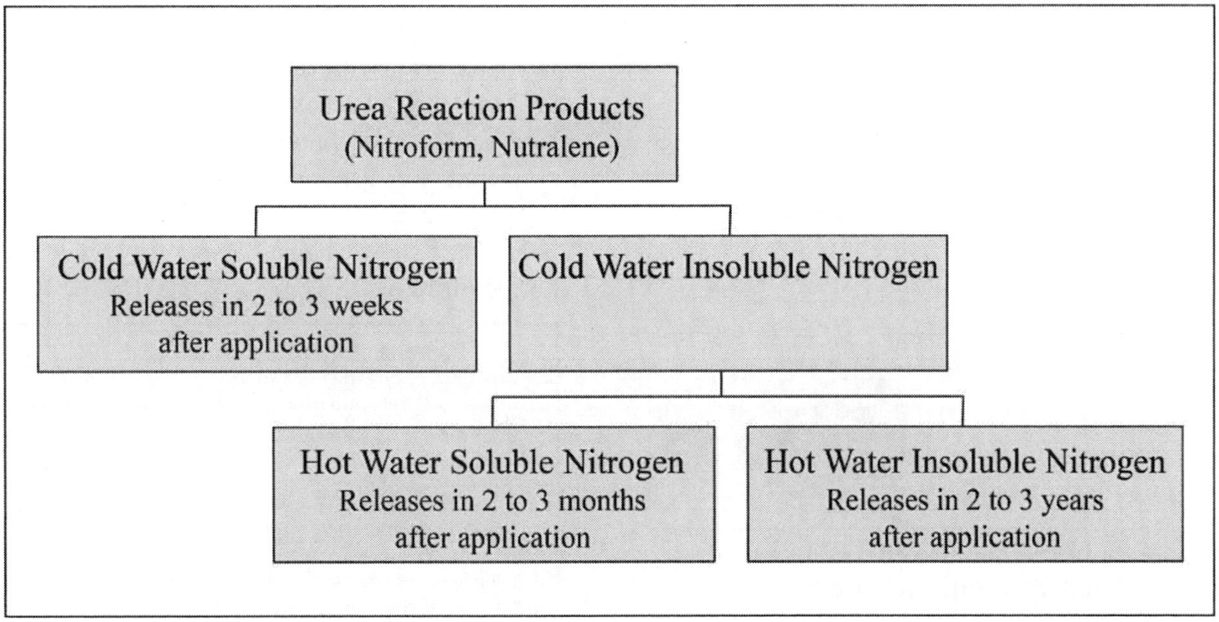

Figure 3. Fertilizer in Nitroform SRF is released through mineralization

There are also polymer coated sulfur coated urea products (PCSCU) such as Poly-S and XCU® - which release via what is known as "catastrophic release" (ruptures, cracks, fractures in the coating) (Figure 4).

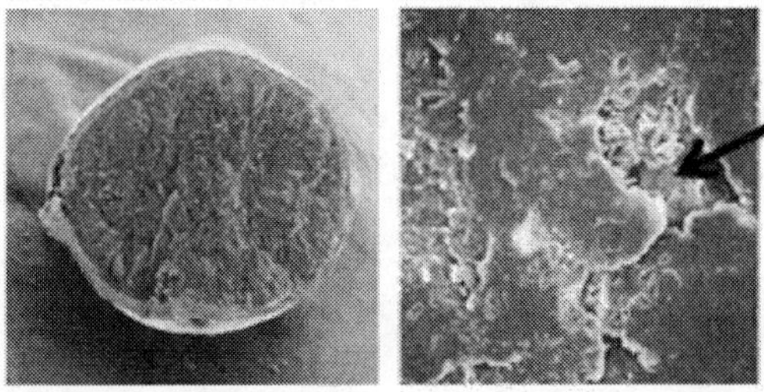

Sulfur-coated urea particle (left) and rupture on the surface of the layers coating urea fertilizer substrate (right, arrow) due to increase in osmotic pressure that enables fertilizer release.

Figure 4. Fertilizer in sulfur coated urea is dispensed via catastrophic release via the rupture/ fissure in the polymer coated layers.

In summary, SRF's have no "controlled release" mechanism. They are also less efficient and are not ideal as the sole source of a fertility program. However, they are great additions in a more balanced approach to complement CRF's.

CONTROLLED-RELEASE FERTILIZERS (CRFS)

CRF's are classified into two different groups based on their coating: **resin coated** (coating comes from plant origins) or **polymer coated** (synthetic coating).

Resin coated

Osmocote® is the only true resin coated product on the market today. A resin is "any of numerous clear to translucent yellow or brown solid, or semi-solid substance of plant origin (organic or natural) such as copal, rosin, and amber used principally in lacquers, varnishes, inks, adhesives, synthetic plastics, and pharmaceuticals". It was seen as the "pioneer" of the CRF market. It was developed based on a floor varnish used in the 1940s (Figure 5). It is manufactured by ICL and releases through the process of osmosis (Figure 6). The prills swell when water enters and once it swells - it cannot contract. Fissures form in the coating, and the water/nutrient solution then exits the prills. The longevity of Osmocote® is based on the coating thickness at an average temperature of 21°C (70°F). Higher temperatures and physical damage to the prills will speed up the fertilizer release.

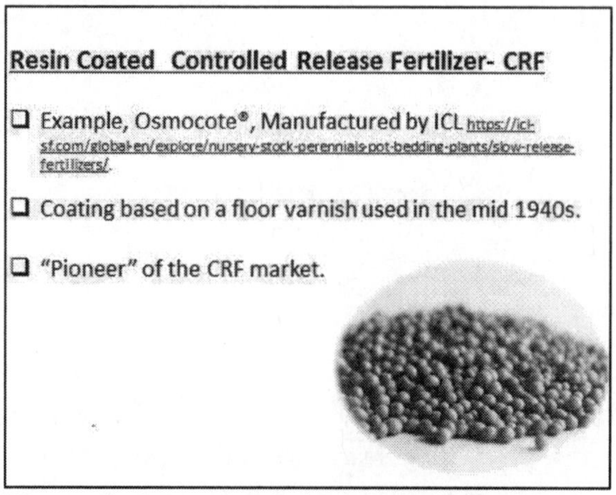

Figure 5. Osmocote® is an example of a resin coated, controlled release fertilizers (CRF).

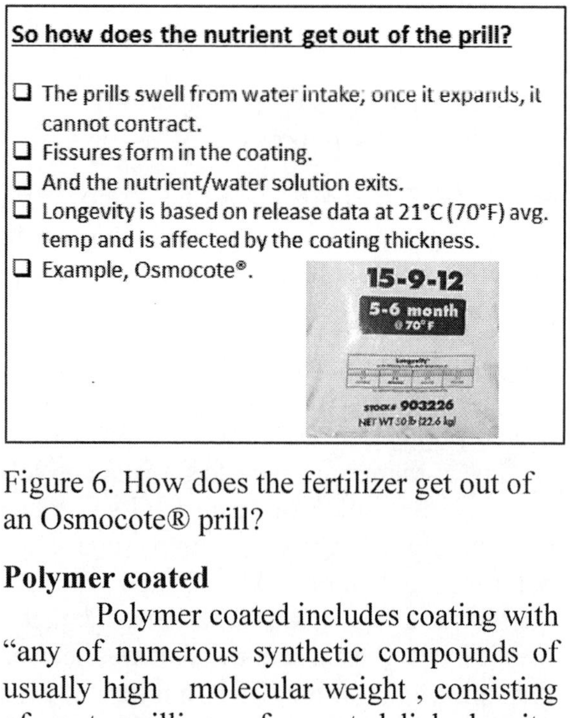

Figure 6. How does the fertilizer get out of an Osmocote® prill?

Polymer coated

Polymer coated includes coating with "any of numerous synthetic compounds of usually high molecular weight , consisting of up to millions of repeated linked units, each a relatively light and simple molecule." An example is Nutricote®, which is manufactured by JCAM in Japan. JCAM creates the raw and finished product in Japan and then ships it to the USA where Florikan® is the exclusive distributor. Nutricote® is unique in that the longevity is not based on the coating thickness; rather, it is the ratio of two polymers (with different release times) that coat the prills. This determines the fertilizer release period. The release pores of the polymers can close in cooler temperatures. The release mechanisms is via the solute concentration gradient of the two polymers coating the prill (Figure 7). The longevity of Nutricote® is based on an average temperature of 25°C (77°F).

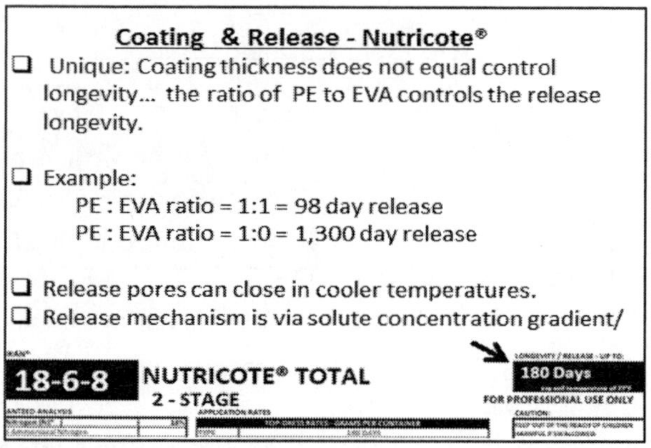

Figure 7. Coating and release characteristics of Nutricote® controlled release fertilizer (CRF).

Polyon® is the other major CRF in the market today. It is manufactured by Koch Agronomic Services in the USA and is exclusively distributed by Harrell's. Polyon® is custom blended and sold directly to the end user. The longevity of Polyon® is related to the coating thickness of the two monomers incorporated with polyurethane (Figure 8). Polyon® also releases via the solute concentration gradient, and its longevity is based on 30°C (86°F).

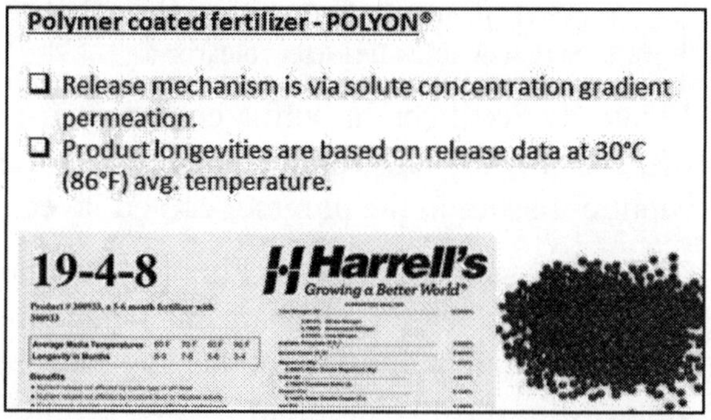

Figure 8. Polymer coated Polyon® controlled release fertilizer (CRF).

Other CRF products mentioned briefly are similar in characteristics to Polyon®. These are, as follows:

- GAL-Xe$^{ONE®}$ —formerly known as "Florikote®" that is owned by Simplot http://simplotgalxeone.com/ and sold nationwide. Release longevities are based on 38°C (100°F)
- Multicote®—Manufactured by Haifa in Israel. Release longevity is based on 21°C (70°F). Its availability is based on temperature and potassium release, not nitrogen.
- Plantacot® —longevity is based on 21°C (70°F).
- Duration®—Former Agrium product, but now is part of Koch Agronomic Sciences technology. Release data is based on 20°C (68°F).
- Purkote® —a new product from Pursell that is still under evaluation. Release data is based on 30°C (86°F).

SUMMARY

I encourage IPPS members to compare more than cost when looking at different CRF products. For instance, a $10 per 50lb bag difference in pricing only equates to $0.04 difference per 3-gal container (assuming the analysis and longevities are the same) (Figs. 9 and 10). Also, be careful to check the temperatures in which the longevities are determined. A "4 month" CRF product based on release data at 21°C (70°F) - is reduced to a "2 month" product at 30°C (86°F).

Cost Analysis

❑ Brand X Fertilizer (8-9 month) $55.00/50lb

		Rate in Grams					
BRAND X 8 month	Size (Gallons)	Low	$/pot	Med	$/pot	High	$/pot
$55.00	1	11	$0.027	13	$0.031	16	$0.039
	2	26	$0.063	31	$0.075	38	$0.092
	3	46	$0.111	56	$0.136	68	$0.165 ←
	5	62	$0.150	75	$0.182	92	$0.223
	7	91	$0.220	112	$0.271	136	$0.330
	10	114	$0.276	139	$0.337	170	$0.412
	15	128	$0.310	156	$0.378	191	$0.463
	20	160	$0.388	195	$0.472	238	$0.577

❑ Brand Y Fertilizer (8-9 month) $45.00/50lb

		Rate in Grams					
BRAND Y 8 month	Size (Gallons)	Low	$/pot	Med	$/pot	High	$/pot
$45.00	1	11	$0.020	13	$0.024	16	$0.030
	2	26	$0.048	31	$0.057	38	$0.070
	3	46	$0.085	56	$0.104	68	$0.126 ←
	5	62	$0.115	75	$0.139	92	$0.170
	7	91	$0.168	112	$0.207	136	$0.252
	10	114	$0.211	139	$0.257	170	$0.315
	15	128	$0.237	156	$0.289	191	$0.353

Figure 9. Cost analysis of a CRFs based on unit prices

Cost Analysis

❑ Brand Y Fertilizer (8-9 month) $45.00/50lb.

❑ Brand X Fertilizer (8-9 month) $55.00/50lb.

Take Home Message

❑ $10/bag difference between Brand Y & X.

❑ On a 3 gal pot the difference is only $0.04.

❑ Remember to compare more than price.

❑ A 4-month product at 21°C (70°F) is equivalent to a 2-month product at 30°C (86°F).

Figure 10. Factors in selecting a CRF should not be based just on the unit price – but rather the usage/ longevity cost under local, commercial production conditions.

Products

Nature Safe https://www.naturesafe.com/

Nitroform Blue-Chip https://kochturf.com/Products/SlowRelease/Nitroform/

Poly-S https://icl-sf.com/uk-en/technologies/poly-s/

XCU® https://www.greenindustrypros.com/lawn-care-renovation/product/10156380/koch-agronomic-services-kas-xcu-slowrelease-fertilizer

Osmocote® https://www.iclfertilizers.com/

Nutricote® http://www.jcam-agri.co.jp/

Florikan® https://www.florikan.com/about

Polyon® https://www.harrells.com/

Combined Proceedings IPPS 68:319-325. 2018.

Nursery-Scale Production of Native Azaleas from Seed

Emily Driskill
Carolina Native Nursery, 1639 Price's Creek Rd, Burnsville, North Carolina 28714, USA

Emily@CarolinaNativeNursery.com

Keywords: Seed collection, seed propagation, *Rhododendron* species, *R. periclymenoides, R. atlanticum, R. canescens, R. austrinum,* IPM – integrated pest management.

INTRODUCTION

At Carolina Native Nursery in mountainous western North Carolina, we grow 12 species of deciduous azaleas that are native to the southeastern United States. Over the past 15 years, we have honed our seed-growing protocol for this special group of ornamentals, native shrubs. We start from seeds and eventually finish in 3-gal containers. What follows is a summary of our methods.

SEED COLLECTION

We collect seed from several locations with the appropriate permission. Interspecific hybrids occur in nature, so be aware of the potential for crossing where species are growing close together. This type of genetic exchange is entirely natural and happens all the time, so we generally do not go to great lengths to avoid cross-pollinated seed. For clients who would prefer a uniform bloom color or stature, we offer tissue-cultured hybrids and varieties. However, for those who are interested in planting native ornamentals for the benefit of local eco-systems - genetically diverse, seed-grown plants are where it is at.

It is advantageous to gather the widest sample of available genetic diversity, so we collect from as many plants as possible. Of course, whenever collecting any type of material from the wild, it is essential to tread lightly and use ethical standards on how much to take. We never collect any more than 20% of the seed from an individual plant, or from the community at large. Please do not collect from the wild if you are unwilling to follow sustainable harvest standards.

Seed collection happens in October and November in our climate, and will vary depending on species, microclimate, and the weather in a given year. It is helpful to first visit the stands where you plan to collect seed during the bloom season. Of course, the plants are much easier to find at that time and seeing the flowers will help you identify what species you are working with. Seeds of *Rhododendron* species are mature when the seed pods have turned brown and are just beginning to split. This is the best time to collect. However, pods can be collected when still green (if at their full size), and then carefully dried to allow seeds to fully mature.

Once collected, leave mature pods to dry in a single layer in a cool place with adequate air flow and no direct sunlight. If seeds are mature when picked, one week of drying time will be sufficient before cleaning. If working with green pods, a longer drying time will probably be required. Once completely dry, seeds not needed for sowing can be stored in the freezer indefinitely. These seeds need no stratification and can be sown immediately if desired.

SOWING

We sow cleaned azalea seeds into community flats [25 x 51 x 5 cm (10 x 20 x2 in.)] in mid to late-November. The flat is filled with about 4 cm (1.5 in.) of our standard composted pine bark medium, which also contains dolomitic limc and a humic acid product. On top of that, we place 0.6 cm (.25 in.) of pre-moistened ground peat moss, which is tamped down with a tool to make an even planting surface. Seeds are sprinkled by hand on top of the peat. The goal is to end up with 700-1000 seedlings per flat. This is a heavy seed rate, but we will transplant seedlings into cell trays starting in March, avoiding issues of fungal diseases developing because of insufficient air flow in the thick stand of the seedlings. We have not found it necessary to cover the seeds with more peat. We then use a hose-end mist nozzle to fully wet the seeds, which also functions to push them down, ensuring full contact with the peat.

Community flats are placed on bottom heat inside of a 3-mm clear poly humidity tent (or with individual humidity domes on each flat) within the greenhouse and monitored carefully for moisture (Figure 1). Flats should not dry out on top, nor should they be soaking wet. A thermostat is set using a soil sensor to keep the root zone at about 21°C (70°F). During extremely cold weather, we may turn the bottom heat setting up around 24°C (75°F) to ensure the tent stays at or above 10°C (50°F) overnight. A continuous photoperiod is provided using 600w HPS lights that come on at dusk and turn off at dawn. Germination should occur within about 3-4 weeks; at which time we begin a weekly maintenance schedule to provide fertility and avoid pest and disease threats.

Figure 1. Humidity tent (left) and inside the humidity tent with bottom-heated bench (right).

SEEDLING MAINTENANCE

When germination has clearly begun, we begin our maintenance schedule which involves weekly drenching with products to provide fertility and protect against pests and diseases (Figure 2). Fungus gnats and fungal diseases are the main pests that we worry about during this stage.

Figure 2. *Rhododendron periclymenoides* seedlings in flats in early January after being sown in November 29, 2017.

We use a checklist format for keeping track of this schedule, which is an easy way to ensure we are using a proper fungicide rotation (to avoid pathogens gaining resistance) and to make sure we get the work done on schedule (Figure 3). We use water-soluble fertilizer with minor nutrients included to provide fertility. The checklist includes the analyses of the two fertilizers we rotate. Following the checklist, we step-up the fertilizer rate gradually as the plants grow. We start out using ¼ of the label rate on tiny seedlings, and step-up to ½ rate when seedlings have developed 2 true leaves, and again to ¾ or full rate when they have 4-6 true leaves.

We follow this maintenance procedure until the seedlings are transplanted in March-May. Once seedlings have leaves, make sure to open up the humidity structure to vent on days when the sun is out. It is surprising how quickly the clear poly can warm up on a sunny day even in January. The seedlings need sufficient air flow during these times. Bottom heat and supplemental lighting are turned off in March, at least two weeks before we plan to transplant.

December 22	**December 29**	**January 5**	**January 12**
• 20-20-20 ¼ tsp/gal • Zerotol ½ TBS/gal • Captan 1 TBS/gal • Change sticky cards, scout for pests	• 12-48-8 ¼ tsp/gal • Gnatrol ½ tsp/gal	• 20-20-20 ¼ tsp/gal • Actinovate 1tsp/gal • change sticky cards, scout for pests	• 12-48-8 ¼ tsp/gal • Essential 2 oz/gal • Gnatrol ½ tsp/gal
January 19	**January 26**	**February 9**	**February 16**
• 20-20-20 ¼ tsp/gal • Zerotol ½ TBS/gal • Change sticky cards, scout for pests	• 12-48-8 ½ tsp/gal • Gnatrol ½ tsp/gal	• 20-20-20 ½ tsp/gal • Triathlon BA 2 tsp/gal • Change sticky cards, scout for pests	• 12-48-8 ½ tsp/gal • Essential 2 oz/gal • Gnatrol ½ tsp/gal

Figure 3. Azalea seedling care checklist, winter 2018.

TRANSPLANTING SEEDLINGS

Once seedling flats have been properly hardened off, we are ready to transplant the azaleas for the first time. We use our composted pine bark medium in RootMaker 18-cell flats, which are placed on benches in a cold frame with a 30% shade cloth.

Seedlings are taken from the flats in chunks and carefully separated to preserve the roots. We have built ourselves a dibble board with wooden dowels, which we press onto the flats to make the necessary holes in each cell to plant into. Transplanting seedlings at the correct depth requires some skill (Figure 4). Each tiny plant needs to be placed no deeper than it was originally growing in the flat, but not so high in the cell that the small plant flops over when watered. We generally assign this task to our most skilled workers - because so much of the quality of our future crop depends upon this step. Once each flat is complete, we sprinkle 20 grams of Harrell's Polyon® 16-6-11+ micros-controlled release fertilizer (CRF) evenly over the entire flat, and water thoroughly by hand.

Figure 4. Freshly transplanted *Rhododendron canescens* seedlings, April 1, 2018

Once a block of seedling transplanting is complete, we drench with Actinovate (*Streptomyces lydicus* strain WYEC 108) or a similar biological fungicide. If transplanted by the first of May, seedlings can be expected to fully root into cells before the first of September (Figure 5). We use a hedge trimmer to prune the seedlings back to a consistent height about two months after transplanting, when lots of active growth is present. We repeat this trimming 2-3 more times throughout the season to encourage branching.

Figure 5. Azalea liners in propagation house, September 2018.

Seedlings are overwintered in cold frames with 30% shade and no supplemental heat. Sufficient air flow provided by fans and/or opening doors on clear and sunny days is important for preventing development of fungal diseases and liverwort/moss invasion.

SUBSEQUENT TRANSPLANTING AND MAINTENANCE

The young azaleas are ready to be transplanted again the next spring. Around early April, we pull them from the cell trays, gently loosen the root mass, and transplant to a 1-gallon pot (Figures 6 and 7). Planting at the proper depth (root crown must be no lower than it was in the cell) is just as important at this stage. However, it is easier to do since you are working with larger material. We then topdress each pot with 10 grams of the Polyon® 16-6-11 CRF, and cover that with a 2.5 cm (1-in.) layer of rice hulls, which acts as a mulch to keep down weeds and maintain a more constant moisture level within the container. One-gallons are set on the ground (gravel) in cold frames with 3mm opaque overwintering poly covering. We vent these cold frames by opening the doors on any day with temps above about 3°C (38°F). Around the time the leaves have expanded on the local trees, we take down the poly covering and replace with 30% shade cloth.

We do not currently space 1-gal containers, but it may be advantageous to do - because it leads to a fuller-shaped plant, and results in greater airflow, which can prevent fungal infection. Leaf rust is something to keep an eye out for at all stages of growth. Azaleas can be expected to fully root into the 1-gal containers by August or early September (Figure 8). One-gal azaleas are hedge trimmed 3-4 times throughout the season, taking just an inch or two off the tops of the plants to maintain height consistency and encourage more branching.

The next spring, azaleas are transplanted in the same manner to 3-gal containers. We hedge trim the blocks one more time before preparing plants for transplant. We have used both squat pots and standard 3-gal with success. Three-gal containers receive a topdressing of 38g Polyon® 16-6-11 CRF, and again 2.5 cm (1 in.) of rice hulls. The pots are spaced about a foot apart on all sides.

Figure 6. *Rhododendron austrinum* liners, Sept 4, 2018. Transplanted on April 20, 2018.

Figure 7. Bed of 1-gal *Rhododendron atlanticum*, June 12, 2018. Seedlings were transplanted March 22, 2018.

Three-gal plants may be hedge trimmed once after blooming, if needed. We make a pass through 3-gal crops in June or July to hand-prune branches growing at strange angles.

Fertility is supplemented throughout the season with water-soluble fertilizer and other sources of nutrition (like kelp extract) as a foliar spray - depending on how the growth is looking and what our electro-conductivity (EC) readings are for the containers. With the exception of hand-watering just after transplant, containers are watered via overhead irrigation in a cyclical pattern (i.e., watering for two cycles with an interval in between, rather than one long cycle). This increases absorption capacity and decreases runoff in our fast-draining medium.

Azaleas transplanted into three-gallon containers in March or April can be expected to be fully rooted and sellable by mid-August through September (Fig. 9). In this manner, we can produce a finished crop of three-gal azaleas from seed in three growing seasons.

Figure 8. Growth of 1-gal *Rhododendron atlanticum* on June 6 (left) and September 28 (middle); seedlings were transplanted on March 22, 2018. Growth of 3-gal *Rhododendron atlanticum* on 26 September 2018 (right); seedlings were transplanted on 15 March 2018.

INTEGRATED PEST MANAGEMENT (IPM)

During all stages of growth, we practice Integrated Pest Management (IPM). We continually scout, identify, and strive to prevent pest and disease outbreaks. We have found that working to prevent pest problems through regular scouting, encouraging beneficial predators, using cultural strategies like spacing, and preventative spraying - can reduce the time needed to manage these problems. We also strive to reduce the costs of chemical inputs. Once we see a growing presence of our common pests such as western flower thrips in the spring - we begin a bi-weekly spray program intended to prevent pest populations from reaching the threshold of economic damage.

Our rotation this season included neem oil, Mainspring (cyantraniliprole), and a tank mix of azadirachtin and the entomopathogenic fungus *Beauveria bassiana.* We aim to incorporate a fungicide into the rotation about every 6 weeks, but this interval may decrease if there is an acute problem present.

We rotate through many fungicides including biological fungicides and some traditional classes, such as thiophanate methyl and strobilurin. Tank mixes may also include foliar fertilizers as appropriate; be sure to check that tank mix component are compatible. We continue our bi-weekly spraying until pest pressure has died down. Sticky cards placed throughout the nursery help us monitor insect presence and abundance throughout the season.

Propagating azaleas from seed is not a new topic to IPPS. We would like to credit J.P. Jackson and Lindy Johnson (and all those who came before), who authored the paper in the IPPS annals which we have used as a jumping-off point for our protocol (Jackson and Johnson, 2013). Hopefully we have been able to add a few nuggets of azalea wisdom to the collective consciousness by hashing over the details of our own standard operating procedure.

For further information, please check out our YouTube channel where you will find a three-part video on propagating native azaleas.
https://www.youtube.com/watch?v=phOO-YS8taA

Literature Cited

Jackson, J.P. and Johnson, L. (2013). Growing Native Azaleas from Seed. Comb. Proc. Intl. Plant Prop. Soc. *63*: 441-442.

Combined Proceedings IPPS 68:326-330. 2018.

Designing with Hosta

Ben Ford
Garden Guru Services, 234 Morrell Road, Suite 151, Knoxville, Tennessee 37919, USA

benford2@hotmail.com

Keywords: Garden site design, construction, *Aphelenchoides fragariae,* strawberry crimp nematode, *Hosta virus X*, Hosta stem blight.

INTRODUCTION

Since the fall of 2013, I have been working with Cornelia B. Holland of Franklin, Tennessee, to develop a hosta and shade garden at the University of Tennessee Gardens in Knoxville, Tennessee. To date, over 500 *Hosta* sp., hybrids, *Rohdea* and other Asian origin plants, shrubs and trees have been dug from Cornelia's garden and transported to Knoxville to establish the *Tranquility – Cornelia B. Holland Hosta Garden* at the University of Tennessee Gardens.

SITE CONDITIONS

The half-acre garden site was a basically a blank canvas. There are nine existing bald cypress (*Taxodium distichum*) and three post oak (*Quercus stellata*) to work with and incorporate the garden around. The site is also the low point of the garden. Standing water was a huge problem in the site, so raising the soil level became a requirement for the site.

THE DESIGN

The garden is divided into a series of areas separated by paths and surrounded by varying trees and shrubs - that create a sense of enclosure and privacy. Spaces for benches and large boulders provide seating throughout the garden. At the heart of the garden is large open lawn area that will serve as a venue for weddings and events. The area will eventually have a moon gate for guests to enter the room that will face a large pergola to serve as the stage for such events. The lawn area will comfortably hold 175 people. The lawn is surrounded by fragrant hostas near the lawn's edge with larger shrubs behind them to create privacy and enclosure. The garden also features a temple bell and several water features to add a sense of sound to the garden.

The garden location has presented many challenges. To address the issues of standing water, over 31 cm (12-in.) of topsoil and compost were added to the entire garden. The addition of berms raised beds and drainage trenches were needed to push water out of the site and into the nearby drainage swell and wetland.

With a need to create additional shade for the garden for future garden expansion, the design required additional oaks and fast-growing tulip poplar (*Lirodendron tulipifera*) to increase the size of the above canopy (Figure 1).

Small trees, such as dogwood (*Cornus florida*) and various Asian maples (*Acer sp.*) were also planted throughout the garden. *Hosta sp.* are the main focus of the garden. Hostas provide a wide variety of colors and textures, as well as offer a diversity of habits and sizes.

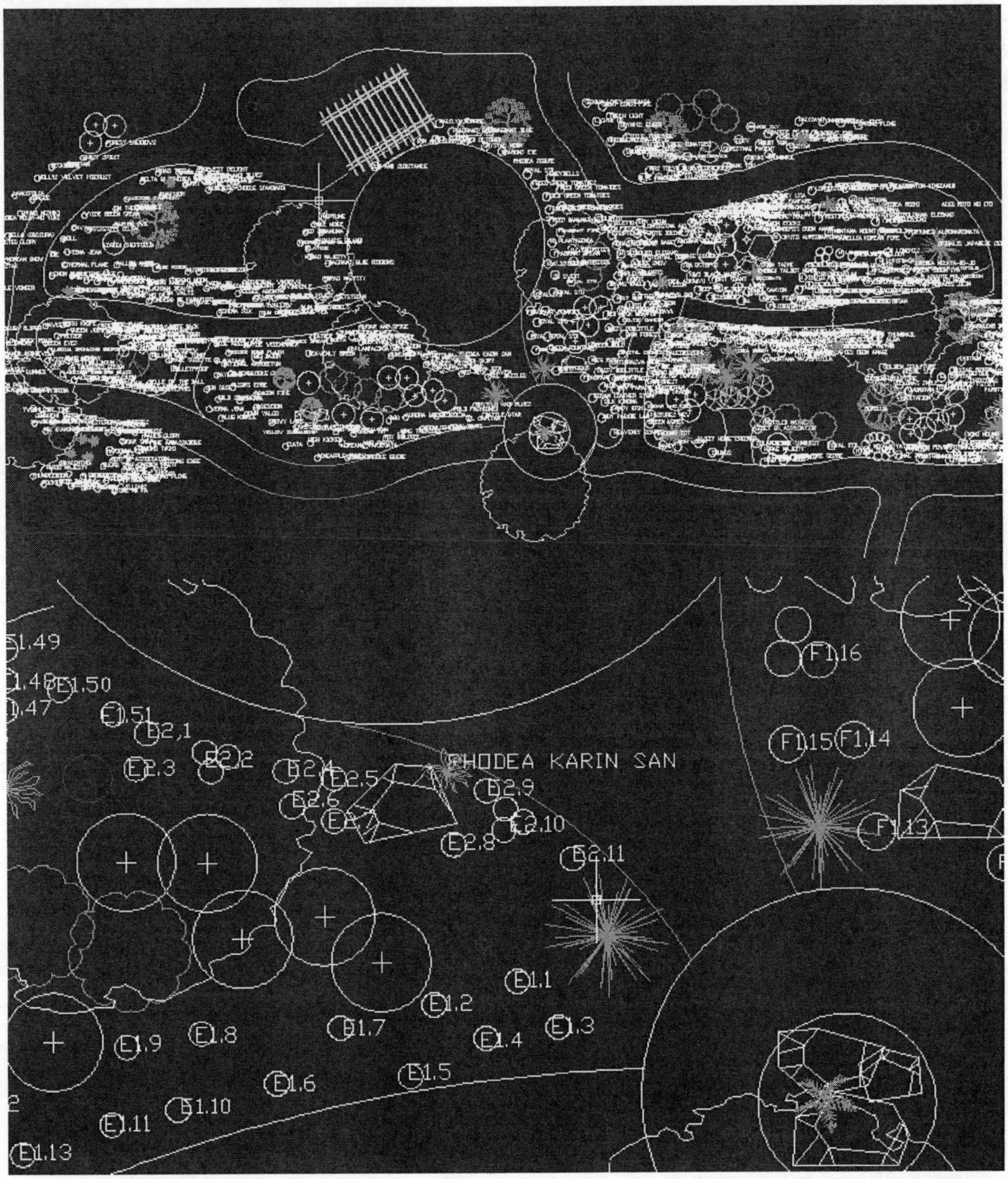

Figure 1. The design of the "*Tranquility – Cornelia B. Holland Hosta Garden*" at the University of Tennessee Gardens, Knoxville, Tennessee.

VARIOUS HOSTA

Designing with a collection of plants can be quite the challenge (Figure 2). The hostas were incorporated into the design by color, size and habit. For example, hostas with red petioles were planted so visitors could see and enjoy the red stems. Fragrant hostas were planted in masse in a large common area of the garden where future events could be held.

Figure 2. The "*Tranquility – Cornelia B. Holland Hosta Garden*" at the University of Tennessee Gardens, Knoxville, Tennessee.

- Fragrant hosta and yellow hosta tend to be capable of tolerating more sun, whereas hosta with strong variegation tend to require dense shade. Hosta that feature blue foliage also require more shade to increase the time that the foliage remains blue. The hosta can sustain the blue hues in more shade but as the summer heat intensifies, the blue color gradually fades away to green.
- Empress Wu hosta - a giant dark green hosta that can reach a width of 1.2 m (4 ft) and a height of 76 cm (30-in.) (Fig. 3).
- Marylin Monroe hosta - a medium size hosta that can reach a width 30.5 cm (12-in.) and a height of 46 cm (18-in.). This plant features wavy leaves with white undersides that tend to grow erect. (Figure 3).
- Fire Island hosta - another medium size hosta that reaches a height of 31 cm (12-in.) and a width 71 cm (28-in.). This hosta features brilliant yellow leaves, which are extended from red petioles (a rare stem color in hosta).

Figure 3. 'Empress Wu' (top) and 'Marylin Monroe' (bottom) hosta.

- Jetstream hosta- a large hosta that grows can reach a height of 66 cm (26-in.) and a width of more than 102 cm (40-in.). This hosta creates a deep blue clump and features leaves that are heart shaped.

- Guacamole hosta- another large hosta that grows to a height of 61 cm (24-in.) and a width of 1.2 m (3.9 ft) (Figure 4). Light green centers bordered by dark green edges are a common characteristic of this hosta. This hosta is also fragrant!

The garden will also include a large selection of the original hosta species, which are the mother plants for today's vast selection of hybrids. Many of these hostas are rare and almost impossible to find. This large portion of the garden will be designated as a species area for housing and preserving the parents of today's hybrid varieties.

Figure 4. 'Guacamole' and 'First Frost' hostas.

A large collection of *Rohdea* (Sacred Lily) are also planted throughout the garden. In Japan, *Rohdea* are known as "o moto," which means large leaf base. *Rohdea* are rarely available in the plant industry. A handful of growers throughout the entire United States preserve these rare plants from Japan. The collection in the University of Tennessee Gardens will help educate others about these plants and help to preserve some of the rare cultivars that were donated to the collection.

RESEARCH AREA

An area to research foliar nematodes will also be designated in the garden. The primary goal of the research in the Knoxville garden will be to determine the *Aphelenchoides fragariae* (strawberry crimp nematode) lifecycle, and their seasonal movement among infected plants. The research will give a more detailed understanding of nematode movement and determining the best time to treat for nematodes. Integrated pest management

(IMP) techniques can then be implemented into management plans to help prevent nematode damage. In order to develop adequate management strategies for this nematode on hostas, cultivar susceptibility and existing chemical and cultural control options can be tested to determine more efficient ways to use and manage hosta in an environment.

In order to study nematodes, infected plant material will be required for examination. A quarantined area designated only for researching nematode infected hosta will be required. It will be best to water hosta with drip irrigation and prevent spread throughout the entire collection.

THE GARDEN'S MAINTENANCE

General Notes. Lightly use mulch. Never pile a large amount over the crown of the plant. Too much mulch will promote and set the stage for Southern stem blight (*Hosta virus* X). Symptoms of the disease include collapsing of hosta leaves. The staff should be aware of and know how to recognize. Test plants and remove if infected.

February. Apply 3-month Osmocote.

April. Spray a 20% ammonia 80% water solution on hosta shoots as they emerge to reduce snail and foliar nematode damage. If slugs are an issue, scatter a light layer of sand throughout the garden to help keep the slugs away.

April, May and early June. Before temperatures reach above 32°C (90°F) - spray 1 or 2 times with Miracle Grow tomato food. Use per instruction on label. This will help increase the vigor of the hosta and provide magnesium.

June. Apply 3-month Osmocote.

July and August. Divide hosta every 3 to 4 years to maintain plant vigor.

October, November and December. Tree leaves should be removed to prevent slugs. Winter cleanup. Do not have to cut the hosta leaves. Let leaves die back and remove them.

CONCLUSION

The garden itself is an exciting opportunity for the University of Tennessee Gardens to further the research of nematodes. Many rare plants will be protected and preserved as well as fulfilling the University of Tennessee Garden's mission to teach people about plants.

HOSTA INFORMATION SOURCES

- Hosta Library
  http://www.hostalibrary.org/
- American Hosta Society
  http://www.americanhostasociety.org/index.htm
- American Hosta Society – *Hosta Virus* X
  http://www.americanhostasociety.org/Education/HostaVirusX.htm
- American Hosta Society – Diseases
  http://www.americanhostasociety.org/Education/Diseases.htm

Combined Proceedings IPPS 68:331-335. 2018.

Not My Father's Nursery

William Hackney
Hackney Nursery Inc., 3690 Juniper Creek Road, Quincy, Florida 32351, USA

williamhackney@hackneynursery.com

Keywords: Sales, automation, robots, pods, labor, propagation, potting, herbicides, shipping.

INTRODUCTION

Hackney Nursery was started by my father, George Hackney, in 1991. The nursery was built around cheap land, cheap freight, and cheap labor. None of these attributes exist anymore. Much has changed since 1991 - not only in the way we sell, ship, and grow our plants - but also, who grows them. These changes have evolved because of regulation, necessity and advancing technology. All this has dramatically revamped the way my brothers, George, Martin, Joseph, and I - manage our roles at the nursery.

SALES

The biggest change to the nursery has been in our sales department. In the early days of the nursery, we would take pictures with a polaroid camera, and snail mail them to customers, along with an availability - and any notes a salesman might want to add. Now days we use our smart phones to email or text a picture to our customers in real time. This constant contact has helped us be more efficient in our sales, without having to take and show samples to customers. Along with technology, the sales team has also had to adapt to a growing plant palette with the introduction of many new varieties and brands of plants that are introduced each year. When my father started the nursery, the focus was to grow large numbers of a few varieties, whereas now we are not only growing a larger number of varieties, but we also have many different sizes to adhere to the varying specs on landscape jobs.

LABOR

When my father built the nursery, labor was cheap. The answer to completing many tasks was to just throw more people at it. This style of operation led to hiring a larger number of seasonal workers who were let go at the end of spring shipping season. Now, with healthcare regulations, a decrease in worker availability, and rising wages - we have adapted to create a year around work force. The key is having people who are cross trained to do many different jobs at the nursery. For instance, a pickup crew leader may also work in pruning or propagation, depending on the job priorities for that particular day.

We have also put a major emphasis on incorporating automation at the nursery. Originally our employees were skeptical about automation taking their jobs, but they have now realized that the automation is there to help them make their jobs easier; it frees them up to allow them to do other tasks. We currently see labor as the single limiting factor on the future growth of our nursery and are continually looking for ways to alleviate this issue.

PROPAGATION

Propagation was done outside when the nursery was started. We have now built 36 high tunnels for propagation (Figure 1). The high tunnels are 30.5 m x 7.6 m (100 ft x 25 ft) with a 1.8 m (6 ft) sidewall. By covering with white plastic, clear plastic, or shade cloth - and having the ability to roll the walls up - we have better control over light, temperature, and water exposure. In the early years we were able to get by with a big outdoor mist system. Now with the many different varieties we propagate in smaller, more controlled areas.

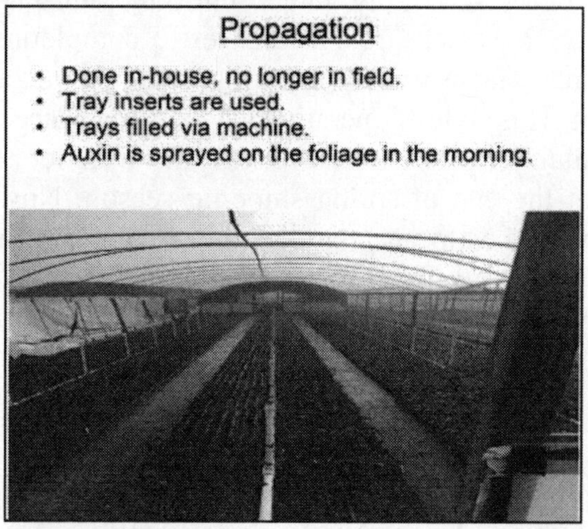

Figure 1. Propagation at Hackney Nursery in protected high tunnels.

All liners are stuck into 7.6 cm (3 in.) disposable inserts - instead of individual reusable cups. We found it easier and quicker filling the trays with inserts. The trays are filled on an Ellis potting machine, stacked on pallets, and wrapped in plastic for storage. When a house is emptied - new trays are set in it and then watered-in for preparation in sticking cuttings. The propagation crew will take cutting in the field and then stick them into the trays on the floor in the house. If the variety of plants the crew is sticking requires auxins, it is applied over the top with a backpack sprayer the morning after being stuck. The auxin application is applied in the morning because the stomates are open in the mornings. It is also safer for the propagation crew, since fewer workers come in contact with applying auxin. We learned this technology through the IPPS.

Once a rooted liner is ready for planting - they are moved to one of the other three locations where they will be potted-up. In the past we have planted multiple 7.7 cm (3 in.) liners into a pot. We are currently running trails with sticking multiple cuttings in a quart pot. We believe that the larger liner will be able to grow a more consistent plant, in a shorter amount of time.

POTTING

When my father started the nursery, he had a full-time potting crew of 15 people. In the past five years we have tried to get away from having one big crew that only pots. Instead, we have invested in 2 EZ Potter potting machines from Ellis Products. These machines allow us to fill propagation trays, 1-,2-,3-,5-, and 7-gal pots. When the machine is accompanied with the drill, we will plant liners to 1-, 2- and 3-gal pots. We also shift 1-gal containers up to 3-gal and 7-gal containers. We prefer to do our potting with one crew of 6 to 7 people - but it is not their full-time job. This crew also works in shipping, consolidation, and bed repair at the nursery.

Once we have potted a plant, we have 3 different configurations in which we set the plants down (Figure 2). The first is 4 solid lines consisting of 3 or 4 pots wide. The reason for the 4 solid lines is we use four Harvest Automation Robots space all of our 2-, 3-, and 5-gal pots. These lines create a constant distance between plants needed to be spaced, and plants the robots have already spaced. For trade gal and full gal pots, we set these down in 4 solid lines of 6 pots. We then us a accordion style spacing device that will pick up one line of 6 pots and slide out to give each pot the correct space it needs.

Figure 2. Potting is facilitated with potting machines and robots.

The second way in which we set down is in what we have termed "pods". Pods are blocks of 3-gal pots that are 9 pots wide and 6 pots deep. This configuration matches up with our Agrinomix Trike Forklifts We can pick up the entire pod and move it to another bed, where there is room to space it. With the pods, we can jam about two and half beds worth of spaced material into one bed. All our 7- and 15-gal material is set down spaced - and generally will not be moved until it is sold. Using this automation, has not only reduced our labor needs from the old way, but we are finding that we are able to grow plants more efficiently. Tasks are now preformed as needed; much more lean flow.

HERBICIDE

Previously, the nursery never used herbicides. Instead, employees had assigned areas and they were awarded bonuses for finishing their areas in a week. However, increased labor cost has caused us to introduce herbicide into the rotation (Figure 3). We continue to use the bonus system and have two weeder teams of whose goal is 10 acres a week. We are currently on a 5-week rotation, where each plant on the nursery will be weeded and herbicide applied every fifth week. We are using mostly liquid herbicides at a lower rate because of the interval. There are instances where we are forced to use granular herbicides. On more sensitive species, like hydrangeas, we have begun using sawdust or rice hulls as a weed barrier.

Figure 3. Weed control with liquid herbicide application.

FERTILIZER

Originally the nursery used liquid fertilizer but shifted to all granular to comply with Florida Best Management Practices for nurseries. We have now created a program, where we are using both Liquid and Granular to offer the plant the correct nutrients during different times of the year. All new plantings are top-dressed with a 6-month fertilizer, and then we will reapply again in 5 months cycles as the plant grows. These applications are performed by two men who use the provided fertilizer spoons, as well as automated Fertileeze fertilizer dispensers. We are also in talks with Harvest Automation to work on designing a fertilizer attachment for their robot that would fertilize the plant when it was spaced. Liquid fertilization is used when we are in optimal growing weather, and the plant needs more nutrients than the granular is providing. We also use the liquid fertilizer to "hold" a plant once it has reached the saleable stage but has not yet been sold.

PESTICIDE APPLICATION

In the beginning, my father's nursery used a converted tomato sprayer to apply pesticides and fungicides. We recently added some new technology as well as some out-of-the box thinking to our spray regimen. Two years ago, we purchased an AirTek air blast sprayer with an ionization system on the boom (Figure 4). We have seen a much better coverage rate, and better pest control by using this technology. Last year, we built a drenching machine, consisting of a 3785 L (1000 gal) tank and a 5-hp pressure pump. The main objective was to use this when hand drenching crops. However, we discovered that with the pressure pump we could design a valve system where we can tie into individual shade and propagation hoses - and subsequently spray a fungicide through the existing sprinkler heads. This drastically reduces the amount of time it takes to do the job.

Figure 4. Pest management using Air-Tek Sprayer for liquid application.

SHIPPING AND FREIGHT

The way we ship our finished product and how much it costs to ship have also greatly changed since the nursery opened. It used to be that we would pick up with wagons and stack all our material with conveyer belts. Now we use racks and pick-up containers on pallets (Figure 5).

Figure 5. A pulled order for shipping using pallets to more efficiently load and unload.

We made this change for two reasons: firstly, it is less damaging to the product and secondly, it decreased loading and unloading time for drivers using electronic logbooks https://eldfacts.com/eld-facts/. We try to have loads racked and staged on the dock before a driver arrives - especially on loads carried by outside drivers.

This allows us to use as little of their drive time as possible. The quicker unloading time with the racks also allows drivers to get two or three loads a week instead of one or two. Also, with the electronic loads, we cannot guarantee your order will be there in one or two days of shipping. The driver could be 40 miles away at lunch but must shut down for the day. This means the customer will not get their product until they open the next day. This has caused us to only ship local trucks after Wednesday to guarantee that our customers will not have to take a Saturday delivery.

The down time caused by the electronic logbooks has caused our freight rates to increase. The local owner operator trucks we use have increased their rates by twenty percent a mile and a 20 percent increase on flat rates in loads to Atlanta and Charlotte.

SUMMARY

The nursery industry has drastically changed since our father opened Hackney Nursery. We, as a family, have had to adapt to increased production cost and regulations. We have adapted with a more skilled trained workforce, more automation and new production practices. With these changes we hope to keep the nursery running well into the future.

Combined Proceedings IPPS 68:336-340. 2018.

Mobile Technology for the Nursery Industry

James Harden
Mortellaro's Nursery, Ltd. 16946 IH 35 North, Schertz, Texas 78154, USA

jamesharden@mortellaro.com

Keywords: Mac/Apple, IPads, IPhones, IOS applications, Evernote, Numbers, Pages, and IOS Calendar, Snappii IOS Platform, efficiency, utility.

INTRODUCTION

Mortellaro's Nursery Ltd. is a wholesale grower with two locations of approximately 40.5 ha (100 acres) total which are 97 km (60 mi) apart. The primary location is 28 ha (70 ac) production and sales area and the second location of approximately 18 ha (45 acres) is strictly a growing area. We employ between 80-120 employees between these two locations. This division of our operations required changes in management methods to ensure both locations were consistent in all operations.

Mortellaro's Nursery has fully embraced the use of mobile technology to allow communication and co-ordination between managers, employees, and locations. Our system has evolved over five years. In adapting new technology, one should not expect instant results. It is important to allow your employees to help implement new uses – and be open to suggestions and advice. Mortellaro's Nursery is currently using 20 IPads and 10 IPhones along with three Bluetooth enabled printers (Figure 1). The IPads are used by management – and designated supervisors in production, propagation, pest control, shop mechanic, HR/Safety, Shipping, and drivers. We use a combination of native IOS applications, purchased off the shelf applications, and custom written applications on a Snappii platform. A native mobile app is a smart-phone application that is coded in a specific programming language, such as Objective C for iOS or Java for Android operating systems. Native mobile apps provide fast performance and a high degree of reliability.

Mobile Technology Equipment Used at Mortellaro's Nursery

- 20 IPads, IPad 2 to IPad pro, IPad mini 1 and 2
- 10 IPhones, IPhone 6 to current
- 3 Bluetooth and Wi-Fi enabled HP printers- We have printers in our field office, shop, and Poteet operation. Two of the locations do not have any type of internet connection.
- We have eliminated 2 fax lines and fax machines with the use of IPads and Wi-Fi enabled printers.
- Laptops vs desktop. A laptop can be used anywhere needed rather than only in a networked building. A laptop can be carried anywhere the same as an IPad.

Figure 1. Mobile technology equipment used at Mortellaro's Nursery.

With the multiple programs and apps that we use and the methods we have developed over time, we are able to effortlessly cover for a missing manager or supervisor (Figure 2). Every manager and supervisor's daily jobs, scheduled crew duties, and other info is accessible by any other manager or supervisor. We use Evernote, Numbers, Pages, and IOS Calendar to organize duties, information, and schedules. These four unmodified programs/apps are used extensively in day to day operations. We use custom programs on a Snappii IOS Platform for many specialized inspections and reports.

Software Used at Mortellaro's Nursery

- Snappii-Custom written Applications
- Evernote
- Dropbox
- Microsoft Office
- Find my Friends
- Google Maps
- Accuweather, My Radar, Radar Pro, WunderStation, Live Lightening, Lightening Alarm.
- JJ Keller ELD software for Drivers

Figure 2. Software used at Montellaro's Nursery.

SNAPPII

Snappii is the most developed program/application that we use. Snappii is a program that works off a remote computer and you either use premade programs available from Snappii or write your own programs to run on their server. We use the IPad's in a way similar to a terminal logging into their system. This allows any mobile device to be used, not just the newer IPads or IPhones. Snappii offers hundreds of readymade template applications for all industries. They also offer the option of using their software to create your own custom application.

I chose to create our own applications after looking at the provided applications for ideas. My programs are written to be very lean and able to be completed in a very short time using drop down lists, checkboxes, and minimal data entry. (Figure 3). I went for efficiency and utility over creative displays. Two of the programs I wrote take less than a minute to input the data and send it to the relevant employees.

At this point I have created the following applications that we use on a daily basis:

- Scouting Report-Quick Reporting of weed, insect, or disease issues. Reports are sent to managers, order assembly, sales and pest control automatically.

- Verbal Warning "Paper Trail" and notification to management of any employee issues. Reports are automatically sent to management and HR.

- Repairs Needed Tickets for all equipment. Reports are automatically sent to management and mechanics.

- DOT Truck Inspection-monthly inspection of highway vehicles. Reports sent to shipping manager.

- Air Compressor Inspection-Monthly Inspection. Reports sent to HR/Safety.

- Fire Extinguisher and Fire Exit Inspection-Monthly Inspection. Reports sent to HR/Safety.

- Off Road Equipment Inspection - Inspections and maintenance checklist. Reports sent to HR/Safety.

- Eye Wash Station, Eye Wash Bottle, and Safety Shower Inspection. Reports sent to HR/Safety.

Figure 3. Snappii program examples.

Each of these modules of the application were written so we could quickly document and inform the relevant people of problems/ issues. The Scouting Report, Verbal Warning Report, and Repairs Needed Report, etc. takes less than a minute to fill out. When the information has been entered, the save/send button sends the information to preselected managers or employees. This allows quick distribution of information from anywhere in the field to multiple people. The user does not need to be concerned with recipients, only entering the relevant data.

EVERNOTE AND DROPBOX

Evernote is an application that allows you to save documents, lists, pictures, pdf's or any other type of information. It can be considered as a notebook with an infinite number of tabs. We use Evernote in a multitude of different ways. We create job lists, equipment info and pdf manuals, production records, production schedules, customer delivery info, employee info, etc.

Dropbox is used as one of two backup methods for all data on our IPads. ICloud is the other backup option we use. All info is backed up onto ICloud when the IPads are charged daily.

WEATHER APPS

Weather is a very important element in our industry. It can affect all aspects of growing and also sales. We not only use forecasting applications, but also radar applications for reporting wind and lightening.

- Accuweather. We use Accuweather for forecasting. It has been shown to be more accurate for colder temperatures than for rain or heat.
- MyRadar and Radar Pro. MyRadar and Radar Pro are used for close-in prediction of rain or storms affecting our locations and also for delivery scheduling during winter storms. My preference is MyRadar for a larger picture or severe weather. But Radar Pro often shows lighter intensity rain that may affect chemical applications.

- Lightning and Lightning Alarm. Both programs are used for lightning strike notification. Lightning will send a text warning to you after you have set distance parameters in which to be notified. It also works for multiple locations to have their own distance parameters. Lightening Alarm includes a circular radius around your central area so that you can see lightning strikes within these areas. Lightening Alarm has to be reset for each location you want to view.
- Weather Underground and Weatherlink. Both of these programs are used to monitor on site weather stations. We use Weather Underground to monitor our ambient weather station at the Schertz location. Weatherlink is used for our Davis weather station at our Poteet facility. We are using two different stations due to internet connection at one location and to test the longevity and quality of the two different brands. We are able see live data of both locations and also keep weather history of both locations. We post current weather conditions on our website, and also allow employees to access the same data.

SHIPPING AND ELOGS

For shipping we use Google Maps, Apple's Find My Friends, Evernote, and JJ Keller Elog software.

Google Maps enables:

- Calculating mileage for freight
- Scheduling deliveries or reroute due to traffic
- Staying on top of developing traffic issues

Find My Friends (Figure 4) enables:

- Use of Native Apps on Apple. Native apps can provide optimized performance and take advantage of the latest technology, such as a GPS, compared to web apps or mobile cloud apps developed to be generic across multiple systems.
- Locates every IPad or IPhone with live tracking.

JJ Keller Elog Software enables:

- Each driver to use the same IPad for Elog as for other uses
- Any truck can be added to the fleet system with a snap in module on the vehicle data port.
- Shipping Manager to see at a glance all remaining hours for each driver for the week. He/she can see daily log on any driver.

Figure 4. Tracking employees to know their locations and availability – helps facilitate the management of two locations, 97 km (60 mi) apart.

CALENDARS

We use Calendars for many different departments. We have custom calendars set up for the following uses:

- Vacation and Holiday work crew scheduling
- Scheduled personal and work injury doctor visits
- Hiring and training of new employees
- Deliveries
- Inventories of supplies
- Safety Inspections of Eye Wash, Extinguishers, Ladders, etc.
- Pest Control Preventive Treatments
- Management and Supervisor scheduling for weekends
- Irrigation Crew schedules
- Equipment PM and Inspections

NUMBERS

- List of customer information
- Production lists to Inventory
- Production to Herbicide Applicator
- Production Quota
- Lockout Tagout lists
- Winter Protection Checklist and heater list

CONCLUSION

- In adapting mobile technology for your nursery, only you can decide how fast to implement. We suggest to walk before you run!
- Start with the right employees, rather than the problem areas. Initially use the path of least resistance. Let the idea and benefits sell themselves rather than forcing it on employees.
- Choose your priorities on mobile technology equipment-price, support, reliability and/or security.
- Your software choices will change as you progress, so do not be inflexible to changing software for support or new features.
- Trial programs similar to what you need first then write your own programs. Create custom software in house as needed - many programs are simple to modify.

Keep control of data and backup critical data continually. Restrict employee's abilities to erase or damage data.

Combined Proceedings IPPS 68:341-348. 2018.

Pre-Plant Nitrogen Rates in Alternative Substrates Affect Production of *Impatiens xwalleriana*

Brianna J. Heilsnis, Glenn B. Fain, J. Raymond Kessler, and Daniel E. Wells
Department of Horticulture, Auburn University, Auburn, Alabama 36849 USA

gfain@auburn.edu

Keywords: Wood fiber substrate, greenhouse crops, loblolly pine, *Pinus taeda*.

Abstract

The effect of nitrogen on an alternative forest-based product substrate (FPS) performance was evaluated using rates of Nitroform®, a slow-release urea-based pre-plant fertilizer, and a water-soluble N-P-K starter fertilizer. FPS is manufactured from loblolly pine (*Pinus taeda*) harvested locally within the Southeastern United States. *Impatiens ×walleriana* were grown in 80:20 peat:perlite (by volume) industry standard, 80:20 FPS:peat, or 100% FPS. Nitroform® was incorporated at 0 or 0.59 $kg \cdot m^{-3}$ N, and ammonium nitrate was incorporated at 0.06, 0.12, or 0.18 $kg \cdot m^{-3}$ N. Nitroform® increased plant size regardless of substrate. Plants grown in 80:20 peat:perlite had higher size indices (SI) than plants in either the 80:20 FPS:peat or 100% FPS regardless of Nitroform® rate. In both FPS substrates, size index increased with increasing N rate, while in the 80:20 peat:perlite substrate, size index decreased with increasing N rate. The results indicate that substrates containing high FPS (up to 100%) have potential in greenhouse substrates with the addition of adequate types and amounts of pre-plant incorporated nitrogen.

INTRODUCTION

Peat-based substrates have been the greenhouse industry standard since their introduction in the 1970's as Cornell peat-lite mixes (Boodley and Sheldrake, 1972; Jackson et al., 2008). Peat moss imports increased from $157 to $312 million from 2000 to 2015 and costs increased from $200 to $271 per ton, respectively (Jasinski 2000; Apodaca, 2015). Due to increased demand for greenhouse media and rising costs of peat (Jasinski 2000), growers are looking for ways to lower their overhead. Approaches include augmenting peat with potentially less costly alternatives such as wood fiber. There was significant research in the last decade on alternative wood fiber substrate components (Fain et al., 2008; Domeno et al., 2010; Gaches et al, 2011, Jackson et al., 2008).

HydraFiber® (Profile® Products LLC, Buffalo, IL) is a wood fiber substrate component currently available in the U.S. Growers and professional blenders are using HydraFiber® at rates of 20– 50% (by volume). Benefits of a manufactured product include product consistency and uniformity, which are crucial for commercial production. Additionally, when substrates are manufactured using resources located near growers, transport costs are reduced, resulting in a more cost-competitive product. Given the abundance of pine trees (*Pinus* spp.) in the Southeastern U.S., wood fiber substrates are a strong candidate for peat alternatives in greenhouse production. Optimum nutrition for FBS has not been determined and is explored in this study.

In this study, we evaluated a forest product-based substrate (FPS) made from loblolly pine (*Pinus taeda*), alone and in combination with peat, and starter fertilizer rates in the production of a greenhouse annual. The objective of this study was to evaluate the performance of FPS with various nitrogen treatments and peat amendments.

MATERIALS AND METHODS

The study was conducted at the Paterson Greenhouse Facility, Auburn University, Alabama. Substrates were blended on June 13, 2018. The treatment design was a 3-way factorial with 3 substrates, 2 Nitroform® rates, and 3 starter nitrogen fertilizer rates. Each of the 18 treatments had 12 single-container experimental units having one seedling from a 200-count plug tray. Treatments were arranged in a randomized complete block design. The study was blocked for greenhouse temperature variation, and each block consisted of one experimental unit per treatment in a single row. Substrate treatments in the study were: 80:20 peat:perlite (by volume) the industry standard, 80:20 FPS:peat, or 100% FPS. Raw peat moss was used in the peat-lite blend and in the peat fraction added to the wood fiber substrate. Nitroform® (39-0-0 powdered slow release, Koch Agronomic Services, LLC., Wichita, KS) treatments were at 0 $kg{\cdot}m^{-3}$ or 0.59 $kg{\cdot}m^{-3}$. Preplant starter fertilizer treatments were at 0.06-0.03-0.06, 0.12-0.03-0.06, or 0.18-0.03-0.06 $kg{\cdot}m^{-3}$ N-P-K, respectively. Nitrogen immobilization was expected in wood-based substrates and therefore, required higher fertilizer rates to compensate (Witcher et al., 2009).

Water soluble ammonium nitrate, potassium sulphate, and potassium phosphate were blended into the ratios listed above. Preplant starter fertilizers were dissolved in 2 L (67.6 oz) of water and applied as a spray to each substrate at mixing. All substrates had the following amendments added at mixing: dolomitic limestone (2.97 $kg{\cdot}m^{-3}$ for peat-lite, 0.59 $kg{\cdot}m^{-3}$ for 80:20 FPS:peat, or 0.30 $kg{\cdot}m^{-3}$ for 100% FPS), 0.59 $kg{\cdot}m^{-3}$ gypsum, 0.30 $kg{\cdot}m^{-3}$ Micromax®(ICL Fertilizers, Dublin, OH), and 3.6 $kg{\cdot}m^{-3}$ Conductor® substrate surfactant (Aquatrols®, Paulsboro, NJ). Limestone varied by substrate based on previous literature (Boyer et al., 2007; Fain et al., 2008; Jackson et al., 2009)

Impatiens ×*walleriana* 'Xtreme White' were acquired from Young's Plant Farm, Inc. (Auburn, AL) on June 25, 2018, transplanted into containers, and watered on June 29, 2018. Containers [Shuttle Container® SS325, 473 cm^3, East Jordan Plastics Inc., East Jordan, Michigan] were filled based on a pre-determined target weight per container calculated from the density of each substrate. Containers were placed in flats and covered with plastic to prevent evaporation until all containers were filled. Visually uniform plugs were chosen randomly from the flats. Holes in the center of each container were dibbled before plugs were transplanted by block. The finished containers were lightly watered and then fertilizer with 150 ppm N 20-10-20 (Greencare, Kanakee, IL). Plants were produced during the experiment using continuous fertilization at 150 ppm N 20-10-20 on greenhouse benches in full sun and irrigated with clear water as needed to address rising EC.

Initial pH and EC were taken using the press method (Scoggins et al., 2001) and fallow containers were brought to saturation using clear water. Mid-study 14 days after planting (DAP) and at termination (28 DAP), pH and EC were collected using the press method and brought to saturation with 150 ppm N 20-10-20. Prior to both press method extractions, shoot tissue was cut at the soil line, bagged per experimental unit, and placed in a forced air-drying oven at 76°C until dry and weighed. For mid-study data collection, four blocks were picked at random from the 12 for destructive harvest. At termination, the remaining eight blocks were harvested for shoot tissue and four of those eight blocks were randomly picked for destructive analysis of pH and EC. The four blocks remaining were used to analyze substrate shrinkage and final water holding capacity (WHC). To determine substrate shrinkage, the void space was measured using 100% fine grade Profile® porous ceramic (Profile® Products LLC, Buffalo, IL), which has a density of 0.6245 $g \cdot cm^{-3}$. The containers were first brought to saturation and weighed (equation, value A). Then a 30 × 30 cm piece of clear 12.7µ thick plastic wrap was loosely centered over the pot, and lightly pressed into the void until the plastic contacted the substrate. Ceramic was added on top of the plastic to fill the void to the top of the container, the container was then reweighed (equation, value B). The ceramic was removed after weighing by grabbing all four corners of the plastic wrap and lifting it out. The saturated containers were placed in a forced air-drying oven at 76°C until dry to determine WHC (Fonteno and Harden, 2003). Shrinkage was calculated using the following formula:

$$\frac{B-A}{0.6245} \times \frac{1}{container\ volume} = \%\ substrate\ shrinkage$$

Throughout the experiment, the date of first bloom was recorded and the days to bloom was then calculated from the date of planting. At termination, a final bloom count was recorded. Size index [SI=(height + width + perpendicular width)/3] was recorded at study termination on all blocks prior to shoots being harvested.

RESULTS

There were interactive treatment effects on size index (Table 1). In the substrate by Nitroform® interaction, plants grown in 80:20 peat:perlite had higher SI's than plants in either the 80:20 FPS:peat or 100% substrates regardless of Nitroform® treatments . In both substrate by Nitroform® and starter N by Nitroform® interactions, plants grown with Nitroform® had higher SI's than those without Nitroform®. SI increased linearly with increasing starter N rate without Nitroform® indicating that starter N rate had a larger effect on plants without Nitroform® than with; however, plants with Nitroform® and the lowest starter N rate were larger than plants without Nitroform® at the highest starter N rate. The addition of Nitroform® to substrates increased SI, regardless of substrates. In the substrate by starter N interaction, both FPS substrates had linear increases in SI with increasing starter N rate, while the 80:20 peat:perlite decreased linearly.

Mid-study shoot dry weight (SDW) showed an interaction among Nitroform®

Table 1. Effects of substrate and nitrogen on size index of *Impatiens ×walleriana*.[zy]

Nitroform N kg·m^{-3}	Substrate			
	80:20 peat:perlite	80:20 FPS[x]:peat	100% FPS	
0	18.6bA[wv]	14.3bB	15.0bB	
0.59	19.5aA	16.8aB	15.8aC	
Starter N kg·m^{-3}	80:20 peat:perlite	80:20 FPS:peat	100% FPS	
0.06	19.3A	15.2B	14.4B	
0.12	19.1A	15.4B	15.5B	
0.18	18.7A	16.1B	16.3B	
sign.[u]	NS	L*	L***	
	Starter N kg·m^{-3}			
Nitroform N kg·m^{-3}	0.06	0.12	0.18	sign.
0	15.3b	15.7b	16.9ns	L***
0.59	17.3a	17.6a	17.2	NS

[z]The substrate by Nitroform®, substrate by fertilizer, and fertilizer by Nitroform® interactions were significant at $P < 0.05$.
[y]Size Index in cm [(height + width + perpendicular width)/3].
[x]Forest-product substrate FPS)
[w]Least squares means comparisons between Nitroform rates (lower case letters in columns) using F-tests at $P < 0.05$.
ns = not significant.
[v]Least squares means comparisons among substrates (upper case letters in rows) using the simulated method at $P < 0.05$.
[u]Not significant (NS) or significant (Sign.) linear (L) trends using qualitative-quantitative regression models at $P < 0.05$ (*) or 0.001 (***).

N except for the lowest rate of 0.06 N kg·m^{-3} with Nitrogen®, in which 100% FPS plants had significantly higher SDW than 80:20 FPS:peat. Between Nitroform® rates within substrates, 80:20 peat:perlite plants had significantly higher SDW's only at the highest starter N rate containing Nitroform®. Plants at the middle and lowest starter N rates were non-significant between Nitroform® rates. Within 80:20 FPS:peat and 100% FPS, both had significantly higher SDW's in the lowest starter N rate containing Nitroform®.

For final SDW (Table 3), the substrate by starter N rate and Nitroform® by starter N rate interactions were significant. Final SDW increased linearly with increasing starter N rate for both FPS substrates while 80:20 peat:perlite showed no trend. Within each of the starter N rates, 80:20 peat:perlite plants had higher SDW's than either FPS substrates. In the Nitroform® by starter N rate interaction, SDW increased linearly over starter N rate without Nitroform® but there was no trend with Nitroform®. Between Nitroform® rates, plants had a higher SDW at the two lowest starter N rates containing Nitroform® with no trend for the highest starter N rate.

Plants in the peat-lite substrate had the highest bloom count at termination followed by plants in the 80:20 FPS:peat and 100% FPS (data not shown). Bloom count increased linearly with increasing starter N rate across all substrates without Nitroform®

Table 2. Effects of substrate and nitrogen on shoot dry weight at 14 days after planting of *Impatiens* ×*walleriana*.[zy]

	Nitroform N kg·m^{-3}					
	0			0.59		
Starter N kg·m^{-3}	80:20 peat:perlite	80:20 FPS[x]:peat	100% FPS	80:20 peat:perlite	80:20 FPS:peat	100% FPS
0.06	0.543aNS[wv]	0.173bB	0.183bB	0.513a	0.253cA	0.358bA
0.12	0.533aNS	0.175bNS	0.238bNS	0.478a	0.245b	0.295b
0.18	0.418aB	0.275bNS	0.260bNS	0.545aA	0.338b	0.298b
sign.[u]	L**	L*	NS	NS	L*	NS

[z]The substrate by Nitroform by fertilizer interaction was significant at $P < 0.05$.
[y]Plant shoot dry weight in grams.
[x]Forest-product substrate (FPS
[w]Least squares means comparisons among substrates within Nitroform rates (lower case letters in rows) using the simulated method at $P < 0.05$.
[v]Least squares means comparisons between Nitroform rates within substrates (upper case letters in rows) using F-tests at $P < 0.05$. NS = not significant.
[u]Not significant (NS) or significant (Sign.) linear (L) trends using qualitative-quantitative regression models at $P < 0.05$ (*) or 0.01 (**).

Table 3. Effects of substrate and nitrogen on shoot dry weight 28 days after planting of *Impatiens* ×*walleriana*.[zy]

	Substrate				NitroForm N kg·m^{-3}	
Starter N kg·m^{-3}	80:20 peat:perlite	80:20 FPS[x]:Peat	100% FPS	Starter N kg·m^{-3}	0	0.59
0.06	3.91aNS[w]	1.78b	1.85b	0.06	2.04b[v]	2.99a
0.12	4.10a	1.79b	2.03b	0.12	2.28b	3.00a
0.18	3.79a	2.25b	2.34b	0.18	2.83ns	2.76
sign.[u]	NS	L**	L***		L***	NS

[z]The substrate by fertilizer and Nitroform by fertilizer interactions were significant at $P < 0.05$.
[y]Plant shoot dry weight in grams.
[x]Forest-product substrate (FPS)
[w]Least squares means comparisons among substrates (lower case letters in rows) using the simulated method at $P < 0.05$.
[v]Least squares means comparisons between Nitroform rates (lower case letters in rows) using F-tests at $P < 0.05$. ns = not significant.
[u]Not significant (NS) or significant (Sign.) linear (L) trends using qualitative-quantitative regression models at $P < 0.01$ (**) or 0.001 (***).

but decreased linearly with Nitroform®. Substrate pH was within an acceptable range (4.8–6.4) at 0, 14, and 28 DAP across all treatments (data not shown) and the substrate × Nitroform® × starter N rate interaction was significant at each test date. An increased EC was observed in substrates containing Nitroform®. The 80:20 peat:perlite substrate had much higher EC values (2.3-4.8 $mS \cdot cm^{-1}$) than either FPS substrate (0.9-1.9 $mS \cdot cm^{-1}$). EC stayed high in 80:20 peat:perlite (2.5-.8 $mS \cdot cm^{-1}$) from 0 – 14 DAP. At termination, 80:20 peat:perlite EC had dropped to 1.1 $mS \cdot cm^{-1}$. EC in both FPS substrates remained consistent between 0.9 and 2.5 $mS \cdot cm^{-1}$ throughout the study.

DISCUSSION

Results indicate that substrates high in FPS (up to 100%) have potential as greenhouse substrates with the addition of adequate forms and amounts of pre-plant nitrogen. Previous research documented nitrogen loss issues in substrates containing higher amounts of wood fiber. Nitrogen immobilization and microbial respiration were recorded by Boyer et al. (2012) in which a high wood fiber content substrate, clean chip residual (CCR), was incubated in sealed glass and carbon mineralization measured to determine the amount of microbial respiration. Boyer concluded that respiration increased with increasing nitrogen; the more finely processed the wood substrate, the more microbial respiration occurred; and CCR had the lowest available nitrogen. Nitrogen immobilization was documented in pine tree substrate (PTS) in earlier studies by Jackson and Wright (2007). Nitrogen drawdown index (NDI) and substrate CO_2 efflux were recorded to determine the extent of N-immobilization. The authors determined that PTS CO_2 efflux rate was five times as high as peat and twice as high as pine bark. Additionally, NDI results showed that 68% of PTS's available substrate N was immobilized compared to 13% in peat. Therefore, future studies should evaluate N-immobilization in FPS using one or more of these procedures. Further studies should also evaluate higher rates of nitrogen, as well as varying formulations and delivery methods to overcome the effects of N-immobilization on crop production.

Literature Cited

Apodaca, L.E. (2017). Peat, p. 54.1–54.8. In: 2015 minerals yearbook. U.S. Geological Survey.

Boodley, J.W. and Sheldrake, R. Jr. (1972). Cornell peat-lite mixes for commercial plant growing. New York Agr. Expt. Sta. Res. Bul. 43.

Boyer, C.R., Gilliam, C., Fain, G., Sibley, J., Torbert, H., and Gallagher, T. (2007). Lime and micronutrient use in clean chip residual substrate amended with composted poultry litter or peat for use in annual production. Proc. Southern Nursery Assoc. Res. Conf. *52*:77–82.

Boyer, C.R., Torbert, H.A., Gilliam, C.H., Fain, G.B., Gallagher, T.V., Sibley, J.L. (2012). Nitrogen immobilization in plant growth substrates: clean chip residual, pine bark, and peatmoss. Intl. J. Agron. *2012*:1–8.

Domeno, I., Irigoyen, I., and Muro, J. (2010). New wood fiber substrates characterization and evaluation on hydroponic tomato culture. European J. Hort. Sci. *75*:89–94.

Fain, G.B., Gilliam, C.H., Sibley, J.L., and Boyer, C.R. (2008). WholeTree substrate and fertilizer rate in production of greenhouse grown petunia (*Petunia* ×*hybrida* Vilm.) and marigold (*Tagetes patula* L.). HortScience *43*:700–705.

Fonteno, W.C. and Harden, C.T. (2003). Procedures for determining physical properties of horticultural substrates using the NCSU porometer. Hort. Substrates Lab., Raleigh, N.C.

Gaches, W.G., Fain, G.B., Eakes, D.J., Gilliam, C.H., and Sibley, J.L. (2011). Comparison of aged and fresh WholeTree as a substrate component for production of greenhouse-grown annuals. J. Environ. Hort. *29*:39–44.

Jackson, B.E. and Wright, R.D. (2007). Pine tree substrate: fertility requirements for nursery and greenhouse crops. Proc. Southern Nursery Assoc. Res. Conf. 52:523–526.

Jackson, B.E., Wright, R.D., and Barnes, M.C. (2008). Pine tree substrate, nitrogen rate, particle size, and peat amendment affect poinsettia growth and substrate physical properties. HortScience *43*:2155–2161.

Jackson, B.E., Wright, R.D., and Gruda, N. (2009). Container medium pH in a pine tree substrate amended with peat moss and dolomitic limestone affects plant growth. HortScience *44*:1983–1987.

Jasinski, S.M. 2000. Peat, p. 56.1–56.2. In: 2000 minerals yearbook. U.S. Geological Survey.

Scoggins, H.L., Bailey, D.A., and Nelson, P.V. (2001). The press-for plug testing success. Southeastern Floriculture. July/August, p. 24–25.

Witcher, A.L., Fain, G.B., Blythe, E.K., and Spiers, J.M. (2009). The effect of nitrogen form on pH and petunia growth in a WholeTree substrate. Proc. Southern Nursery Assoc. Res. Conf. *54*:428–433.

Combined Proceedings IPPS 68:349-355. 2018.

Substrates Trends for Propagation, Production and Profit

Brian E. Jackson and Paul C. Bartley
North Carolina State University, Department of Horticultural Science, 130 Kilgore Hall, Raleigh, North Carolina, 27695-7609, USA

brian_jackson@ncsu.edu

Keywords: Wood fiber, peat moss, fiber properties, coir; physical, chemical, hydrological, biological properties of wood; *Cannabis*, CAT scan, Computer Assisted Tomography, tomography.

THE REVOLUTION OF WOOD FIBER

Research on growing media (substrates) has been an important facet in the evolution of containerized horticultural crop production for over 50 years. During this time, we have relied on peat moss as the primary substrate component to grow most greenhouse crops. Peat moss is undoubtedly an ideal material based on its excellent physical and chemical properties. Research on substrates continues today as vigorously (maybe more) than ever, despite the successes and familiarity of our traditional peat-based mixes. Much has been reported in recent years about the development and potential of wood-based substrate components in the floriculture, nursery, and edible production industries here in the United States. These trends are equally, maybe more so emphasized, in many European countries and markets. Different regions of the world face different challenges related to horticultural system advancements, labor force issues, evolving consumer preferences and demands, as well as economic concerns and governmental policies and regulations.

Embedded in the discussions and product development in many European and North American companies is the continued interest in wood materials as components in indoor and outdoor crop production. As has been previously reported, there are now even more wood products that are being produced and used successfully (Figure 1). While visually different, these commercial materials have been and seemingly are, being used successfully. The scale to which many of these wood materials are being made has grown rather large over the years to the point that today, the color of peat and bark storage yards and production facilities is becoming more and more "blonde" in color (Figure 2)! The large-scale production of these materials is evidence of the increase in sales and demand for these products.

Figure 1. Wood substrate materials from European and North American manufacturers show variations in particle size and structure.

Figure 2. Success of many commercialized wood substrates has led to large-volume, mass production of materials for use in professional and retail products.

The majority of the wood products being commercialized are primarily made by one of three processes: 1) single or twin-screw extrusion; 2) twin disc refiners; or 3) hammer mills. The first two processes, used extensively throughout Europe, are thermo-mechanical techniques which involve high temperatures and friction to make the products. These technologies also exist here in the U.S. Wood products made with hammer mills are mostly confined to companies and grower operations here in the U.S., even though hammer mills are used for many purposes in the substrate industry throughout Europe for other material processing purposes. The differences among the different wood materials from these three processes include fiber size and thickness, sterility/chemical properties of the end-product, type of wood feedstock used, and varying abilities to be compressed, handled, and blended with other materials. Based on the resulting fiber properties and structure of wood produced from the different techniques, some may be better suited as loose-filled materials (blended with peat, bark, coir, etc.) while others are more capable of being compressed in small or large bales (Figure 3).

Compressed bales may offer unique advantages relative to storage and transport of these materials and are being used for both professional and retail/consumer soil and substrate products. Compared to "loose" fiber materials (not compressed) there is the added step of bale busting/loosening that must occur prior to substrate blending and use.

Figure 3. Some wood fiber materials can be easily compressed in different sized bales to aid in storage and transport.

There has been a tremendous amount of research conducted on wood substrates and substrate components over the past decade. The data and observations generated from those trials has been the foundation to all that we know today about the uses and potential of these substrate materials. While previous data is valid and has answered many questions (while generating many more), the learning curve for how to properly research, evaluate, and characterize these materials has been steep. As we understand more and more about the physical, chemical, hydrological, and biological properties of wood materials we have to continually evolve how we conduct our substrate research.

Based on all that we have learned in the past decade, when conducting trials today we must consider many variables (many of which are potentially confounding) about wood substrate materials before we conduct specific trials to learn more about a specific question. For example, we cannot design an experiment to understand fertility needs/issues (i.e. nitrogen immobilization) without considering and accounting for pH differences, liming adjustments, porosity variations, water/irrigation management, potential toxicities, etc. of the wood materials being tested compared to whatever control is being used. When these variables are known, considered and minimized as much as possible we get closer to comparing "apples to apples" as opposed to "apples to oranges" when comparing plant growth and substrate performance. Anytime data is presented (at trade shows, education sessions, company advertising, marketing propaganda, etc.) that shows plant growth differences or similarities as it relates to substrate performance or comparison to other products - it is important to keep in mind, and ideally ask the person presenting the data - what the conditions were that the crops were grown under (were all variables the same); and if the results were obtained from comparing "apples to apples" or "apples to oranges".

In addition to discussions about wood fiber and other "alternative" substrate materials for our current and future cropping systems, I would also like to applaud the peat industry for all that they are doing in support of continued substrate science and product development as well as their collective extreme awareness and involvement in sustainability and environmental stewardship. There continues to be debate and, in many instances, false narratives about peat and its sustainability in the future. A few things my travels and engagement with the peat industry both in North America and Europe have taught me is that they: 1) are committed to sustainability efforts; 2) are proactive with peatland management and restoration; 3) are adamant about maintaining proper harvesting techniques; 4) invest vast resources and efforts into product consistency and quality assurance; and 5) are willing to evolve as horticultural production needs and challenges arise in the future. Many peat companies are currently among the global leaders in wood fiber substrate development and commercialization - and they are excited to expand their product pallet to offer what growers want and need.

VISUALIZING SUBSTRATES

The opaqueness of containers and substrates have caused researchers to exercise some creativity to overcome their lack of visibility. To quantify total pore space and air space at container capacity, we saturate a substrate-filled container with water, weight it, allow it to drain, and weight it again. To get an idea of the pore structure and how water may move through the system, we incrementally apply pressure to the container and associate the volume of water drained and pressure applied as a function of pore diameter. If root growth data is being looked at, the most common way to do so is by painstakingly handwashing rootballs to carefully separate the roots from the substrate. Instead of going through all of these time consuming and invasive procedures, would it not be nice to simply - *see inside*?

If a doctor ever needed to non-invasively *see inside* you, it's likely that you have had the misfortune of experiencing a CT or CAT scan (Computer Assisted Tomography). Tomography, simply described, is the combination of hundreds or thousands of X-ray images which are reconstructed to digitally render a 3D object. This donut-shaped, claustrophobia-inducing instrument once exclusively utilized in the medical and petroleum industries is finding new uses (and maybe more willing patients) in other research fields. Since the first tomographic research studies in plant and soil relations were conducted in the mid- to late-1980's, it follows that the idea to subject plants and substrates to tomographic imaging is nothing novel. However, since the 1980's, the capabilities of CT instruments and analytical software have improved by several orders of magnitude. Images can be captured at the micrometer and nanometer scale - compared to the millimeter scale. Access to a CT instrument no longer requires a doctor's appointment. In collaboration with the Shared Materials and Instruments Facility at Duke University, we in the Horticultural Substrates Lab at North Carolina State University are finally capable of seeing our substrates and plants *inside and out* in both 2D and 3D.

In preliminary scans, substrate components were packed into 7.6 cm (3-in.) cores and scanned at a resolution of 50 microns. Qualitative differences in the inherent physical structures of each material were apparent (Figure 4).

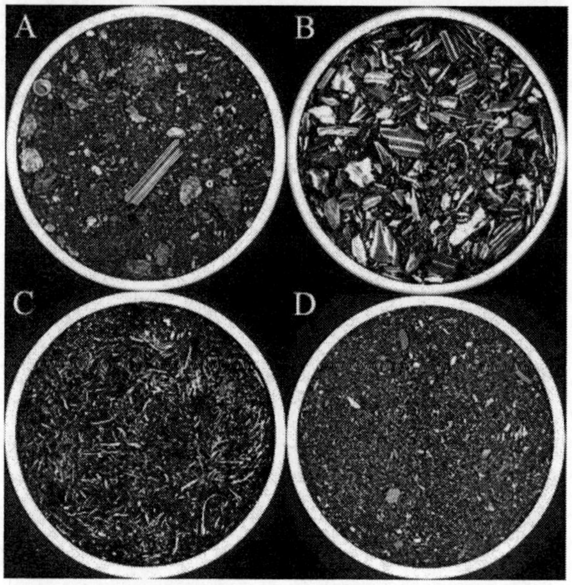

Figure 4. Two-dimensional horizontal slices of peat (A), pine bark (B), wood fiber (C), and coconut coir (D) substrate components packed in 7.6 cm (3-in.) diameter polyacrylic cores.

Peat may be best described as a heterogeneous mixture of fibrous particles and partially decomposed stems. Coir, thought to be very similar in texture to peat, appeared more homogenous and consisted of granular shaped, sponge-like particles. Pine bark appeared the coarsest and contains particles consisting of two layers, the dense periderm layers (brighter white layers) and less dense layers comprised of crushed phloem and expanded parenchyma cells (grey layers). The elongated, fibrous network of wood fiber substrates visually distinguished it from other materials.

To examine the effect of water on CT scans, two pine bark-filled cores were analyzed, one irrigated and another not irrigated. Since the density of water and organic components are similar, it can be difficult to discern what is water and what is pine bark in the irrigated sample. However, what is apparent is the spatial distribution of water in the container. From this orientation, the layer of water held by capillary tension in the irrigated sample, commonly called the "zone of saturation," stands out in bright contrast from the non-irrigated sample (located between 0 and 20 mm; Figure 5).

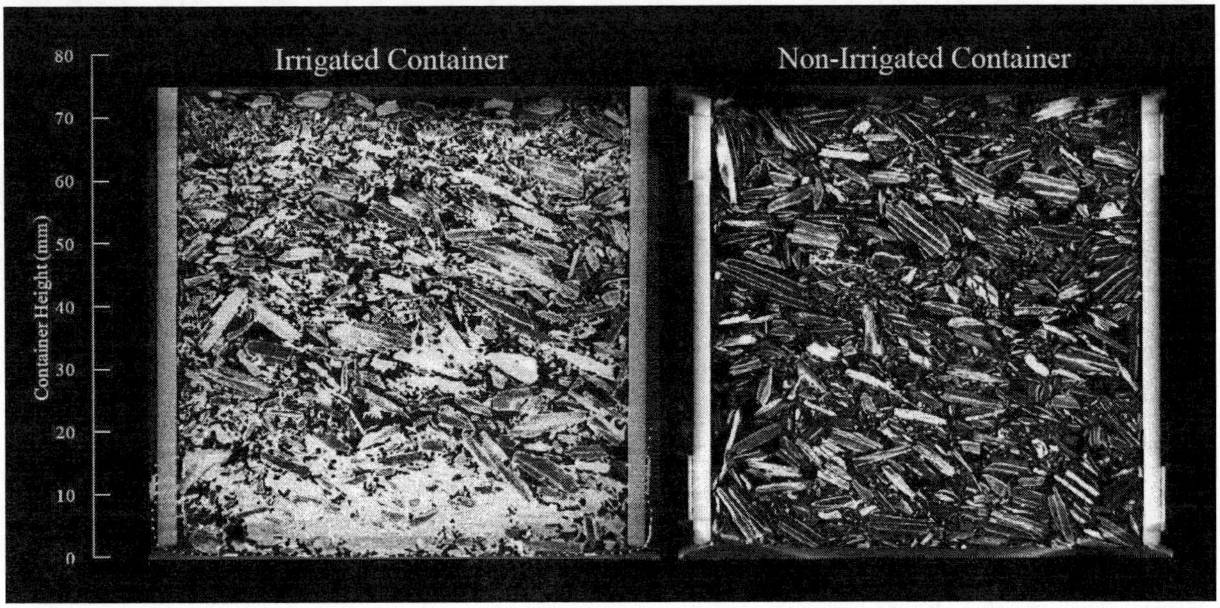

Figure 5. Pine bark cores were analyzed before and after irrigation. White or light grey objects indicate the presence of solids and water. Black spaces indicate air-filled regions.

A valuable relationship to understand is that between substrates and roots. However, separating plant roots and substrates from CT scans can be challenging as both components are organic and comprised of similar elements. In order to isolate a root system, there must be sufficient contrast created between the roots and substrate. Under the appropriate conditions, this contrast can be created and reveal remarkable detail in the plant's root architecture (Figure 6). Similar to substrate characterization, 3D rendered root systems can be characterized by root volume, length, surface area, and diameter.

The spatial distribution of the root system can also be characterized with the same analyses used to generate data for water distributions within a container. There may be no greater dynamic relationship within a container than the relationship between plant roots and the substrate.

How do substrates affect root development? Conversely, how does root development affect substrate physical and hydraulic properties? Aside from substrate/root affects, this technology may allow any abiotic and biotic growth affects, particularly during sensitive stages of plant development, to be non-invasively studied.

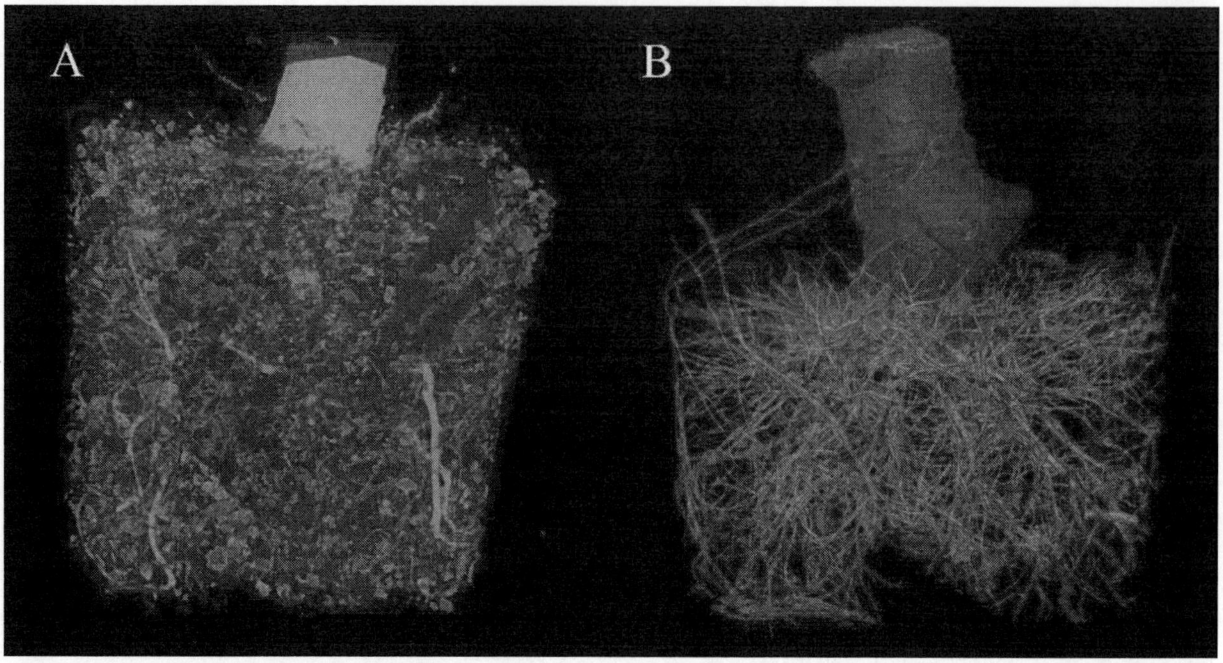

Figure 6. (a) A root geranium cutting is rendered and displayed using a color panel to differentiate materials by their apparent density. (b) Given sufficient contrast, the root system of the plant can be isolated from the substrate and analyzed.

The factors affecting callus tissue development after cutting propagation or grafting could be observed without laying a finger to the plant. The extensive research being conducted to understand the genetic traits responsible for specific root characteristics could be accomplished *in situ*, offering a unique perspective with 3D characterization. There is no doubting the significance future tomographic research could have in the area of plant growth and development.

THE FUTURE IS/OF CANNABIS?!

Is *Cannabis* a horticultural crop? If it is grown indoors under controlled environment conditions, it most certainly is! If grown outdoors (much of the industrial hemp industry) on a large acreage basis - then it may fall more under agronomic jurisdiction depending on who you ask. Regardless the designation, *Cannabis* that is grown in containers requires the use of some substrate (growing media, medium, soilless media, potting soil, etc.) for production. The cannabis industry, due to legal hurdles and crippling stigmas, is somewhat deprived of scientific literature on many production practices and issues. Growers and industry professionals rely heavily on personal experience and information from other industries (i.e. Floriculture/Greenhouse). When personal experience is not enough, they are often forced to online forums, YouTube videos, and decades old handbooks for information. Although hands on experience is an invaluable source of knowledge, having a scientific base of information to rely on can greatly expedite the learning process for growers, both experienced and inexperienced. A specific area in need of information is that of container substrates.

At NC State University, the Horticultural Substrates Laboratory began in the mid 1980's and has since become one of the only laboratories in the world that solely focuses on substrate science to assist growers and retailers/consumers with substrate-related issues and opportunities. Currently, Drs. Brian Jackson (Director) and Bill Fonteno (Founder) operate this lab with responsibilities in conducting grower and industry trials, substrate diagnostic testing, soil/substrate certification, graduate student training, and course instruction.

In 2018, we have received permits to grow/research *Cannabis* (low THC/high CBD) and now are broadening our research pallet to include the needs and opportunities of the ever-growing *Cannabis* industry (Figures 7 and 8).

Figure 7. *Cannabis* trials (CBD Oil Industrial Hemp) at NC State University are beginning to yield some of the first scientific data on substrate-water-plant interactions in controlled environment production.

Current and future understanding of *Cannabis*-substrate interactions will increase drastically as more and more state institutions (and private) are allowed to research and study these crops. Grower trials and experimentation will also continue to provide reliable information about containerized *Cannabis* production. Growers can conduct accurate and reproducible trials at their operations. Research does not have to be conducted in a laboratory! The perfect balance of science and application exists in grower and researcher partnerships which has been the key to success for numerous other horticultural industries.

Figure 8. Indoor production of container-grown *Cannabis* relies on many different organic and inorganic growing media (substrate) components.

Combined Proceedings IPPS 68:356-362. 2018.

Evaluation of Glyphosate Resistant and Susceptible Horseweed

Christina S. Jennings and Anthony L. Witcher
Tennessee State University, Otis L. Floyd Nursery Research Center, 472 Cadillac Lane, McMinnville, Tennessee 37110, USA

christinasjennings@gmail.com

Keywords: Round-up, glufosinate, marestail, *Conyza canadensis*, seed germination, whole-plant assay.

Abstract

Glyphosate resistant horseweed (*Conyza canadensis*) is a significant issue in nursery crop production but has not been studied broadly. Our objective was to identify a glyphosate resistant horseweed population, evaluate seed viability, and determine the effectiveness of glyphosate rate on resistant and susceptible seedlings. Two populations were identified, and individual plants were treated with glyphosate or glufosinate Resistant plants survived glyphosate treatments and died following glufosinate treatments, while all treated susceptible plants died. Seeds were collected from each population and a germination test was used to evaluate viability. Seedlings were used for a whole plant assay, with glyphosate at 0, 2.2, and 11.1 L ha^{-1} (0, 1, and 5 qt ac^{-1}).

Germination test results indicate that resistant plant seeds are more likely to have a higher viability. In the whole plant assay, resistant seedlings treated with glyphosate did not have significantly lower shoot weights than non-treated seedlings. However, with susceptible seedlings there was a significant decrease in shoot weight with increasing glyphosate rates. Results suggest that glyphosate resistance is readily passed down and resistant seedlings are more likely to survive and reproduce.

INTRODUCTION

Nursery crops were valued at over $5.1 billion based on annual sales as of 2012 (USDA, 2012). Maintaining nurseries that are already in production may cost anywhere from $2000 to $3500 per acre annually until harvested; the production period can be three to five years depending on the crop (Halcomb, 2009).

Weed management is a critical component for producing quality nursery crops because weeds can cause competition for light, nutrition, and water - causing a negative impact on production and profit.

Glyphosate is a non-selective post-emergent herbicide that is used to control annual and perennial weeds. Glyphosate is commonly used in field grown nursery crops to control weeds within the row and is typically applied using a shielded sprayer to prevent contact with the crop. Glyphosate is most effective on young, actively growing plant tissue. (Kleinman and Rubin, 2016). The mode of action of glyphosate is the interruption of a plant's shikimate pathway, which is a metabolic pathway important for plant growth. The specific target site for glyphosate is a metabolic enzyme called chloroplastic enzyme 5-enolpyruvyl shikimate-3-phosphate (EPSP) synthase which produces energy for the plant. When EPSP is blocked, the buildup of shikimic acid and metabolic processes failing to function leads to chlorosis and plant death (González-Torralva et al., 2017). Glyphosate resistant crops were introduced in the late 1990's (Shaner et al., 2005). Overreliance of glyphosate has caused glyphosate resistant weeds to develop, hindering the ability of the herbicide to control weeds (Kumar, et al. 2017).

Conyza canadensis (L.) Cronquist, commonly known as horseweed, is an annual weed native to much of North America (Tilley, 2012). Due to its wide native range and adaptability, horseweed has developed glyphosate resistance from the repeated use of glyphosate particularly in agronomic crops. Glyphosate resistant horseweed has been identified in at least 25 U.S. states in some 12 countries (Heap, 2017). Horseweed has several growth stages, which can be identified as the seedling, rosette and bolting stages (Shoup et al., 2012). Horseweed can germinate throughout the year (Kumar, et al. 2017), can produce 50,000 to 250,000 seeds per plant, which are easily spread by wind. As a result, glyphosate resistant horseweed has spread to areas producing other crops, including nurseries. Horseweed also plays host to pests such as tarnished plant bugs (*Lygus lineolaris*), which is a serious horticultural pest (Steckel, 2018).

The impact of horseweed glyphosate resistance in nursery crops has not been extensively evaluated. Identifying glyphosate resistant horseweed and developing management practices to prevent and control these weeds will benefit nursery producers throughout the U.S. The objective of this preliminary research project was to identify a glyphosate resistant population of horseweed, evaluate seed viability of resistant and susceptible plants, and determine the effectiveness of glyphosate rate on resistant and susceptible seedlings.

MATERIALS AND METHODS

Suspected glyphosate resistant and susceptible populations were first identified then screened for resistance. The suspected resistant population was located in a soybean field in Warren County, TN. The field had been planted in glyphosate resistant crops (soybeans) for at least 10 years. The suspected susceptible population was located at the Tennessee State University Otis L. Floyd Nursery Research Center (NRC) in McMinnville, TN, in a field that had not been treated with glyphosate in several years.

Twenty individual plants were flagged in each location and were treated with either glyphosate (1.3 oz/gal; GLY-4; 41% glyphosate; Universal Crop Protection Alliance, LLC, Eagan, MN) or glufosinate (1.3 oz/gal; Finale; 11.33% glufosinate; Bayer, Research Triangle Park, NC). One week after treatment, plants were evaluated for herbicide activity. Plants at the soybean field treated with glyphosate showed no damage or minor damage exhibited as chlorosis of the terminal leaves. At the NRC, all plants were killed (no green tissue remained on the above ground portion of the plant) with both the glyphosate and glufosinate treatments. Seeds were collected (Sept. 22, 2017) from ten non-treated plants within each population. A separate vial (7 ml polypropylene; Evergreen Scientific, Rancho Dominquez, CA) was used to store seeds from each individual plant for a total of twenty vials (ten per population), and vials were placed in the refrigerator.

To evaluate seed viability, a germination test was performed using methods described by Travlos and Chachalis (2013). Seeds were sown in petri dishes (100 x 15 mm; VWR International, Radnor, Pennsylvania) on 2 sheets of filter paper (top layer Whatman 4, bottom layer Whatman 598; Whatman Ltd., Maidstone, England) saturated with deionized water. Ten seeds from an individual plant were placed in each petri dish, and there were ten petri dishes (replications) per plant for a total of 100 seeds per plant. Petri dishes were then placed in a growth chamber at 26/21°C day/night temperatures with a 14 h daylength. After 72 h, petri dishes were removed from the growth chamber. Seeds were determined to have germinated if a radicle and cotyledon were present and germination percentage was calculated. The experiment was conducted twice (Test 1 and Test 2).

To evaluate the effectiveness of glyphosate rate on resistant and susceptible horseweed seedlings, a whole plant assay was conducted based on methods used by Koger and Reddy (2005). Seedlings were grown from the seeds described above. Forty-eight seeds per plant were sown into 72 cell flats (2 seeds per cell; PROP-72-RD; T.O. Plastics Inc. Clearwater, MN) filled with seed starter mix (Morton's #1; McMinnville, TN). The flats were placed in a growth chamber (as described above) and were watered as needed with a handheld pump sprayer.

When seedlings were at the three to five leaf stage, they were transplanted to individual 8.9 cm square containers (SVD 350; T.O Plastics Inc., Clearwater, MN) filled with substrate (1 pine bark: 1 peat moss by volume; amended per cubic yard with 4 lb Nutricote 13-11-11, 5 lb Dolomitic Lime, 1 lb Aqua Gro, 2 lb Plantex 10-5-10 Media Starter, and 0.75 lb Micromax). When seedlings reached the rosette stage (approximate width of 12 cm), they were treated with glyphosate (GLY-4 Plus) applied at 2.2 or 11.1 L ha^{-1} (1 or 5 qt ac^{-1}) using a CO_2 pressurized sprayer calibrated to deliver 33 gpa at 30 psi. At 21 days after treatment, shoot fresh weight was measured for all seedlings.

Data were analyzed using the GLIMMIX procedure of SAS (Version 9.3; SAS Institute, Inc., Cary, NC). Differences between means were determined using the Shaffer-Simulated method ($P < 0.05$).

RESULTS AND DISCUSSION

Seed germination rate varied within plant populations, but overall was higher in the resistant population (Table 1). Within the resistant population germination rates ranged from 56 to 80.3% and 57.1 to 89.9%, respectively, for Tests 1 and 2. Within the susceptible population, germination rates ranged from19.0 to 72.2% and 7.7% to 79.0%, respectively, for Tests 1 and 2. When all plants within a population were averaged, resistant seeds in Test 1 had 70.2% germination compared to 52.2% within the susceptible population. For Test 2, resistant seeds had an overall germination rate of 75.6% compared to 46.4% for the susceptible seeds. In both tests, we observed the five highest germination rates occurred for resistant plants while the five lowest germination rates were from susceptible plants. In a previous study, Travlos and Chachalis (2013) found no differences in seedling vigor between the glyphosate resistant and glyphosate susceptible horseweed in their study.

Table1: Mean germination percentage of seeds from resistant and susceptible horseweed (*Conyza canadensis*) populations. Seeds were collected form 10 plants within each population and the germination test was conducted twice.

	Germination Rate (%)			Germination Rate (%)	
Plant	**Test 1**	**Test 2**	**Plant**	**Test 1**	**Test 2**
Resistant 1	74.1	86.8	Susceptible1	68.1	74.0
Resistant 2	74.0	83.4	Susceptible 2	19.0	13.4
Resistant 3	77.1	79.8	Susceptible 3	30.4	34.9
Resistant 4	56.6	63.9	Susceptible 4	72.2	73.6
Resistant 5	58.5	58.0	Susceptible 5	67.3	79.0
Resistant 6	60.0	77.8	Susceptible 6	65.4	40.2
Resistant 7	75.5	89.9	Susceptible 7	56.5	68.0
Resistant 8	73.8	86.0	Susceptible 8	63.1	7.7
Resistant 9	72.0	57.1	Susceptible 9	55.7	41.6
Resistant 10	80.3	73.0	Susceptible10	24.7	31.3
All Plants	70.2	75.7	All Plants	52.2	46.4

They had optimal germination rates (65 %) at the temperature regime of 15/25°C and intermediate germination rate (31 to 51%) at 20/30°C.

In the whole plant assay, resistant seedlings treated with glyphosate did not have significantly lower shoot fresh weight compared with the non-treated resistant seedlings except for Resistant Plant 4 (Figure 1a). However, susceptible seedlings treated with glyphosate had lower shoot fresh weight compared with non-treated susceptible seedlings (Figure 1b). When shoot fresh weight was averaged for all seedlings within the resistant group, non-treated resistant seedlings had a higher fresh weight compared with glyphosate treated seedlings (Figure 2). However, there was no difference between resistant seedlings treated with 2.2 or 11.1 L ha^{-1} (1 or 5 qt ac^{-1}) glyphosate. Within the susceptible population, the non-treated seedlings also had a higher fresh shoot weight compared with the treated seedlings, however fresh shoot weight significantly decreased with each increase in glyphosate rate.

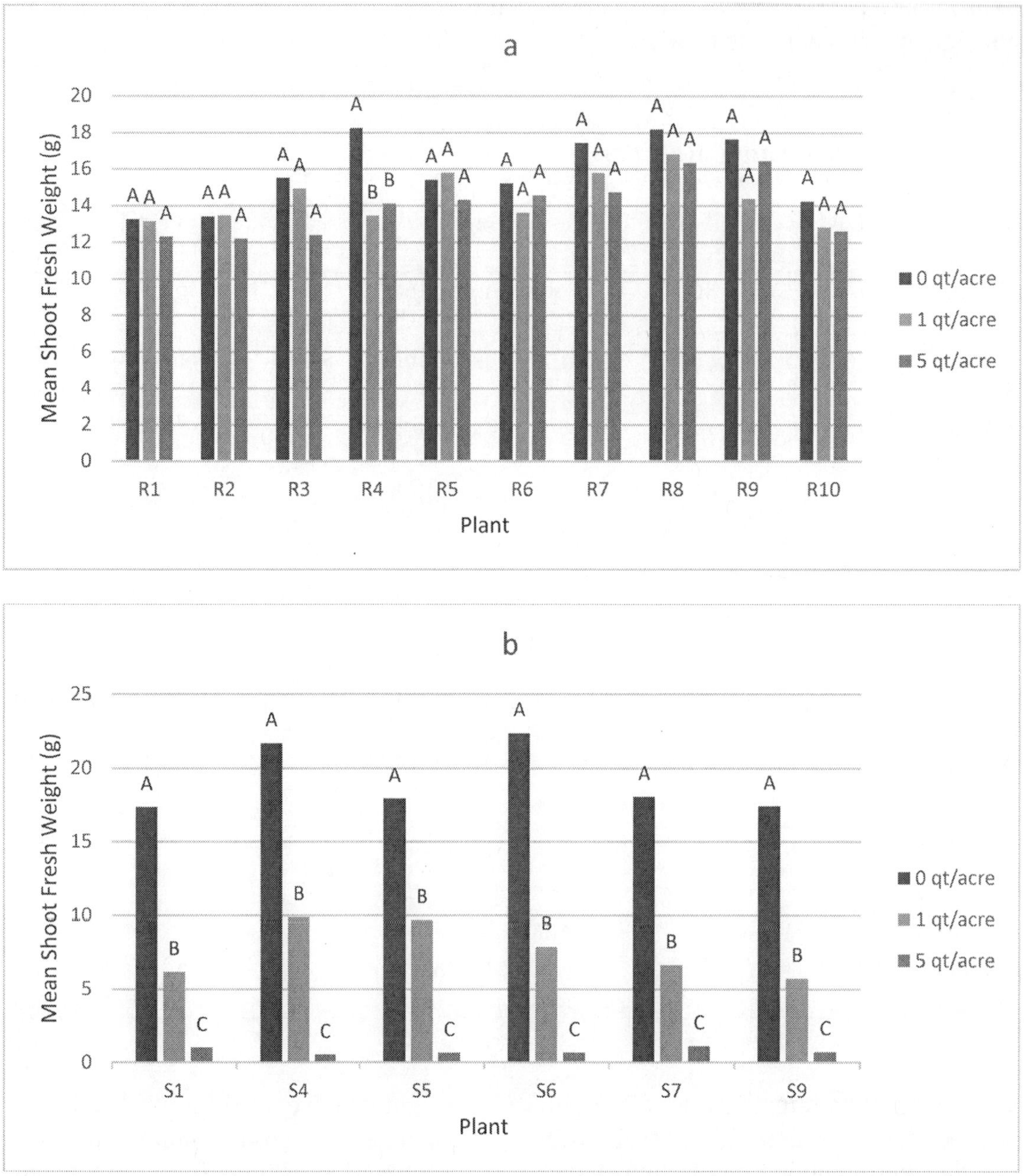

Figure 1: Mean shoot fresh weight (grams) of resistant (a) and susceptible (b) horseweed (*Conyza canadensis*) seedlings from multiple plants 21 days after glyphosate treatments at 0, 2.2, and 11.1 L ha^{-1} (0, 1, and 5 qt ac^{-1}). Columns with same letters within an individual plant are not significant at $P \leq 0.05$.

The germination study results show that resistant plants are more likely to have a higher germination rate than susceptible plants. The results indicate that resistant plant seeds may be more likely to establish due to being able to germinate and survive more readily than susceptible plants. The whole plant assay results suggest that resistance is passed down from parent plant to seedling. This means that resistant plants that are treated with glyphosate can survive the treatment and reproduce. Future studies will include assessing more rapid laboratory methods to confirm glyphosate resistance and evaluating establishment prevention of horseweed in nursery fields.

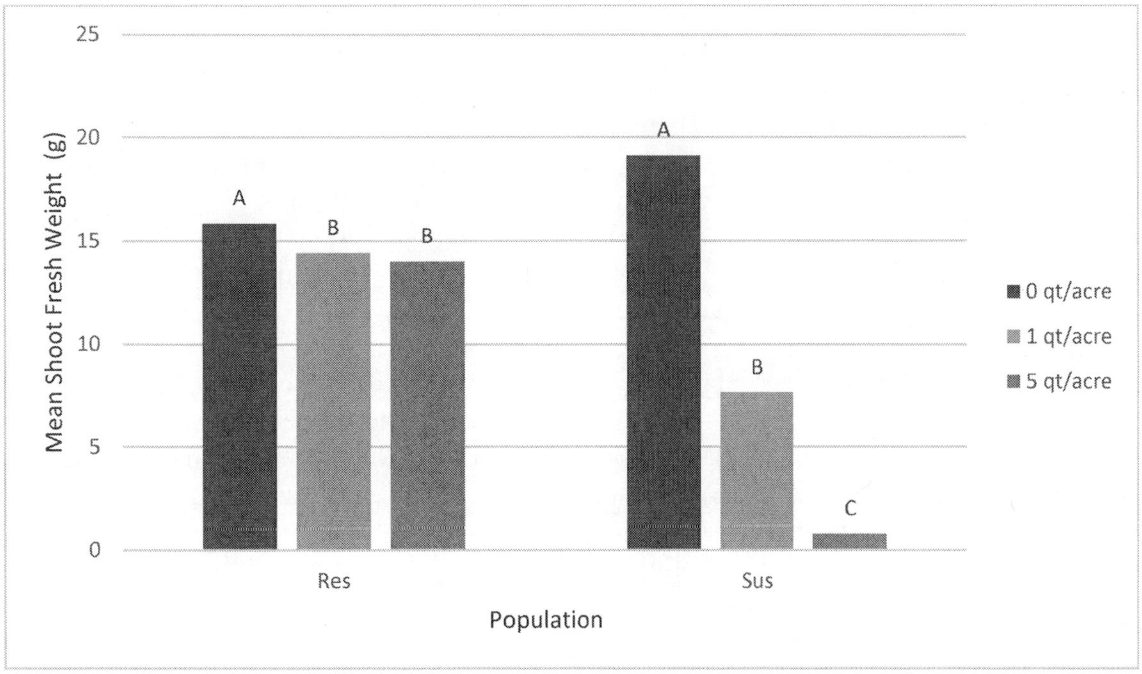

Figure 2: Total mean shoot fresh weight (grams) from all plants of resistant and susceptible horseweed (*Conyza canadensis*) populations 21 days after glyphosate treatments at 0, 2.2, and 11.1 L ha^{-1} (0, 1, and 5 qt ac^{-1}). Columns with different letters within a population are significant at $P \leq 0.05$.

Literature Cited

Altland, J. E., Gilliam, C. H., Wehtje, G. (2003). Weed control in field nurseries. HortTechnology *13*:9-14.

González-Torralva, F., Brown, A. P., Chivasa, S. (2017). Comparative proteomic analysis of horseweed (*Conyza canadensis*) biotypes identifies candidate proteins for glyphosate resistance. Sci. Rep. 7:42565.

Halcomb, M. 2009. Nursery field production. University of Tennessee. https://extension.tennessee.edu/mtnpi/Documents/handouts/Field%20Production/Field_Production_Handout-8-09.pdf . Accessed December 22, 2018.

Heap, I. The international survey of herbicide resistant weeds. December 8, 2017. http://www.weedscience.com/Summary/Species.aspx.

Kleinman, Z., and Rubin, B. (2016). Non-target-site glyphosate resistance in *Conyza bonariensis* is based on modified subcellular distribution of the herbicide. Pest Management Science *73*: 246-253.

Koger, C. H. and Reddy, K. N. (2005). Role of absorption and translocation in the mechanism of glyphosate resistance in horseweed (*Conyza canadensis*). Weed Sci. *53*: 84-89.

Kumar, V., Jha, P., and Jhala, A. J. (2017). Confirmation of glyphosate-resistant horseweed (*Conyza canadensis*) in Montana cereal production and response to post herbicides. Weed Tech. *31*:799-810.

Shaner, D. L., Nadler-Hassar, T., Henry, W. B., and Koger, C. H. (2005). A rapid in vivo shikimate accumulation assay with excised leaf discs. Weed Sci. *53*:769-774.

Shoup, D., Peterson, D. E., Thompson, C. R., and Martin, K. L. (2012) Marestail control in Kansas. Manhattan, Kan.: Agricultural Experiment Station and Cooperative Extension Service, Kansas State University. https://www.bookstore.ksre.ksu.edu/pubs/mf3014.pdf . Accessed December 22, 2018.

Steckel, L. (2005). Horseweed. University of Tennessee Extension. https://extension.tennessee.edu/publications/documents/W106.pdf. Accessed December 22, 2018

Tilly, D. (2012). Ecology and management of Canadian horseweed (*Conyza canadensis*). USDA- Natural Resources Conservation Service. Tenn. Plant Materials *59*:1-5.

Travlos, I.S., and Chachalis, D. (2013). Assessment of glyphosate-resistant horseweed (*Conyza canadensis* L. Cronq.) and fleabane (*Conyza albida* Willd. Ex Spreng) populations from perennial crops in Greece. Intl. J. Plant Prod. *7*: 665-676.

USDA. (2012). Census of Agriculture. December 8, 2017. https://www.nass.usda.gov/Publications/AgCensus/2012/Full_Report/Volume_1,_Chapter_1_US/usv1.pdf Accessed December 22, 2018

Combined Proceedings IPPS 68:363-368. 2018.

An Initial On-Site Analysis of Irrigation Management Practices at Four Alabama Container Nurseries

Claire Krofft, Adam Newby, Jeremy Pickens and Glenn Fain
Auburn University, Dept. of Horticulture, Auburn, Alabama 36849, USA

cek0054@tigermail.auburn.edu

Keywords: Ornamentals, BMPs, irrigation management, survey, leaching, distribution uniformity (DU), efficiency.

Abstract

The ornamental container nursery industry is large contributor to Alabama's green industry. However, it relies on intensive irrigation and fertilization to achieve high economic yields. In this study, observational data including application rate, distribution uniformity, and leaching fraction were collected from four Alabama container nurseries in July 2018 to assess current irrigation management practices. All four nurseries fell short of the recommended distribution uniformity (DU) value of greater than 80%. One nursery exceeded the recommended maximum leaching fraction value of 15%. Improvements can be implemented to improve efficiency, reduce inputs, and decrease environmental impact. Effort should be put forth to increase education and training for irrigation management in container nurseries.

INTRODUCTION

The use of containers in ornamental production has increased over the past 50 years because plants can be produced more efficiently than in-field cultivation (Majsztrik et al., 2011). In the state of Alabama in 2014, container nursery stock generated $91.9 million in sales (USDA, 2014). However, container-cultivated ornamentals rely on intensive irrigation and fertilization practices to produce high yield and economic returns.

Many container-grown nursery crops are irrigated with overhead sprinkler systems (Bilderback, 2002). These systems are a common choice of nursery growers because they have low installation and maintenance needs and cover large areas. The downside is that overhead sprinklers can be inefficient if installed improperly, and the uniformity of a system may be impacted by environmental factors of wind and evaporation (Whitcomb, 1984).

To compensate for uneven distribution, nursery growers often run irrigation until the driest areas are sufficiently irrigated, causing excess water to be applied in other areas (Biernbaum, 1992). Due to the large amount of applied water, growing media in nursery production must be porous to allow proper drainage and avoid pest problems caused by overwatering. Nurseries in Alabama typically use porous pine bark-based growing media which has low nutrient-holding capacity.

Due to the large amounts of irrigation necessary to prevent containers from drying out, loss of nutrients occurs readily (Rathier and Frink, 1989). Hence, high fertilizer application rates are used to offset leaching due to current irrigation practices. Consequently, because of high fertility rates, frequent irrigation is deemed necessary to prevent the buildup of soluble salts. Despite being effective for producing container-grown nursery crops, these cultural practices waste expensive resources and contribute to environmental problems.

Inefficient irrigation systems at container nurseries create a significant amount of runoff water which can transport dissolved substances into the surrounding area (Briggs et al., 1998). Leachates from nursery plants are primarily nitrate fertilizers, but also include soluble phosphates and traces of heavy metals (Biernbaum, 1992). Unutilized nitrates can lead to contamination of surface and groundwater - and have detrimental environmental effects such as excessive growth of algae and death of aquatic species (Howarth, 1988; Kabashima, 1993; Rauschkolb and Hornsby, 1994). Inefficient distribution within irrigation systems may cause plants to be over- or under-watered, causing disease, plant stress, and death. In addition, leaching of nutrients means loss of expensive inputs from the production cycle. Degradation of crops combined with the high inputs needed to produce container ornamentals can lead to economic losses for nursery growers. By implementing better irrigation management practices, growers can prevent leaching of applied nutrients and other chemicals and decrease production costs.

Container nurseries can improve irrigation efficiency by implementing Best Management Practices (BMPs), strategies intended to mitigate environmental impacts of nursery crop production (Southern Nursery Association, 2013). Implementing BMPs can reduce water use and leaching of fertilizers, thereby reducing inputs and environmental impacts.

Inefficient irrigation management practices increase the cost of production and contribute to environmental problems. To begin improving irrigation management practices in Alabama, current practices must be assessed. There has not been an assessment of irrigation practices at container nurseries in Alabama in the last 20 years (Fain et al., 2000). The objective of this study is to collect observational data about irrigation management practices in Alabama container nurseries.

MATERIALS AND METHODS

Visits to four participating nurseries in Alabama began in July 2018. Two nurseries are located in central Alabama, and two are located in southern Alabama on the Gulf Coast. Application rate, distribution uniformity and leaching fraction were assessed on blocks of dwarf yaupon holly (*Ilex vomitoria* 'Nana') in #3 containers under overhead irrigation at each nursery. The area of the block, number of plants, container dimensions, orientation of risers, and sprinkler head type were recorded. Growers were asked to run a typical irrigation cycle for data collection.

Distribution uniformity (DU) is a measure of how uniformly water is applied to an irrigation area. To determine DU, cups were placed in a uniform grid pattern within an irrigation zone. After an irrigation cycle, the volume of water collected in the cups were measured with a graduated cylinder. The average of the lowest quartile of application volumes were divided by the overall average application volume and multiplied by 100.

$$DU = \frac{\begin{array}{c}average\ application\\ of\ lower\ quartile\end{array}}{\begin{array}{c}average\ overall\\ application\end{array}} \times 100$$

According to BMPs, DU values should be greater than 80%, with a percentage lower than 60% indicating a more thorough audit is required to determine design or hardware malfunctions in the system.

Leaching fraction (LF) is the proportion of applied water that leaches from a container. To calculate LF, 8-12 plants (depending on the number of plants within a block) were randomly selected (Figure 1).

Figure 1. Leaching fraction samples (pink flags) within a block.

Each plant was fitted tightly into a 5-liter bucket and "skirted" with a plastic bag so that no irrigation water could enter the bucket between the container and the buckets (Figure 2).

Figure 2: "Skirted" sample for irrigation efficiency analysis

A section of six-inch PVC pipe was placed under the container to ensure adequate space between the bottom of the container and the bottom of the bucket for leachate to collect. The plant with the bucket and plastic bag skirt was then weighed to determine its pre-irrigation weight and placed back in its original location within the block. After running a typical irrigation cycle, the samples were allowed to drain for 30 minutes and weighed again. Total applied irrigation per plant was calculated by subtracting the pre-irrigation weight from the post-irrigation weight. The plants were then removed from the 5-liter buckets, and leachate that had collected in the bottom of the bucket was measured with a graduated cylinder.

To determine the LF, the leachate recovered is divided by total applied irrigation and multiplied by 100.

$$LF = \frac{leacheate\ recovered}{total\ applied\ irrigation} \times 100$$

BMPs state that LF should be 15% or less to avoid excessive irrigation. If the percent LF is over 20%, irrigation run time should be reduced.

Size index of the same plants used to measure LF was recorded as the average of the height, widest width, and width perpendicular to the widest width. Media pH and electrical conductivity (EC) were measured on the same plants using the pour-through nutrient extraction method (Wright, 1986). Leachate pH and EC were measured using a HACH Pocket Pro+ Multi 2 Tester (Hach Co., Loveland, CO).

RESULTS AND DISCUSSION

Results were collected on 20 July at Nursery A and Nursery B, 19 July at Nursery C, and 30 July at Nursery D. Cultural irrigation practices are different for each nursery and vary between irrigation run time, sprinkler arrangement, and sprinkler head type. The average irrigation amount was around 1.3 cm (0.5 in.) for all nurseries. However, DU for all nurseries was below the recommended value of 80%, ranging from 61% to 74% (Table 1). Sprinkler orientation (corner, side, or center) and spray pattern (90⁰, 180⁰ and 360⁰) affect DU. The hollies at Nursery A was irrigated with four 180° sprinklers and six 360° sprinklers and had a DU of 67%. Nursery B was irrigated under sprinklers with a 360° spray pattern and had the highest DU (74%). Nursery C's block was irrigated by two 90° and six 180⁰ sprinklers and had the lowest DU of 61%. Nursery D's block was irrigated with two 180° sprinklers and two 360° sprinklers and had a DU of 66%. None of the nurseries had installed matched precipitation sprinklers, therefore sprinklers applied the same volume of water per minute despite spray pattern.

Table 1. On-site irrigation data collected from four nurseries located in Macon, Mobile, and Montgomery county in Alabama in July of 2018.

Nursery	Irrigation duration (min)	Average irrigation depth (in)	Irrigation depth range (in)	Distribution uniformity[z]	Average leaching fraction[y]	Leaching fraction range
A	30	0.52	0.22–0.95	67%	7.4%	1.8%–15.5%
B	60	0.48	0.33–1.05	74%	14.4%	1.7%–35.2%
C	90	0.53	0.27–1.01	61%	8.5%	0.2%–41.5%
D	120	0.54	0.3–1.05	66%	36.4%	5.0%–67.2%

[z]Distribution uniformity = (average application of lower quartile)/ (average overall application) × 100

[y]Leaching fraction = (leachate recovered/total applied irrigation) × 100

Matched precipitation sprinklers match the flow rate of the nozzle to the spray pattern and can improve the DU. For example, a corner sprinkler spraying a 90⁰ spray pattern would apply half the volume compared to a sprinkler with a 180⁰ spray pattern since it is covering half the area (Bilderback, 2002).

Average LF for nurseries A, B, and C are within the BMP guideline of ≤15%, however poor DU led to a large range of LF values. Nursery B LF ranged from 1.7% to 35.2%, while Nursery C LF ranged from 0.2% to 41.5%. Average LF at nursery D exceeded the recommended LF with an average of 36.4% and ranged from 5.0% to 67.2%. Nursery A had the lowest average LF at 7.4% and ranged from 1.8% to 15.5%. Splitting irrigation into multiple applications throughout the day, a recommended BMP known as cyclic irrigation, can reduce the LF.

Distribution uniformity is the biggest limitation to efficient overhead irrigation systems (Southern Nursery Association, 2013; Fare et al., 1994). If DU were improved and irrigation was applied more evenly throughout the area, irrigation duration and therefore volume can be reduced. Additionally, LF range and LF averages would improve as well.

Decreasing LF reduces the amount of nutrients lost to leaching and decreases environmental impact. Using matched precipitation sprinklers can significantly improve DU. Regular inspection and maintenance of irrigation systems is also an important step to identify and resolve DU issues.

The initial results of this study are encouraging for Alabama growers, but small changes could be implemented to improve irrigation management practices, reduce inputs, and decrease the impact in the environment. Effort should be put forth to increase education and training for irrigation management in container nurseries. Irrigation assessments at several Alabama nurseries will continue through 2019, and results will be used to develop irrigation BMP training workshops for nursery growers through the Alabama Extension Service.

Literature Cited

Biernbaum, J.A. (1992). Root-zone management of greenhouse container-grown crops to control water and fertilizer. HortTech. *2*:127–312.

Bilderback, T.E.(200)2. Water management is key in reducing nutrient runoff from container nurseries. HortTech. *12*:541–544.

Briggs, J.A., Riley , M.B., Whitwell, T. (1998). Quantification and remediation of pesticides in runoff water from containerized plant production. J. Environ. Qual. *27*:814–820.

Fain, G.B., Gilliam, C. H., Tilt, K.M, Olive, J.W. and Wallace, B. (2000). Survey of best management practices in container production nurseries. J. Environ. Hort. 18:142–144.

Fare, D.C., Gilliam, C.H., Keever, G.J. (1992). Monitoring irrigation at container nurseries. HortTech. *2*:75–78.

Fare, D.C., Gilliam, C.H., Keever, G.J., Olive, J.W. (1994). Cyclic irrigation reduces container leachate nitrate-nitrogen concentration. HortScience. *29*:1514–1517.

Howarth, R.W. (1988). Nutrient limitation of net primary production in marine ecosystems. Ann. Rev. Ecol. *19*:89–110.

Kabashima, J.N. (1993). Innovative irrigation techniques in nursery production to reduce water usage. HortScience *28*:291–293.

Majsztrik, J.C., Ristvey, A.G., and Lea-Cox, J.D. (2011). Water and nutrient management in the production of container - grown ornamentals. Hort. Rev. *38*:253-97.

Rathier, T.M. and Frink, C.R. (1989). Nitrate in runoff water from container grown juniper and Alberta spruce under different irrigation and N fertilization. J. Environ. Hort. *7*:32–35.

Rauschkolb, R.S. and Hornsby, A.G. (1994). Nitrogen management in irrigated agriculture. Oxford Univ. Press, New York.

Southern Nursery Association. (2013). Best management practices guide: Guide for producing nursery crops. 3rd ed. Southern Nursery Association. Acworth, Georgia.

U.S. Department of Agriculture. (2014). Census of horticultural specialties. Table 17: Nursery stock sold: 2014. U.S. Dept. Agr., Washington, D.C.

Whitcomb, C.E. (1984). Plant production in containers. Lacebark Pub. Stillwater, OK.

Wright, R.D. (1986). The pour-through nutrient extraction procedure. HortScience *21*:227-229.

Combined Proceedings IPPS 68:369-372. 2018.

Evergreen seed germination technique

Buddy Lee
Transcend Nursery, 52063 Ridgecrest Drive, Independence, Louisiana 70443, USA
Email: buddyazaleas@yahoo.com

Keywords: Seed dormancy, *Rhododendron*, azalea, seed capsules, germination.

INTRODUCTION

Successful evergreen azalea seed germination can be achieved using healthy viable seeds and the appropriate conditions. Evergreen azaleas are in the genus *Rhododendron,* subseries *Obtusum*. The seed capsules begin to form after a complete azalea flower opens and matures and pollen is transferred to the stigma of the pistil. This occurs via physical contact, wind, pollinators or controlled crossing by humans. The pollen moves down the pollen tube to the ovary and with successful pollination - it will develop into a capsule maturing to approximate size of 0.6 cm X 0.6 cm (0.25 in. X 0.25 in.). A mature seed capsule can contain 100-500 azalea seeds (Figure 1).

Seed capsules are usually harvested in the fall of the year. However, care must be taken not to harvest too late because the seed capsule will dry and split open into a star shaped configuration and the seed will be disbursed.

AUGUST 1, 2018

Azalea seed capsules were collected in early August as the result of a controlled pollination on March 15, 2018.

Figure 1. Evergreen azalea seed capsules (arrow).

In observations from the past, seed collected at least ninety days after pollination seem to have many viable seed especially - if the outer portion on the seed capsule is starting to turn brown. The capsules were collected and placed on a folded sheet of paper and moved to a dry area away from direct sunlight. The folded paper helps to remove moisture from the capsules. By August 12, the seed capsules were dry and ready to be cleaned. The dried capsules were then placed on a flat sheet of paper so that the tiny seeds dropping out of the capsule could be collected.

IPPS Vol. 68 – 2018

To facilitate extracting the seeds from inside the capsule, a small pair of forceps was used to gently crack the seed open, starting at the base of the capsule and carefully moving along the capsule to the end tip - breaking the dried seed capsule shell. Once the capsule is cracked open, the seeds start falling into the sheet of paper and can easily be seen (Figure 2).

Figure 2. Evergreen azalea seed capsules and dispersed seed (arrow).

After all the capsules have been cracked opened and seeds released, the old capsule parts and the seeds can then be separated by using a small mesh screen that allows the seeds to pass through - and collect on a sheet of paper under the screen. The old capsule pieces can be discarded.

With the clean seed ready to be sown, seed trays need to be prepared. Evergreen azaleas thrive best in an acidic, well drained, organic material. A pH of around 5.5 is usually best for azaleas. Solid sides and bottom Standard 1020 trays with drain outlets on the bottom of the tray can be used as container flats.

A good media is to mix of 50% fine ground pine bark: 50% PRO-MIX BX general purpose media. This mixture provides the material for the bottom portion of the tray. Fill approximately eighty percent of the tray and level the media. The remaining twenty percent is then filled with moistened peat moss to the top of the tray and leveled off. Using a water spray nozzle, wet the media to settle in the mixture and make sure that the material is thoroughly soaked. Place the tray in a protective greenhouse. To make a smooth and level seed bed - remove any exposed peat stems or other large moss structures from the top of the tray seed bed.

AUGUST 12, 2018

Cleaned azalea seeds were sown on prepared seed germination tray. For sowing seed, I use a sheet of paper and fold it in half. After placing the seed on the inside of the fold, I gently shake the seed down the crease of the paper until they start spilling out onto the peat moss in the seedling trays. Care must be taken to avoid thickly sowing seed. It is best to perform seed sowing in the greenhouse with fans off, and as little wind as possible.

Azalea seeds are very tiny and can be blown is every direction with the slightest puff of wind. After the seeds have been sown, the seed is misted with a very light mist spray. This mist will allow the seed to settle and anchor into the peat most. Place the tray of planted seeds back on the bench and do not allow them to become dry; however, do not let them become overly wet. Use only a light mist to moisten the peat moss. Check the tray at least once a day to make sure the peat does not become dry. The water that I use to irrigate my seedling trays has a pH of 5.2, with no measurable salts. It is my experience that azalea seed germination is slowed or hindered using irrigation water from neutral to a slightly alkaline pH – and higher.

My greenhouse/germination chamber, where the seedling trays are placed, has a dimension of 3 m X 6.1 m (10 ft by 20 ft). The bench that the trays sit on is about 1.2 m (4 ft) off the floor. The top of the greenhouse is covered with hard clear plastic sheeting; during seed germination, 35% shade cloth is doubled and also placed on the top of the greenhouse. The greenhouse faces the East and gets direct morning sun. However, the shade cloth is also in the top front of the greenhouse and blocks the direct sun. In the afternoon, shade from trees help keep the greenhouse cooler. The bottom sides of the greenhouse are open during the summer and late fall. This allows better air flow which keeps the area cooler than the outside temps. The cooling fan thermostat is set to turn on at 35°C (95°F). The temp inside the greenhouse very seldom exceeds 35° C (90°F) during August and September. In August, the average ambient temperature for this area of Southeast Louisiana is a high of 33°C (92°F) and a low of 23°C (73°F). In September, the average high is 31°C (88°F) and a low of 21°C (69°F).

Later in fall/winter, when the weather gets cooler - the sides are covered with plastic sheeting. Heat is then provided to keep temp from dropping below 4.4C° (40°F) during the winter. The methods that I use for this greenhouse to create a favorable environment for seed germination has evolved over years of growing azalea seedlings. It works well for me at this location. However, at other locations, you will need adjust your greenhouse and environmental conditions.

SEPTEMBER 3, 2018

In September, the tiny azalea seeds show visible signs of germination. Once they start to germinate, it is very important to monitor them daily. Make sure the peat moss remains moist. At this time, I start applying Gnatrol® Biological Larvicide at the recommended rate. It is applied weekly for protective control against fungus gnat larvae. Fungus gnat larvae can eat away at small azalea seedling roots and destroy the seedlings. I closely monitor for fungus gnats and other pests throughout the process because it doesn't take very long for a pest to destroy many seedlings in a very short time.

OCTOBER 15, 2018

By mid-October, he little azalea seedlings are growing well and are approximately 0.3 to 0.6 cm (1/8 to ¼ in.) long. As the seedlings continue to grow, they are monitored for signs of stress and disease, which can be corrected quickly (Figure 3).

Figure 3. Two-month-old evergreen azalea seedlings.

No fertilizer is needed at this time because the peat moss has enough nutrients to sustain the seedlings. When the seedlings are five to six months old, I usually start applying liquid fertilizer at 25% or less of the recommended rate. A fertilizer for acidic plants is best for azaleas. Be very careful with fertilizer. It's always best to try a trial application on a small area of seedlings and wait a few days to make sure that there is no damage. When I am sure that the fertilizer rate is safe, I will usually apply the fertilizer two to three times a week. I use a handheld two-gallon size spray to mist over the top of the seedlings and allowing a small amount of fertilizer to soak into the soil. It does not take a lot of fertilizer to make the young azalea seedlings thrive and look healthy and green.

When the seedlings reach approximately 2.5 cm (1 in.) in height they can be successfully transplanted into small individual cell plugs. I use fifty cell size that fit in a standard nursery tray (Figure 4).

A mix of 75% fine grounded pine bark and 25% medium perlite is used for the growing media. Once the seedlings are well rooted in the fifty cells, a light application of glandular nursery fertilizer can be applied. As the seedlings grow and the root system becomes well developed, they then came be potted up into larger containers and grown and maintained under usual nursery production routine.

Figure 4. Evergreen azalea seedlings transplanted into 50-cell nursery seedling trays.

Combined Proceedings IPPS 68:373-377. 2018.

Rethinking Weed Control in the Nursery

Chris Marble
University of Florida/IFAS Mid-Florida Research and Education Center, Apopka, FL 32703, USA

marblesc@ufl.edu

Keywords: Herbicides, hand weeding, non-chemical weed control, sanitation, top-dressing, dibbling, incorporating, chemical rotation, pre- and post-emergence, solarization.

INTRODUCTION

The only weed management tools in container plant production (other than a few herbicides for grass weed control) are preemergence herbicides and hand weeding. Labor for hand weeding is becoming more and more expensive and difficult to find. It may be more important now than ever to find ways to manage weeds efficiently. Recent estimates show that weed control costs can exceed $10,000 per ha ($4,000 per ac.) (Mathers, 2003). These costs typically include herbicide costs, application costs, and hand weeding costs. These estimates often do not include the opportunity cost that is lost when labor is diverged from profit generating tasks (potting, propagating, etc.) to hand weeding - a profit reducing task.

Here I will describe five common themes I have observed in the most consistently clean nurseries I have visited. These methods are supported by research, but more importantly, have all been proven successful (and economical) in the real world. You should reconsider some aspects of your current weed program.

HANDWEEDING FREQUENTLY IS SUPERIOR TO HANDWEEDING LESS

The goal of improving a weed control program should be to reduce hand weeding. However, the goal should be to reduce the time spent weeding, not the frequency. Research has shown that weeding every 2 weeks reduces labor costs by an average of 36% (and some much more). This was compared with weeding only every 8 weeks, or just before herbicides are reapplied (Barker and Neal, 2016). Several nurseries have adopted this approach and significantly reduced their hand-weeding costs. More frequent weeding works in several ways to reduce the total time (and cost) of weeding overall. First, and most importantly, weeds do not have time to produce seed in two weeks. This reduces the overall weed pressure throughout the rest of the season. This allows subsequent herbicide applications to be more effective because the "seed bank" is very low. Weeding this frequently also allows crews to skip over very small weeds and only focus on weeding plants that may go to seed within the next two weeks. More information on this practice including how to implement this method has been previously published (Barker and Neal, 2016).

ENFORCE STRICT SANITATION

Proper sanitation is discussed in many different Integrated Pest Management (IPM) guides. In terms of weed management, the main principles to follow include:

- Hand weed frequently.
- Do not recycle potting media. Many growers do this to reduce costs but savings on potting media can easily be lost by heavy weed pressure (and possibly disease issues). If someone is determined to recycle soil, solarization is one way to significantly reduce weed pressure (Steed, 2015). If that is not an option or is not feasible, at least do not use this potting soil on herbicide sensitive plants because preemergence herbicides will be needed and needed often, and you do not want to limit your options. Using this soil in larger containers may also allow the use of directed postemergence herbicide applications that will be useful as weed pressure is likely to be very high. This is not a recommended practice but if someone must use old or recycled soil, these steps may help mitigate some of the negative impacts.
- Spot-spray and utilize preemergence herbicides in non-crop areas (walkways, aisles, ditches, etc.). Weeds in non-crop areas will inevitably move into the crop if they are not controlled. Non-selective herbicides such as glyphosate (or possibly glufosinate if glyphosate resistant weeds are present) can be used to control most weeds. Ultra-low volume sprayers can be utilized for applying glyphosate in these areas. These sprayers reduce the amount of water needed to make the application and allow workers to cover more ground quickly. Preemergence herbicides such as flumioxazin or indaziflam are also effective in these areas and provide broad-spectrum weed control.
- Clean pots before reusing them. Pressurized water, heated water or steam can be used to reduce weed seed presence and/or viability.
- Moving liners and other plants in the greenhouse off the ground. Tables are expensive and not feasible in all cases, but simply moving plants in the greenhouse up on benches can help prevent weed growth in the crop. This also allows the use of postemergence herbicides or indaziflam (a preemergence herbicide) to be used on the floor underneath the benches.
- Start with weed-free liners and keep liners weed free prior to potting.
- Prevent pot blow-over. Blown over pots can create several issues – first, the herbicide barrier in the top portion of the soil is broken, so weeds have greater opportunity to take over after a pot has been blown over. Secondly, spilled soil in production areas is an ideal place for weed seeds to germinate.

USE NON-CHEMICAL CONTROLS WHERE NEEDED

Some common ornamental species are notoriously sensitive to preemergence herbicide applications (e.g. hydrangea, herbaceous perennials, etc.). Crops that cannot be treated with preemergence herbicides often become "hot spots" or areas where weed growth is concentrated and prolific. Loose-fill organic mulch is much more expensive than herbicides, but much cheaper than hand weeding. Large particle, well-drained mulch materials such as pine bark nuggets, rice hulls, and other can be used to significantly reduce weed growth (Altland et al., 2016; Richardson et al., 2012). In many cases, these mulch materials can outperform herbicides and require only one application.

Fertilizer placement has also been used as a very effective way to manage weeds without herbicides. Fertilizers are most often top-dressed or incorporated, and while dibbling is effective for weed management, dibbling is associated with some crop safety concerns and is not as commonly practiced. Subdressing or layering fertilizers is a method used by some growers to eliminate the negative effects of dibbling, but also utilize the advantages of strategic fertilizer placement. When fertilizers are subdressed or layered, the pot is filled halfway at potting, fertilizer is applied in a single layer, and then the rest of the potting soil is added along with the liner. This results in a 2 to 4 in. (or greater) layer of potting soil that contains no fertilizer; thus no fertilizer is available to weeds germinating on the soil surface. In research at the University of Florida, growth and seed production of common container weed species was reduced by over 90% in several studies and other research has shown no negative impacts on crop growth (Broschat and Moore, 2003).

SELECT THE BEST HERBICIDES AND ROTATE THROUGHOUT THE YEAR

Currently there are around 25+ preemergence herbicide products labeled for over-the-top use in container plant production. Unfortunately, these herbicides are different combinations with just a few different modes of action (MOA) - including Weed Science Society of America's herbicide MOA groups 3, 14, 15, 21, and 29 (Senseman, 2007).

With most herbicides, growers are limited to around 2 to 3 applications per year, resulting in herbicide applications every 8 to 10 weeks throughout the year (or growing season, dependent upon climate). Hence, multiple herbicides will be needed. There are no "magic bullet" herbicides and all herbicides have weak areas - thus, several herbicides will be needed in a preemergence herbicide program. These products should be rotated throughout the year, utilizing herbicides with different MOA, when they are most effective, and achieving the best return on investment.

Avoiding back-to-back or sequential applications of the same herbicide can reduce chances of phytotoxicity and improve weed control. A list of common preemergence herbicides including their MOA and general weeds they control are included in Table 1. Weed efficacy guides, such as the 2017 Southeastern U.S. Pest Management Guide for Nursery Crops (Neal et al., 2017) can be used to reference which herbicides are most effective on different weeds throughout the year and determine which crops these herbicides can be applied.

The first priority (assuming the herbicide is safe for the crop) should be selecting a herbicide that is highly effective on the most troublesome weed during a particular time of year.

Table 1. Common Preemergence Herbicides Utilized in Ornamental Plant Production.			
Active ingredient	**Trade name[z]**	**MOA Group[y]**	**Weeds controlled[x]**
dithiopyr	Dimension 2EW	3	Grasses and a few broadleaves
oryzalin	Oryzalin Pro, Surflan	3	Grasses and a few broadleaves
pendimethalin	Corral, Pendulum 2G, Pendulum AquaCap, Pin-dee, etc.	3	Grasses and a few broadleaves
prodiamine	Barricade, Prodiamine, RegalKade	3	Grasses and a few broadleaves
trifluralin	Treflan, Trifluralin, Preen	3	Grasses and a few broadleaves
flumioxazin	Broadstar, SureGuard	14	Broadleaves and grasses
oxadiazon	Oxadiazon, Ronstar	14	Broadleaves and grasses
oxyfluorfen	Galigan, Goal	14	Broadleaves and grasses
dimethenamid-p	Tower	15	Grasses, sedge suppression, broadleaves
s-metolachlor	Pennant Magnum	15	Grasses, sedge suppression, some broadleaves
isoxaben	Gallery	21	Broadleaves and a few grasses
indaziflam	Marengo SC, Marengo G	29	Broadleaves and grasses
benefin + oryzalin	XL 2G	3 + 3	Grasses and some broadleaves
pendimethalin + dimethenamid-p	FreeHand	3 + 15	Grasses and broadleaves
trifluralin + isoxaben	Quali-Pro TI, Snapshot	3 + 21	Grasses and broadleaves
prodiamine + isoxaben	Gemini, Gemini G	3 + 21	Grasses and broadleaves
dithiopyr + isoxaben	Fortress	3 + 21	Grasses and broadleaves
oxadiazon + prodiamine	RegalStar II	14 + 3	Grasses and broadleaves
oxyfluorfen + oryzalin	Rout	14 + 3	Grasses and broadleaves
oxyfluorfen + pendimethalin	OH2	14 + 3	Grasses and broadleaves
oxyflourfen + prodiamine	Biathlon	14 + 3	Grasses and broadleaves
oxyflourfen + trifluralin	Granular Herbicide 75	14 + 3	Grasses and broadleaves
oxyflourfen + oxadiazon	OO-Herbicide, Two OX E-Pro, Regal OO	14 + 14	Grasses and broadleaves

[z]Trade names are included for educational purposes only and do not indicate an endorsement. Similar products may be available with the same active ingredient and/or formulation and would be suitable.

[y]MOA = mode of action group. Modes of action were adopted from Senseman, 2007.

[x]Weeds controlled only shows broad classifications of weeds. Detailed information is available on product labels and in pest management guides (Neal et al. 2017).

As several herbicides are likely effective, the next step would be to select the best option that controls both the most troublesome weed and the second or third most troublesome weed(s). For the next application, a herbicide with similar efficacy but a different MOA could be chosen. This process is repeated throughout the year until a preemergence herbicide program is developed for the year. Some of the most successful nurseries may utilize 4 or 5 (or more) different preemergence herbicide products, using each to take advantage of the products strengths such as crop safety or efficacy of key weed species during different times of year. Example rotations have been published previously (Neal et al., 2017) but many different herbicide combinations can be used successfully depending upon local conditions.

MAKE WEED CONTROL A PRIORITY

The most consistently clean nurseries all make weed management a top priority. They understand how much weed control failures cost and take necessary steps to prevent failures from happening.

Many common weed species in nurseries can begin producing seeds within three weeks. So, within a month, a second generation of weeds are beginning to germinate. Weed populations will increase exponentially until action is taken. As there are few options, prevention is always the best course of action.

Literature Cited

Altland, J.E., Boldt, J.K., and Krause, C.R. (2016). Rice hull mulch affects germination of bittercress and creeping woodsorrel in container plant culture. Amer. J. Plant Sci. 7:2359-2375.

Barker, A. and Neal, J.C. (2016). Frequent hand weeding saves money. Accessed 10 Oct. 2018. https://content.ces.ncsu.edu/frequent-hand-weeding-saves-money

Broschat, T. and Moore, K.K. (2003). Influence of fertilizer placement on plant quality, root distribution, and weed growth in container-grown tropical ornamental plants. HortTech. *13*:305-308.

Mathers, H. (2003). Novel methods of weed control in containers. HortTech. 13:28-31.

Neal, J.C., Chong, J.C., and Williams-Woodward, J. (eds.). (2017). The 2017 Southeast pest management guide for nursery crops and landscape plantings. Southern IPM Center. Accessed 9 Sept. 2018. https://content.ces.ncsu.edu/southeastern-us-pest-control-guide-for-nursery-crops-and-landscape-plantings

Richardson, B., Gilliam, C.H., Fain, G.B., and Wehtje, G.R. (2008). Container nursery weed control with pinebark mini-nuggets. J. Environ. Hort. *26*:144-148.

Senseman, S.A.(2007). Herbicide Handbook, 9th edition. Weed Science Society of America. Lawrence, KS. 458 p.

Steed, S. (2013). Large scale recycling of used potting media with solarization. SARE Project OS13-075 Final Report. Accessed 10 Oct. 2018. https://projects.sare.org/sare_project/os13-075/

Combined Proceedings IPPS 68:378-384. 2018.

IPPS European Exchange 2017

Lis Meyer
Department of Horticultural Science, North Carolina State University, Raleigh, North Carolina 27695-7609, USA.

lis_meyer@ncsu.edu

Keywords: IPPS-SRNA Early-Career Exchange Program, European green industry.

INTRODUCTION

As a young faculty member at North Carolina State University, IPPS has always been a valuable resource for me to see the concepts that I teach on campus applied in a real-world setting. It is one thing to discuss plant propagation in a lecture and to practice the techniques in a laboratory classroom—it is another to see those techniques and principles at work in the industry and hear them discussed by professionals who have been using them years to run their businesses. I have always appreciated the perspective that the annual conference for the IPPS-Southern Region of North America (IPPS-SRNA) allows me to gain. In July 2017, I had an amazing opportunity to broaden the scope of my knowledge of our industry even farther. I participated in the IPPS-SRNA Early Career Exchange Program with the European Region, which allowed me to tour sites of interest in England, Belgium, and Holland - as well as attend the European Region annual conference.

One of my mentors in the department of Horticultural Science at NC State once told me that if you can get one good photo to use for class on a trip, then that is all you need to make that trip worth the cost. According to that principle, the number of photos, videos, and relevant information for the classes that I teach made this trip invaluable for me. I have had many excellent resources of this nature passed down to me from retired faculty members at NC State. But to have content that I personally have gathered, and stories that I can tell students about the concepts we are learning that I have experienced - makes my lectures more meaningful to the students in my classes.

ENGLAND

My trip began in England, where after experiencing rather stringent questioning by customs, I made my way through Gatwick airport and boarded a train towards West Dean Gardens in West Sussex. On my arrival, I was awed by the history behind the gardens there—this one having existed on site since 1622. I was both captivated by and envious of the lush, thriving plant life after watching the water and heat stressed gardens of home succumb to the climate of eastern North Carolina in July. A highlight of this garden for me was seeing their partially underground glass cold frames where they overwintered tropical and annuals (Figure 1).

Figure 1. A cold frame containing annuals at West Dean Gardens in England.

BELGIUM

The next day, our group boarded the ferry in Calais to cross over into Belgium and continue our tour. Perhaps one of the most impressive stops of the whole trip was the Solitair Nursery, which specializes in large specimen trees. This nursery was reminiscent of an art gallery, with each of their trees displayed as a masterpiece to be considered for purchase (Figures 2 and 3).

Figure 2. Rows of trees and shrubs on display at Solitair Nursery.

Figure 3. IPPS members admiring trees at Solitair.

I was amazed that trees of this size could be successfully moved (Fig 4). I learned that years of careful irrigation and root pruning were involved in preparing the trees for eventually being transplanted.

Figure 4. I stand in front of a ball and burlapped tree at Solitair for scale.

I also noticed at Solitair what a theme for the rest of the trip would be—the entire nursery was immaculate. There was not a single weed to be seen, the boxwoods were completely free of blight, and even the buildings where equipment and machinery was stored gleamed with the obvious care that been taken to keep everything clean and orderly.

HOLLAND

Just down the road from Solitair but across the border into Holland was one of my personal favorite nurseries from the trip, Piet Vergeldt Boomkwekerij. This family-owned nursery produces around 100,000 grafted trees a year with only the five members of the family doing the grafting. In addition to a number of unique conifers, this nursery also specialized in grafting Japanese maples, magnolias, and crabapples (Figure 5).

Figure 5. Japanese maples at Piet Vergeldt Boomkwekerij.

While observing their field-grafted trees, I noticed with surprise that the leaves and branches of the rootstocks were being left on the plants long after the graft union had healed (Figures 6 and 7). When I asked about the reason for this practice, I learned that it assisted in making sure the rootstock trunks of the graft plants grew sufficiently in girth to support the scion growth. In my classes at NC State, we have experienced a problem with a similar graft that we perform with the rootstocks not being thick enough to support the vigorously growing scions. I am eager to apply Piet Vergeldt Boomkwekerij's solution to this class activity and see the results.

Figure 6. Grafted trees in the field with rootstock branches still growing.

Figure 7. A witch's broom on a conifer at Piet Vergeldt Boomkwekerij, and a photo of a miniature conifer grafted from that witch's broom.

From Piet Vergeldt Boomkwekerij, our group made a stop at Delta Works to see the Maeslantkering storm surge barrier. In 1953, a massive flood from the North Sea in Holland resulted in such a high death toll and loss of agricultural land, crops, and livestock. Hence, the Delta Works project was implemented to find and engineering solution that would prevent such a catastrophe from happening again. Our tour guides informed us that the massive gates of Maeslantkering are made up of an amount of steel equivalent to four Eiffel Towers (Figure 8). Holland has 20% of its land below sea level and 50% of it no more than three feet above sea level. As someone who was born and grew up in eastern North Carolina, specifically in an area frequently hit hard by flooding resulting from hurricanes, I was very interested in how the Dutch have addressed this issue. I hope to one day see solutions like this at work in my own home state, rather than the heartbreaking destruction and rebuilding from scratch that seems to happen again and again.

Figure 8. The Maeslantkering storm surge barrier.

The scale of the horticultural industry in Holland continued to amaze me for the duration of the trip. I watched mile upon mile of state-of-the-art glasshouses pass by my window as I rode the tour bus between stops. Each stop showed me plants in greater quantities than I had ever seen before, and those plants were always of the highest quality. In eastern North Carolina, Venus flytraps are native. At Corn.BAK nursery in Holland, there are more Venus flytraps in their greenhouses than probably in my whole state (Figure 9)!

Figure 9. A greenhouse full of Venus flytraps at Corn.BAK.

In North Carolina, roses and lilacs struggle to thrive in the heat and humidity. At the nursery we visited that markets the "Parfume of Nature" line of plants - the roses and lilacs saturated the air with their fragrances. And I saw species and cultivars that I never even knew existed. It would be impossible to describe with words or even capture in pictures the scale of the cut flower industry on display at Aalsmeer Flower Market (Figure 10). Flowers from all over the world are stored in the world's largest refrigerator during the auction and conveyed through the most complicated system of human and machine transport to buyers from all over the world.

Additionally, this trip allowed me to see the most modern nursery and propagation technologies right next to the some of the oldest roots of the horticulture industry. At Deliflor Chrysanthemum, I watched a robot sticking cuttings right next to a room of women doing the same task (Figure 11).

Figure 10. Carts of flowers being transferred for pickup by buyers on the floor of Aalsmeer Flower Market.

Figure 11. A robot and women workers sticking cuttings at Deliflor Chrysanthemum.

Our hosts described the pros and cons of each system—the women were faster than the robot, but the robot could work longer hours. Once the chrysanthemum cuttings were stuck, they were not touched by human hands again. I even witnessed the young plants being placed on their own small train to transport them out to their place in the greenhouses. This nursery, which was the most technologically advanced I had ever witnessed, was less than an hour away from the historic nursery district of Boskoop.

At Boskoop, I was astonished to see nurseries placed directly along bodies of water without fear of flooding as I toured by boat (Figure 12). This area has been involved in nursery production since 1222!

Figure 12. One of the many nurseries of the Boskoop District.

There was even a museum dedicated to the nursery industry (Figure 13). This museum and the Boskoop district as a whole stood out to me as a contrast to the United States. Horticulture in Holland is such an integral and highly visible part of their history, culture, and economy.

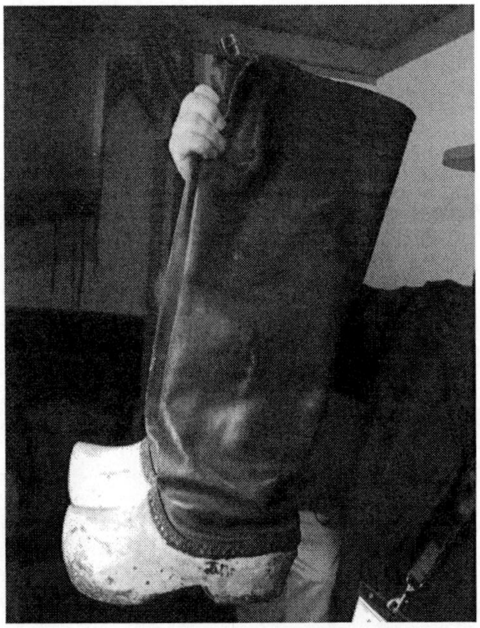

Figure 13. Traditional Dutch clog style work boots at the tree nursery museum in Boskoop.

At home, and especially in the university, we are constantly discussing how to increase the visibility and awareness of horticulture in a place where many people are not even aware of its existence.

A GREAT OPPORTUNITY

As with all IPPS trips, though, it would be a mistake to only focus on the plants. The experiences and conversations that I shared with IPPS members from all over the world made this experience truly memorable. I was able to strengthen relationships with members from my own region and to make new ones with members from regions in Europe, Australia, and New Zealand. I made new friends who I will look forward to seeing throughout the years to come - as we "seek and share" together in this amazing industry. I am extremely grateful for my experiences with the IPPS-SRNA Early-Career Exchange Program, and I will be sure to recommend it to my students and other young professionals.

Combined Proceedings IPPS 68:385-395. 2018.

Breeding Powdery Mildew Resistant Dogwoods and More at Rutgers University

Thomas J. Molnar
Department of Plant Biology and Pathology, The School of Environmental and Biological Sciences, Rutgers University, 59 Dudley Road, New Brunswick, New Jersey 08901, USA.

molnar@aesop.rutgers.edu

Keywords: Holly, *Ilex* sp., *Cornus* sp., hazelnuts, *Corylus* sp., *Erysiphe pulchra*, woody ornamental breeding, clonal propagation, budding, seedlings, molecular tools, novel cultivars.

INTRODUCTION

The Rutgers University Woody Ornamental Breeding Program began in 1960 under the direction of Dr. Elwin Orton. The early focus of the program was the breeding of hollies (*Ilex* sp.), with work on big-bracted dogwoods (*Cornus* sp.) starting in the 1970s. The program continues today with the addition of hazelnuts (*Corylus* sp.) for nut production and ornamentals. Over 40 cultivars have been released since the initiation of the program, and a number have become widely grown in the nursery and landscape trade. A list of releases can be found in Molnar and Capik (2013). The most noteworthy Rutgers introductions, at least from the perspective of plants propagated and sold thus far, are likely the hybrid dogwoods. These were largely the results of crossing *Cornus kousa* with *C. florida* to create a series of unique plants (subsequently named *Cornus* × *rutgersensis* [Mattera et al., 2015]) that combined traits of both parental species to create attractive, high-value landscape specimens. The hybrids generally exhibit increased vigor over their parental species and better drought tolerance than *C. florida.* However, their most important attribute is likely their resistance or increased tolerance to diseases such as dogwood anthracnose (*Discula destructiva*) and powdery mildew [(PM), *Erysiphe pulchra*) (Hibben and Daughtrey, 1998; Ranney et al., 1995)]. They also tend to get significantly fewer stem borers, extending their lifespan and reducing maintenance needs in the landscape.

Seven F_1 hybrids were released from the program, mostly from 1990-1991 (Orton, 1993), with the most popular and widely grown being 'Rutgan' Stellar Pink® (Figure 1). Each cultivar differs in its floral attributes and growth habits and exhibits sterility,

Figure 1. *Cornus* × *rutgersensis* 'Rutgan' Stellar Pink® hybrid dogwood.

avoiding any concerns of naturalization in the landscape while also making them a dead-end for further breeding. While only the sterile selections were originally released, Dr. Orton did develop a small number of semi-fertile interspecific hybrids (trees that produced a very small number of fertile seeds) that formed the basis for further hybrid breeding.

POWDERY MILDEW IMPACTS ON *CORNUS FLORIDA*

In the 1990s, introduction of the PM pathogen *Erisyphe pulchra* caused major issues for *C. florida*, which is generally very susceptible to the disease (Li et al., 2009; Ranney et al., 1995). It causes a white bloom of mycelium on the leaves and can cause unattractive leaf distortion and red and brown coloration, leading to a significant reduction in overall aesthetic value, plant health, and vigor (Klingeman et al., 2004; Figure 2). This damage is especially evident in nursery operations where plants are grown in dense rows in close proximity (Figures 3 and 4). Individual trees in the landscape can vary in their reaction to PM, as environmental factors like wind and rainfall influence the severity of the disease.

Figure 2. Powdery mildew caused by *Erisyphe pulchra* on leaf of *Cornus florida.*

However, in general, susceptible cultivars lose leaf quality and health especially later in the summer, and the weakened trees tend to decline in vigor over subsequent years and succumb to stress-related disorders like stem borers, which ultimately can kill the trees.

Prior to the 1990s, in the absence of the disease, the breeding and selection for resistance to PM was not an objective of the Rutgers breeding program. The introduction of PM directly impacted several cultivars that were developed and released but which ultimately proved too susceptible for commercial production.

An example includes *C. florida* 'D-376-15' Red Beauty®, which is a very attractive, dark-red bracted specimen tree (Figures 5 and 6) that gets too much PM in liner production to be commercially viable (despite being patented it never became commercially available). A similar story could almost be told for 'Rutnut' Red Pygmy®. Fortunately, its unique dwarf form and deep red floral bracts make it novel enough to still be grown on a small scale despite its high susceptibility to PM.

However, the disease generally makes the tree quite unattractive later in the summer, and likely reduces its lifespan in the landscape.

Figure 3. Powdery mildew can cause significant damage, including twisting of leaves and young stems, to susceptible cultivars grown under nursery conditions without the use of fungicides.

Figure 4. *Cornus florida var. forma rubra* (center) and 'Cherokee Brave' (right) showing differential expression to powdery mildew caused by *Erisyphe pulchra*. Picture taken at Hidden Hollow Nursery, Belvidere, TN, October 25, 2018.

Figure 5. Close up of floral bracts of *Cornus florida* 'D-376-15' Red Beauty®, a cultivar developed at Rutgers University but never commercialized due to susceptibility to powdery mildew.

Figure 6. *Cornus florida* 'D-376-15' Red Beauty® at a distance.

In 2006, with Dr. Orton's impeding retirement in 2008 approaching, the Rutgers ornamental program was bolstered by the knowledge that a new breeder (Dr. Tom Molnar) would be hired to continue the project. Working with Dr. Orton, we began by growing thousands of new seedlings derived from open-pollinated (OP) seed originating from a crossing block of unique breeding selections developed and amassed over the previous 40 years. This collection was largely made up of fertile advanced-generation interspecific hybrids, with genetic backgrounds spanning *C. kousa*, *C. florida*, and *C. nuttallii* in a variety of combinations (some known and others somewhat unknown due to open-pollination events in their pedigrees). Based on the out-crossing nature of *Cornus* and the wide diversity of plant material intercrossed, significant variation in many traits was expected in the offspring. The hopes were that some new cultivars might come from this process, or at least some interesting new breeding material to help continue improvement efforts.

As part of this new program, trees germinated from the OP seed were "screened" for tolerance to PM in the greenhouse, where this disease can become quite severe. This allowed for susceptible seedlings to be discarded prior to being field planted for further evaluation (about 25% of the seedlings were discarded at this stage). Several thousand OP trees were thus screened then field planted during 2006 to 2008.

The highest priority breeding goal at the time was to develop the coveted "dark pink" kousa or hybrid dogwood that Dr. Orton worked toward for much of his career (Orton, 1985). Fortunately, we had some very exciting developments on that front in this first generation of new seedlings, exemplified by the release of *C. kousa* 'Rutpink' Scarlet Fire®, which is described in more detail in Molnar (2017) and Molnar et al. (2017). Unexpectedly recovering many dark pink-bracted seedlings in the 2006-2008 planted OP populations reduced the attention put on breeding improved PM-resistant *C. florida*. These efforts were already impacted by the relatively limited germplasm base of PM resistant/tolerant *C. florida* available at Rutgers (Dr. Orton had not worked on *C. florida* since the early 1990s). These limitations, combined with the known rarity of PM resistance in the species (Windham and Witte, 1998), made developing a concerted effort to breed improved PM-resistant *C. florida* daunting and not likely achievable in the near term.

UNEXPECTED NEW SOURCE OF PM RESISTANCE

In 2011, 11 trees from an OP seed lot recorded to be from KF95-1, a *C. kousa* × *C. florida* breeding accession backcrossed to *C. florida* (BC_1), bloomed for the first time. By appearance, these trees looked 100% *C. florida*, despite a pedigree showing them as interspecific hybrids. They were pretty trees (mostly with white bracts, but a few with blush-pink) but not outstanding at first glance in comparison to other *C. florida* selections in the nursery trade. The trees were not given a lot of attention, but since breeding records indicated they came from a hybrid mother tree (representing a possible BC_2 generation assuming their pollen parents were *C. florida*), they were not cut down and were allowed to grow on for further evaluation.

A very bad PM year in 2013 highlighted this particular group of seedlings. Many trees in our program, including known hybrids and even some *C. kousa* selections expected to be highly tolerant to PM, showed disease symptoms that year. Amazingly, several seedlings in the KF95-1 OP seed lot remained free of PM and several others appeared highly tolerant (PM was found on only a relatively small percentage of their leaves). This was unexpected based on their pure *C. florida* phenotypes and the high disease pressure apparent in our fields. Their freedom from PM and the known rarity of PM resistance in the species indicated they might hold some previously unnoticed value for breeding.

Our first step was to clonally propagate the most attractive seedling that was free of PM, today known as selection Rutgers 15-25 (Figures 7, 8, and 9), to see if the resistance held up in replication in our plots and at a different location. This was done with the help of Alex Neubauer at Hidden Hollow Nursery in Belvidere, TN. Multiple trees of Rutgers 15-25 were budded in the summers of 2014, 2015, and 2016 and evaluated for response to PM the following years in the nursery under standard practices and conditions generally suitable for high PM pressure.

Figure 7. Close up image of leaves of Rutgers breeding selection Rutgers 15-25 free of powdery mildew during high disease pressure year.

To our surprise, each year the trees came through evaluations with no signs or symptoms of PM, whereas the standard cultivars propagated at the nursery had their typical amounts of PM and seedling understocks for next year's budding needed significant fungicide applications to remain healthy. The resulting budded trees were dug and bare-rooted each year and sent back to Rutgers where they were lined out in the field or planted in containers for further evaluation. Here, our results showed they also remained free of PM in the greenhouse and field under what we considered high disease pressure, with no preventative fungicide applications.

Figure 8. Grafted tree of Rutgers 15-25 free of powdery mildew in summer 2017, a year of very high disease pressure.

It should be noted that the original tree, Rutgers 15-25, also remains free of PM to date (October 2018) in the field, now spanning 11 growing seasons with no signs or symptoms of disease. Collectively, these trials provided a high level of confidence that this source of resistance was worthy of further exploration.

Looking to capitalize on the new source of PM resistance in breeding, we made a small number of controlled crosses with Rutgers 15-25 and some of its siblings in 2014, which were subsequently evaluated for PM response in the field. While the first populations were too small to examine genetic control, we did recover a significant number of resistant offspring, suggesting that the resistance is heritable and can be used in further breeding.

Figure 9. Floral bracts of Rutgers 15-25.

In more recent years, many additional controlled crosses have been made and large populations are now growing in the field and greenhouse for evaluation and study.

WHERE DID THIS RESISTANCE COME FROM?

Breeding records indicate that Rutgers 15-25 and its siblings originated from seed collected from KF95-1, which is believed to be a backcross hybrid (BC_1) to *C. florida*. Interestingly, KF95-1 is the OP offspring of KF45-29 (Figure 10), which is a full sibling of Stellar Pink®, whose shared parents are *C. kousa* K2 × *C. florida* 'Sweetwater Red' (Mattera et al., 2015).

Figure 10. Blush pink floral bracts of KF45-29 dogwood.

KF45-29 is unique in the fact that it is semi-fertile (Stellar Pink® is sterile). While most fruits are empty, a very small percentage (much less than 1%) set a viable seed. Dr. Orton took advantage of this fact and collected thousands of fruits over many years to search for seed. When he found them, he grew out the rare seedlings to add diversity to his breeding pool.

As a side note and point of reference, 'KF111-1' Hyperion® also originated from seed collected from KF45-29. Hyperion® is an extremely vigorous, white-bracted hybrid dogwood patented and released in 2011 (Figure 11).

Figure 11. Floral bracts of *Cornus* × *rutgersensis* 'KF111-1' Hyperion® dogwood.

It is believed to be a backcross hybrid to *C. kousa*, based on records that suggest KF45-29 bloomed later in the season the year the OP seed was harvested and would have been pollinated by *C. kousa* in the landscape at the research farm (Rutgers Gardens, New Brunswick, NJ). Note that *C. florida* and *C. kousa* bloom about a month apart, and the F_1 hybrids typically bloom at a date between the two parental species. Interestingly, Hyperion® is relatively fertile and can yield fertile offspring. Its fruit, especially when it contains a seed, clearly displays its hybrid background, expressing attributes of both *C. florida* and *C. kousa* (Figure 12).

Figure 12. Fruit of *Cornus* × *rutgersensis* 'KF111-1' Hyperion® dogwood containing viable seed represented as bulging fruitlet.

Figure 13. Fruit of KF95-1 dogwood containing viable seed.

KF95-1, the recorded female parent of our source of PM resistance, is a half sibling of Hyperion®. As mentioned previously, it too is derived from an open pollination event of KF45-29. However, breeding records indicate flowering was during a year where KF45-29 bloomed early and during the late end of the *C. florida* bloom period. Thus, if the records are correct, KF95-1 is a BC_1 hybrid with *C. florida*. The phenotype of KF95-1 suggested this is the case, including its early bloom time closer to that of pure *C. florida*. Unfortunately, KF95-1 was cut down a few years ago before we realized its potential value, and a picture of the bracts are not available. However, an image of its fruit with viable seed was available that clearly shows its hybrid nature, similar in form to that of Hyperion® (Figure 13).

Since the KF95-1 bloom period in 2006 overlapped that of the later end of *C. florida*, the OP seed collected are expected to be the result of pollination by *C. florida* and represent a possible BC_2 generation. With a significant portion of their genome now coming from *C. florida* (the genomic contribution of BC_2 hybrids would be approximately 87.5 % *C. florida*), we did not consider it unrealistic that the plants would look very much like *C. florida* in phenotype.

As such, we operated under the assumption that the records were correct. However, considering that the PM resistance may have been derived from *C. kousa* (which would be of substantial breeding value since *C. kousa* is generally very resistant to PM), we decided to investigate further. Thus, we turned to molecular tools to examine this possibility with more certainty.

MOLECULAR TOOLS PROVIDE INSIGHT INTO THE ORIGIN OF PM RESISTANCE

The first glimpse of the genetic background of the PM resistant seedlings was completed as part of a Master's degree thesis by Robert Mattera (2016). In his study, 11 simple sequence repeat (SSR) markers were used to fingerprint 337 accessions, which included over 50 *C. florida,* as well as many *C. kousa* and interspecific hybrids of diverse genetic backgrounds and combinations held in the Rutgers collection with some contributed by the University of Tennessee. While the primary goal was to examine the genetic diversity present in the Rutgers and University of Tennessee collections, the inclusion of the OP seedlings of KF95-1 was useful for shedding light on their interrelationships and genetic background. Results showed that the seedlings were grouped tightly together in one clade, suggesting a common parent and supporting the hypothesis that they came from the same seed lot. This was an important finding—they most likely came from a single mother tree. However, the clade in which they were placed was clearly nested in the well-defined *C. florida* subgroup of the study, which was strongly supported and contained plants only known to be of *C. florida* species background. None of the known *C. kousa* or hybrids in the study were placed in the *C. florida* group (alternatively, and as even greater support for the groupings, no known *C. florida* accessions were place exterior to the *C. florida* group). This finding for the KF95-1 seedlings was unexpected based on their breeding history and was our first indication that the records could be wrong. Note that KF95-1 was unfortunately cut down and DNA not available for the SSR study.

While the SSR results made us question the origin of the KF95-1 seedlings and their reported connection to *C. kousa*, 11 SSR markers cannot clearly define the species background of a potential interspecific hybrid, especially those with a very high percentage of genetic composition from one parental species, such as in BC_2 hybrids. More work would need to be done to confirm their true origin. Regardless of this fact, one useful point was that the seedlings were shown to be genetically distinct from a number of the PM resistant and tolerant *C. florida* cultivars and selections included in the study (e.g., Appalachian Mist) that were developed at the University of Tennessee (Windham et al., 2003). The Tennessee-sourced plants were generally all placed in Subgroup 2 of the *C. florida* group in the dendrogram. This provided good evidence that the "KF95-1" seedlings are at least from a different, distinct genetic background and may possibly carry an unrelated source of PM resistance (Mattera, 2016).

Next, we considered a different approach to test the hypothesis that there are *C. kousa* genes present in the PM resistant seedlings. Knowing that chloroplasts are maternally inherited in most higher plants, we set out to identify if the chloroplasts of the seedlings were derived from *C. kousa* (the maternal line of KF95-1 is *C. kousa*). To do this, Muehlbauer et al. (2018) sequenced three conserved chloroplast genes in 11 of the PM resistant seedlings, as well as a panel of known *C. florida* and *C. kousa* cultivars and their reported mother KF95-1 (a dry herbarium specimen was identified and used to extract chloroplast DNA of KF95-1) to identify their maternal species lineage. The chloroplast genes included were *matK*, *rbcL*, and *ycf1*, which have been used previously in phylogenetic analysis of the Cornaceae family (Xiang et. al., 1998, 2002). The genes were successfully sequenced, and BLAST searches including the consensus sequence search revealed that each of the PM-resistant seedlings showed very close alignment with *C. florida* (again in contrast to our breeding records).

However, there were some confounding sequencing results for the *C. kousa* controls where they did not explicitly discern *C. kousa* from *C. florida* cultivars in all cases, leaving some level of ambiguity. Further, gene sequence identities were also ambiguous for KF95-1, which made it hard to draw conclusions on its maternal origin (*C. kousa* or *C. florida*). However, since the results of KF95-1 were different from that of the 11 seedlings, the results provided some further evidence that they were not carrying a similar chloroplast origin. Unfortunately, due to the somewhat unclear results with *C. kousa*, a deeper study with higher resolution would be needed to fully elucidate their maternal origin.

Fortuitously, in the research lab of Dr. Robert Trigiano at the University of Tennessee, Nowicki et al. (2018) was working on a study to analyze the chloroplast DNA diversity of *C. florida* and *C. kousa* using a high-resolution chlorotyping system developed for *C. florida* (Call et al., 2015). They agreed to include Rutgers 15-25 in their study, which included 332 accessions in total. Their recently published results clearly showed significant differences in chlorotype frequencies between the two species and placed Rutgers 15-25 with the *C. florida* accessions (Nowicki et al., 2018). Thus, the chloroplast in Rutgers 15-25, and likely all of its siblings, was derived from *C. florida*. This new, very well-supported evidence, combined with the previous two studies, make it clear that the seedlings do not fit their pedigree records. They are not the offspring of KF95-1 and most likely are pure *C. florida*. At some point, there must have been a mislabeling in the greenhouse or field. The results are humbling and a reminder of the value of clear notetaking and record keeping, as well as the value of molecular tools in clarifying and supporting breeding records.

PM RESISTANCE FROM UNCERTAIN ORIGIN, BUT IT STILL WORKS!

While the resistance might not come from the origin that was first thought, it appears to be extremely useful resistance that has held up across multiple locations and multiple years and is transmitted to its offspring. Very few other sources of PM resistance exist can claim to have held up completely free of PM in propagation nurseries in TN (most purported sources of resistance are in fact only tolerant; they get some PM in some years). While further testing is needed across a wider area, results to date are very promising. Now the question arises, where did this source of resistance come from? The Rutgers collection of *C. florida* was very small at the time the seed was collected, and none of the trees in the collection remain free of PM each year. Our SSR data tells us that the seedlings were likely collected from the same mother plant (or related plants), but we have no additional information on the source of the seed and no obvious suspects in our germplasm collection.

Regardless of its unknown origin, we are now using this source of resistance in a large-scale breeding program with the goal of developing a series of improved PM resistant plants with a variety of attractive plant types and bract colors. The current research objective is to study inheritance in a more systematic manner and characterize the gene or genes for resistance, similar to the work described by Parikh et al. (2016, 2017). We are also developing a large, full-sibling pseudo-F_2 mapping population to develop a genome-by-sequencing based genetic linkage map to hopefully identify quantitative trait loci associated with resistance. A more comprehensive diversity study, along with the creation of molecular tools to differentiate hybrid species composition, is also underway to help understand the hybrid plant material now being used in breeding at Rutgers. When used alongside the study of other genes for PM resistance (e.g., those discussed in Parikh et al. [2016, 2017]), molecular tools should help us design populations suitable for pyramiding genes and ultimately release improved, novel cultivars that express durable forms of resistance to PM.

CONCLUSIONS

A new source of PM resistance has been identified in the *Cornus* germplasm base held at Rutgers University. Breeding records indicated that the plants were of interspecific origin, but plant phenotypes and molecular studies strongly suggest otherwise.

Although humbling, by using molecular tools, we have been able to better understand the species origin of the plant material in question and can now make better informed decisions about its use in breeding. Regardless of our records, resistance to PM is exceedingly rare in *C. florida*; thus, this new heritable source may hold significant value in breeding improved dogwood cultivars. Systematic breeding is now in progress alongside genetic studies to better understand the source of resistance with the goal of using it to develop a series of new, high-value dogwood plants that require reduced chemical inputs in the nursery - and live longer, healthier - and more attractive lives once planted into landscapes.

Literature Cited

Call, A., Sun, X.Y., Yu, Y, Pearman, P.B., Thomas, D.T., Trigiano, R.N., Carbone, I. and Xiang, Q. (2015). Genetic structure and post-glacial expansion of *Cornus florida* L. (Cornaceae): integrative evidence from phylogeography, population demographic history, and species distribution modeling. J. Systematics Evol. *54*:136–51.

Hibben, C.R. and Daughtrey, M.L. (1998). Dogwood anthracnose in the northeastern United States. Plant Dis. *72*:199-203.

Klingeman, W.E., Eastwood, D.B., Brooker, J.R., Hall, C.R., Behe, B.K. and Knight, P.R. (2004). Consumer survey identifies plant management awareness and added value of dogwood powdery mildew resistance. HortTech. *14*:275–282.

Li, Y., Mmbaga, M.T., Windham, A.S, Windham, M.T., and Trigiano, R.N. (2009). Powdery mildew of dogwoods: current status and future prospects. Plant Dis. *93*:1084-1092.

Mattera, R. (2016). The Rutgers hybrid dogwood: naming and genetic diversity analysis. Master thesis (New Brunswick, USA: Rutgers, The State University of New Jersey), p.1–195. https://rucore.libraries.rutgers.edu/rutgers-lib/50049/PDF/1/play/

Mattera, R., Molnar, T., and Struwe, L. (2015). *Cornus* × *elwinortonii* and *Cornus* × *rutgersensis* (Cornaceae), new names for two artificially produced hybrids of big-bracted dogwoods. PhytoKeys 55: 93–111. https://doi.org/10.3897/phytokeys.55.9112

Molnar, T.J. and Capik, J.M. (2013). The Rutgers University woody ornamentals breeding program: past, present, and future. Acta Hort. *990*:271–280.

Molnar. T.J., Muehlbauer, M., Wadl, P., and Capik, J. (2017). *Cornus kousa* 'Rutpink' (Scarlet Fire®) dogwood. HortScience. *52*:1438–1442.

Molnar, T.J. (2017). Going nuts: continuing a 40-year-old woody ornamental breeding program. Acta Hort. 1174: 305-312 (Comb. Proc. Intern. Plant Prop. Soc. *67*:305-312; DOI: 10.17660/ActaHortic.2017.1174.62).

Muehlbauer, M., Honig, J., and Molnar, T. (2018). The use of chloroplast gene sequences to confirm maternal backgrounds of powdery mildew resistant interspecific hybrid dogwood (*Cornus kousa* × *C. florida*). Acta Hort. *1191*:17-26. (DOI 10.17660/ActaHortic.2018.1191.3)

Nowicki, M., Boggess, S.L., Saxton, A.M., Hadziabdic,D., Xiang, Q-YJ., Molnar,T., Huff, M.L., Staton, M.E., Zhao,Y. and R. N. Trigiano. (2018). Haplotyping of *Cornus florida* and *C. kousa* chloroplasts: insights into species-level differences and patterns of plastic DNA variation in cultivars. PLoS ONE 13(10): e0205407. https://doi.org/10.1371/journal.pone.0205407

Orton, E.R. (1985). Interspecific hybridization among *Cornus florida*, *C. kousa*, and *C. nuttallii*. Comb. Proc. Int. Plant Prop. Soc. *35*:655-661.

Orton, E.R. (1993). New large-bracted dogwoods from Rutgers University. Comb. Proc. Int. Plant Prop. Soc. *43*:51-54.

Parikh, L.P., Mmbaga, M.T., Kodati,S., Zhang, G. (2016). Estimation of narrow sense heritability of powdery mildew resistance in pseudo-F_2 (F_1) population of flowering dogwoods (*Cornus florida* L.). Eur. J. Plant. Pathol. *145*:17–25.

Parikh, L.P., Mmbaga, M.T., Kodati, S., Blair, M., Hui, D., and Meru, G. (2017). Broad-sense heritability and genetic gain for powdery mildew resistance in multiple pseudo-F_2 populations of flowering dogwoods (*Cornus florida* L.). Scient. Hort. *213*:216–221.

Ranney, T.G., Grand, L.F., and Knighten, J.L. (1995). Susceptibility of cultivars and hybrids of kousa dogwood to dogwood anthracnose and powdery mildew. J. Arboriculture *21*:11-16.

Windham, M.T., and Witte, W.T. (1998). Naturally occurring resistance to powdery mildew in seedlings of *Cornus florida*. J. Environ. Hort. *16*:173–175.

Windham M.T., Witte, W.T., and Trigiano, R.N. (2003). Three white bracted cultivars of *Cornus florida* resistant to powdery mildew. HortScience *38*:1253-1255.

Xiang, Q., Soltis,D., and Soltis, P. (1998). Phylogenetic relationships of Cornaceae and close relatives inferred from matK and rbcL sequences. Amer. J. Bot. *85*:285. https://doi.org/10.2307/2446317

Xiang, J.Q., Moody, M.L., Soltis, D.E., Fan,C., and Soltis, P.S. (2002). Relationships within Cornales and circumscription of Cornaceae-matK and rbcL sequence data and effects of outgroups and long branches. Mol. Phylogenet. Evol. *24*:35–57. https://doi.org/10.1016/S1055-7903(02)00267-1

Combined Proceedings IPPS 68:396-401. 2018.

Crop Efficacy and Weed Control Evaluation of a New Herbicide for Herbaceous Ornamentals

John R. Penney, Adam F. Newby, and J. Raymond Kessler
Auburn University, Dept. of Horticulture, Auburn, Alabama 36849, USA

jrp0042@auburn.edu

Keywords: Container nurseries, sensitive crops, Fortress, preemergence.

Abstract

There is a need for preemergence herbicides in the container nursery industry due to the high cost of hand weeding, but few preemergence herbicides are labeled for sensitive herbaceous ornamental crops. Currently, Snapshot (isoxaben + trifluralin) is the only granular formulated preemergence herbicide labeled for many sensitive herbaceous ornamental crops. Fortress (isoxaben + dithiopyr) is a new granular preemergence herbicide made by OHP, Inc. for use on sensitive herbaceous ornamental plants. In this experiment, four species of ornamental herbaceous crops in #1 containers were treated with Fortress at 150, 300, and 600 lbs/A, a spray combination of Gallery (isoxaben) plus Dimension (dithiopyr) at 0.75+0.38 lbs ai/A, and Snapshot (isoxaben + trifluralin) at 150 lbs/A. Also, #1 containers filled with amended 6 pine bark : 1 sand substrate were treated with Fortress at 100, 150, and 200 lbs/A, Gallery (isoxaben) plus Dimension (dithiopyr) at 0.75+0.38 lbs ai/A, and Snapshot (isoxaben + trifluralin) at 150 lbs/A and then overseeded with 25 seeds of either oxalis, bittercress, eclipta, phyllanthus, spurge, or crabgrass. Fortress had no effect on size index and caused no significant phytotoxicity of crops tested. Fortress had excellent control of bittercress and oxalis 30 and 60 DAT, and significantly better control than other treatments 90 DAT. Fortress controlled eclipta well 30 and 60 DAT. It provided good control of spurge 30 DAT, but almost none 90 DAT. Fortress provided poor phyllanthus control 60 DAT while Snapshot provided excellent control. All herbicide treatments provided excellent crabgrass control 90 DAT. Fortress had no effect on size index and caused no significant phytotoxicity of crops tested.

INTRODUCTION

A major challenge for container nurseries is the high cost of labor for hand weeding (Gilliam et al., 1990). Labor costs can be reduced drastically when preemergence herbicides are used to control weeds compared to hand weeding. In addition to cost, another problem for container nurseries is that there are limited management strategies for specific ornamental plants (Fausey, 2003). Weed management in herbaceous ornamental crops is difficult because there are few herbicides that can be tolerated by these plants (Case et al., 2005). Injuries from preemergence herbicides to herbaceous ornamental crops can include leaf burning and stunted growth (Derr, 1994).

Snapshot TG (isoxaben + trifluralin) has been shown to be one of the safest preemergence herbicides on a variety of herbaceous ornamental crops that controls a broad spectrum of grass and broadleaf weeds (Fain et al., 2006; Mervosh and Ahrens, 1998; Porter, 1996; Thetford et al., 1995). OHP, Inc. has developed a new broad spectrum granular preemergence herbicide, Fortress, for sensitive herbaceous ornamental crops in the nursery industry. Fortress contains the active ingredients isoxaben and dithiopyr. These active ingredients have been available separately in liquid sprays, but they have not been previously available in a combined granular product. The objective of this research is to evaluate Fortress for crop safety on several herbaceous ornamental crops for over-the-top application and efficacy on six common nursery weeds.

MATERIALS AND METHODS

On 30 March 2017, #1 containers were filled with a 6 pine bark : 1 sand (by volume) substrate amended with 4 lbs 4 oz dolomitic lime/yd^3, 14 lbs 7 oz/yd^3 15-9-12 with 12 to 14-month Osmocote controlled release fertilizer, and 1 lb 5 oz/yd^3 MicroMax micro-nutrient package. On 12 April 2017, 20 pots each were treated with Fortress at 100 lbs/A, 150 lbs/A, and 200 lbs/A, Gallery (isoxaben, DOW Chemical, Midland, MI) plus Dimension (dithiopyr, DOW Chemical, Midland, MI) spray at 0.75 + 0.38 lbs ai/A, Snapshot 2.5TG (isoxaben + trifluralin, DOW Chemical, Midland, MI) at 150 lbs/A, or left untreated (control). On 13 April 2017, ten pots of each treatment were seeded with 25 oxalis (*Oxalis stricta*), and ten pots of each treatment were seeded with bittercress (*Cardamine hirsuta*). Pots were placed by weed species in a randomized complete block design with ten blocks and one pot per block under a retractable roof greenhouse with open sides and closed roof and under overhead irrigation. Percent coverage ratings were recorded 30, 60, and 90 days after treatment (DAT).

On 13 June 2017, #1 containers were filled with a 6 pine bark : 1 sand (by volume) substrate amended with 4 lbs 4 oz dolomitic lime/yd^3, 14 lbs 7 oz/yd^3 15-9-12 with 12 to 14-month Osmocote controlled release fertilizer, and 1 lb 5 oz/yd^3 MicroMax micronutrient package. On 14 June 2017, pots were treated with the same herbicide treatments as those treated on 13 April 2017. On 15 June 2017, ten pots of each treatment were seeded with 25 seeds of eclipta (*Eclipta prostrata*), longstalk phyllanthus (*Phyllanthus tenellus*), spurge (*Euphorbia maculata*), or crabgrass (*Digitaria ciliaris*) per pot. Pots were placed by weed species in a randomized complete block design with ten blocks and one pot per block in full sun under overhead irrigation. Percent coverage ratings were recorded 30, 60, and 90 DAT.

On 11 April 2017, #1 containers were filled with a 6 pine bark : 1 sand (by volume) substrate amended with 4 lbs 4 oz dolomitic lime/yd^3, 14 lbs 7 oz/yd^3 15-9-12 12 to 14-

month Osmocote controlled release fertilizer, and 1 lb 5 oz/yd^3 MicroMax micro-nutrient package. Liners of tickseed (*Coreopsis grandiflora* 'Baby Sun'), purple coneflower (*Echinacea purpurea*), switchgrass (*Panicum virgatum*), and black-eyed Susan (*Rudbeckia fulgida* var. *sullivantii* 'Goldsturm') were transplanted one per container into 48 containers per species. On 12 April 2017, ten plants each plants were treated with Fortress at 150 lbs/A, 300 lbs/A, and 600 lbs/A, Gallery (isoxaben, DOW Chemical, Midland, MI) plus Dimension (dithiopyr, DOW Chemical, Midland, MI) spray at 0.75 + 0.38 lbs ai/A, Snapshot 2.5TG (isoxaben + trifluralin, DOW Chemical, Midland, MI) at 150 lbs/A, or left untreated (control). Plants were placed in a completely randomized block design by species with ten blocks and one plant per block under overhead irrigation. Plants were placed in full sun. Plants were evaluated for phytotoxicity 30, 60, and 90 DAT. Size indices and flower counts were recorded 90 DAT.

The treatment design for all experiments was a 2-way factorial of herbicide treatment and DAT. Differences among herbicide treatment means were determined using the simulated method. Linear and quadratic trends over DAT were tested using model regression.

RESULTS AND DISCUSSION

For all weed species, the herbicide treatment by DAT interaction was significant. All herbicide treatments provided excellent oxalis control 30 and 60 DAT, but by 90 DAT Fortress at 100 and 200 lbs/A provided significantly better control than Gallery plus Dimension or Snapshot (Table 1).

Table 1. Herbicide efficacy on oxalis percent coverage 30, 60, and 90 DAT.[z]

Treatment	Rate	30 DAT	60 DAT	90 DAT	Sign.[y]
Fortress	100 lbs/A	0.0b[x]	0.0	8.2d	NS
Fortress	150 lbs/A	0.0b	0.0b	13.7cd	NS
Fortress	200 lbs/A	0.0b	0.0b	3.6d	NS
Gallery + Dimension	0.75+0.38 lbs ai/A	0.0b	0.0b	36.0b	Q***
Snapshot 2.5TG	150 lbs/A	0.0b	0.5b	26.6bc	Q**
Control		62.0a	83.1a	97.5a	L***

[z]The treatment by days after treatment interaction was significant for percent coverage at $P < 0.05$.
[y]Not significant (NS) or significant linear (L) or quadratic (Q) trends of percent coverage across DAT using qualitative-quantitative model regression at $P < 0.01$ (**) or $P < 0.001$ (***).
[x]Least squares means comparisons among treatments (lower case in columns) using the simulated method at $P < 0.05$.

Bittercress control among all herbicide treatments was also excellent 30 and 60 DAT (Table 2). By 90 DAT, Fortress at 150 lbs/A and 200 lbs/A provided significantly better control than Gallery plus Dimension or Snapshot.

All herbicide treatments provided good eclipta control 30 DAT, but control 60

DAT was moderate (Table 3). The only treatment with significantly better control than the control 90 DAT was Fortress at 200 lbs/A.

All herbicide treatments provided good spurge control 30 DAT and moderate control 60 DAT (Table 4). However, percent coverage in pots of all herbicide treatments was similar to the control 90 DAT.

Table 2. Herbicide efficacy on bittercress percent coverage 30, 60, and 90 DAT.[z]

Treatment	Rate	30 DAT	60 DAT	90 DAT	Sign.[y]
Fortress	100 lbs/A	0.0ns[x]	0.0b	46.5ab	Q**
Fortress	150 lbs/A	0.0	0.0b	35.1b	Q**
Fortress	200 lbs/A	0.0	0.0b	12.4c	Q*
Gallery + Dimension	0.75+0.38 lbs ai/A	0.0	0.0b	69.5a	Q***
Snapshot 2.5TG	150 lbs/A	0.0	0.0b	69.5a	Q***
Control		30.5	50.5b	81.5a	L**

[z]The treatment by days after treatment interaction was significant for percent coverage at $P < 0.05$.

[y]Linear (L) or quadratic (Q) trends of percent coverage across DAT using qualitative-quantitative model regression at $P < 0.05$ (*), $P < 0.01$ (**) or 0.001 (***).

[x]Least squares means comparisons among treatments (lower case in columns) using the simulated method at $P < 0.05$.

Table 3. Herbicide efficacy on eclipta percent coverage 30, 60, and 90 DAT.[z]

Treatment	Rate	30 DAT	60 DAT	90 DAT	Sign.[y]
Fortress	100 lbs/A	4.3ns[x]	49.1ab	60.5ab	L**
Fortress	150 lbs/A	2.0	29.1b	81.0ab	L***
Fortress	200 lbs/A	0.1	6.3b	46.1b	L***
Gallery + Dimension	0.75+0.38 lbs ai/A	0.3	11.0b	67.0ab	L***
Snapshot 2.5TG	150 lbs/A	1.2	39.0b	68.5ab	L***
Control		26.6	91.0a	91.5a	Q**

[z]The treatment by days after treatment interaction was significant for percent coverage at $P < 0.05$.

[y]Linear (L) or quadratic (Q) trends of percent coverage across DAT using qualitative-quantitative model regression at $P < 0.01$ (**) or 0.001 (***).

[x]Least squares means comparisons among treatments (lower case in columns) using the simulated method at $P < 0.05$.

Phyllanthus control by herbicide treatments was good 30 DAT, although percent coverage of pots treated with herbicide, which ranged from 0.4% to 1.4%, was similar to the non-treated control pots at 15.6% coverage (Table 5). Percent coverage among pots treated with herbicides ranged from 9.6% to 25% 60 DAT and was lower than percent coverage in the control pots. Percent coverage increased significantly in pots treated with herbicide by 90 DAT but were all significantly lower than percent coverage in the control pots.

Table 4. Herbicide efficacy on spurge percent coverage 30, 60, and 90 DAT.[z]

Treatment	Rate	30 DAT	60 DAT	90 DAT	Sign.[y]
Fortress	100 lbs/A	1.6b[x]	37.5b	97.0ns	L***
Fortress	150 lbs/A	2.5b	44.5b	97.0	L***
Fortress	200 lbs/A	0.6b	12.1b	85.6	Q**
Gallery + Dimension	0.75+0.38 lbs ai/A	0.0b	18.0b	97.5	Q**
Snapshot 2.5TG	150 lbs/A	1.1b	25.5b	97.0	Q*
Control		75.5a	99.5a	100.0	L***

[z]The treatment by days after treatment interaction was significant for percent coverage at $P < 0.05$.

[y]Linear (L) or quadratic (Q) trends of percent coverage across DAT using qualitative-quantitative model regression at $P < 0.05$ (*), $P < 0.01$ (**) or 0.001 (***).

[x]Least squares means comparisons among treatments (lower case in columns) using the simulated method at $P < 0.05$.

All herbicide treatments provided excellent control of crabgrass 30, 60, and 90 DAT, but it is important to note that all Fortress treatments had some amount of weed germination while the others did not (Table 6).

Table 6. Herbicide efficacy on crabgrass percent coverage 30, 60, and 90 DAT.[z]

Treatment	Rate	30 DAT	60 DAT	90 DAT	Sign.[y]
Fortress	100 lbs/A	0.1b[x]	9.1b	9.5b	NS
Fortress	150 lbs/A	0.0b	0.0b	0.5b	NS
Fortress	200 lbs/A	0.0b	0.0b	7.5b	NS
Gallery + Dimension	0.75+0.38 lbs ai/A	0.0b	0.0b	0.0b	NS
Snapshot 2.5TG	150 lbs/A	0.0b	0.0b	0.0b	NS
Control		68.5a	99a	100.0a	Q**

[z]The treatment by days after treatment interaction was significant for percent coverage at $P < 0.05$.

[y]Not significant (NS) or quadratic (Q) trends of percent coverage across DAT using qualitative-quantitative model regression at $P < 0.01$ (**).

[x]Least squares means comparisons among treatments (lower case in columns) using the simulated method at $P < 0.05$.

No phytotoxicity occurred on any of the herbaceous crops tested. Tickseed was the only herbaceous crop species that that was affected by herbicide treatments (Table 7). Fortress 100 and 150 lbs/A had significantly greater size index 90 DAT, while the plants in the Gallery plus Dimension treatment produced the largest number of flowers 90 DAT.

Overall, Fortress treatments provided equal or better weed control than Snapshot or Gallery plus Dimension. Fortress particularly provided longer residual control of oxalis compared to Snapshot or Gallery plus Dimension.

Table 7. Size index and flower count of tickseed 90 DAT.

Treatment	Rate	Size Index	Flower Count
Fortress	100 lbs/A	58.2a[z]	5.9b
Fortress	150 lbs/A	61.4a	5.4bc
Fortress	200 lbs/A	53.3ab	5.0bc
Gallery + Dimension	0.75+0.38 lbs ai/A	56.1ab	10.9a
Snapshot 2.5TG	150 lbs/A	46.6ab	2.8c
Control		51.2b	3.4bc

[z]Least squares means comparisons among treatments (lower case in column) using the simulated method at $P < 0.05$.

Literature Cited

Case, L.T., Mathers, H.M., and Senesac, A.F. (2005). A review of weed control practices in container nurseries. HortTech. 15:535-545.

Derr, J.F. 1994. Weed control in container-grown herbaceous perennials. HortScience *29*:95–97.

Fain, G.B., Gilliam, C.H., Keever, G.J. (2006). Tolerance of hardy ferns to selected preemergence herbicides. HortTech. *16*:605-609.

Fausey, J.C. 2003. Controlling liverwort and moss now and in the future. HortTech. *13*:35–38.

Gilliam, C.H., Foster, W.J., Adrain, J.L. Shumack, R.L. (1990). A survey of weed control costs and strategies in container production nurseries. J. Environ. Hort. *8*:133–135.

Mervosh, T.L. and Ahrens, J.F. (1998). Preemergence herbicides for container-grown perennials. Proc. Northeastern Weed Sci. Soc. *52*:131.

Porter, W.C. (1996). Isoxaben and Isoxaben Combinations for Weed Control in Container-Grown Herbaceous Flowering Perennials. J. Environ. Hort. *14*:27-30.

Thetford, M., Gilliam, C.H., and Williams, J.D. (1995). Granular Preemergence Applied Herbicides Influence Annual Bedding Plant Growth. J. Environ. Hort. *13*:97-103.

Combined Proceedings IPPS 68:402-407. 2018.

Developing Crapemyrtle Pollen Sampling Methods for a Neonicotinoid Pathway Study

Asija Rice[1] and Yan Chen[2]
[1]Southern University and A&M College, Urban Forestry Program, J.W. Fisher Hall Rm.102 Baton Rouge, Louisiana 70813, USA
[2]Louisiana State University Agricultural Center, Hammond Research Station, 21549 Old Covington Highway, Hammond, Louisiana 70403, USA

Asija97@gmail.com

Keywords: *Lagerstroemia*, crapemyrtle bark scale (CMBS), *Acanthococcus lagerstroemiae*, Kuwana, chemical control.

Abstract

Crapemyrtle bark scale is an exotic pest infesting crapemyrtles and has become a major concern because of its potential impact on other economically or ecologically important plant species. The systemic neonicotinoid insecticides are the most effective chemical class controlling this scale. However, pollinator safety has to be considered when developing IPM strategies, since crapemyrtle pollen is a significant food source for many pollinators. Pollen collection is the first step in analyzing for neonicotinoid concentration. In this study, the flowering phenology and the amount of pollen that can be collected from flowers in a timely manner were evaluated to determine the best timing and method for pollen collection. The first week of blooming for a total of 60 crapemyrtle cultivars were recorded to determine early-, mid-, and late-season blooming cultivars. Then, the clusters of crapemyrtle flowers of ten cultivars representing early- and mid-season bloomers were grouped as new, full, and spent blooms based on the percentage of the cluster that showed color. Results indicated that there was a significant difference in flowering phenology among crapemyrtle cultivars, which affected pollen availability and thus sampling period for each cultivar. Three pollen collection trials were conducted to determine the amount of pollen that could be collected from 'White Natchez' crapemyrtle in a particularly timely manner. Results indicate that it is very challenging to collect the amount of pollen currently required by the laboratories for neonicotinoid concentration analysis. An improved laboratory procedure or analyzing methodology, requiring smaller amount of pollen will be critical for successful crapemyrtle pollen neonicotinoid pathway research.

IPPS Vol. 68 – 2018

INTRODUCTION

Crapemyrtles are popular ornamental shrubs or trees in the genus *Lagerstroemia* that have over 400 cultivars in the ornamental trade in the U.S. Their multiple aesthetic attributes include flower color, trunk color, fall foliage color, and long blooming time, which have made them the No. 1 summer blooming shrub or tree in the Southeastern U.S. Nursery production of crapemyrtles had a wholesale value of $66M/yr in 2014 (USDA 2015 Nursery Crop Census).

Crapemyrtles have few diseases and pests. However, since 2005, an exotic scale has become a major concern (Wang et al., 2016). The crape myrtle bark scale [(CMBS), *Acanthococcus lagerstroemiae* Kuwana) infestation is now confirmed in Alabama, Arkansas, Georgia, Louisiana, Arkansas, Alabama, Mississippi, New Mexico, South Carolina, Tennessee, Texas, Virginia and Washington, D.C. In Louisiana, severe infestations were reported from major cities such as Baton Rouge, Covington, Hammond, Mandeville, New Orleans, and Shreveport. With severe infestation, the scale forms a white layer on stems and trunks and causes unsightly black sooty mold from fungi grows on scales exudes, which covers the leaves and the ground underneath the infested trees. Heavy infestation can cause stem dieback, reduction in flowering, stunt growth, and may kill young trees. Infestations are more severe on young and stressed trees.

Currently the most effective chemical control against CMBS is provided by foliar or drench applications of insecticides belonging to the neonicotinoid class, such as imidacloprid, dinotefuran, thiamethoxam, and clothianidin. Neonicotinoids are an important tool for an integrated pest management (IPM) strategy, however, they are systemic insecticides that may potentially be transferred to nectar and pollen, causing concerns that they may create a hazard for beneficial insects and pollinators (Stoner and Eitzer, 2012). Neonicotinoids and their metabolites also have long residual periods in plant tissue, from months up to years. There is a critical need to develop a better understanding of their pathway by analyzing their concentrations in pollen, flower, and leaf tissue of popular crapemyrtle cultivars.

Crapemyrtles do not have nectar but their pollen is an important food source to pollinators. However, current laboratory analysis methodology requires a minimum of 4 grams of pollen per sample for analyzing one neonicotinoid insecticide (i.e., imidacloprid) and two of its biologically active metabolites (i.e., imidacloprid-olefin and its 6-hydroxy acid). Pollen availability varies during the blooming period of a flowering plant, and collection methods also affect the amount and quality of pollen samples.

Therefore, the objectives of this study were to: 1) determine best pollen sampling time by recording flowering phenology of selected cultivars, and 2) determine the amount of pollen that can be collected in a timely manner from 'White Natchez' crapemyrtles.

MATERIALS AND METHODS

This study was conducted with a new crapemyrtle collection at the Louisiana State University Agriculture Center, Hammond Research Station, that was established during 2013 and 2014. A total of 60 cultivars, three to five trees of each, were observed for their first week of blooming from Week 20 (May 14) to Week 30 (July 23rd). Observation was made on every Monday. A total of ten cultivars representing early- and mid-season bloomers were then selected for flowering phenology observations. We purposely selected certain cultivars in the Black Diamond and Delta series because of their popularity in the ornamental trade. Considering each flower cluster as a "bloom", the numbers of new bloom (less than 50% of the cluster showed color and there were no seed pods), full bloom (more than 50% of the cluster showed color), and spent bloom (less than 50% of the cluster showed color and seed pods start to form) were recorded every week for these cultivars from Week 25 (June 18) to Week 30 (July 23rd).

Three trials of pollen collection were conducted with 'White Natchez' during week 24 (June 11) to week 26 (June 25). Different flower collection methods were tested in three trials; Trial 1- individual flowers collected the day before pollen sampling and stored in small containers; Trial 2 -whole clusters of flowers collected the day of pollen sampling, and Trial 3 -the day before sampling. Pollen was collected using a compact vacuum connected to a 1.5 cm long clear tube 'chamber' (blocked with cigarette cotton filter). This was then connected to an Eppendorf pipet tip (Figure 1). Pollen release time was observed daily from 6:30 to 15:00 over a week period of time.

Figure 1. Pollen was collected with a mini vacuum connected to a pipet tip (a and c), and crapemyrtle pollen are only available for a short time of period after being released (d) because of pollinator activities (b).

RESULTS

As presented in Fig. 2, the first week of blooming was significantly different among cultivars ($p < 0.0001$). We then grouped the 60 cultivars as early-, mid-, and late-season bloomers using Week 22 (May 28) and 26 (June 25) as cut-off dates.

Cultivars bloomed before Week 22 are considered early-season bloomers, and cultivars bloomed between Week 22 and 26 are considered mid-season bloomers.

Ten cultivars were selected, including five from the Black Diamond (BD) series

('Lavender Lace', 'Crimson Red', 'Pure White', 'Red Hot', and 'Shell Pink'), and three from the Delta series ('Moonlight', 'Eclipse', and 'Breeze') which are all mid-season bloomers; also included were 'Plum Magic' and 'Red Rooster'. Among these cultivars, 'Lavender Lace' and 'Plum Magic' are early-season bloomers with the rest of the cultivars being mid-season bloomers.

Numbers of full blooms (flower clusters) per plant were significantly different among the ten selected cultivars ($p < 0.0001$). 'Plum Magic' had more full bloom clusters than other early-bloomers at their peak blooming weeks (Week 25 to 27, Figure 2).

Mid-season bloomers generally peaked in full bloom during Week 26 to 28. From Week 29, all cultivars had less than 10 flower clusters that were in full bloom. Number of total blooms (flower clusters) is the sum of new, full and spent blooms on a tree observed each week over a period of six weeks (Figure 3). There was a significant difference among cultivars, and 'Plum Magic', 'Red Rooster', 'Delta Moonlight' and 'Delta Eclipse' had a greater number of total blooms than other cultivars during the majority time of the observation period. Other than data presented here, we observed that individual flowers only open for one day, and pollen release time were about 10:00 AM to 12:00 PM, and visitation from pollinators (bees, bumblebees, beetles, thrips, etc.) significantly reduce pollen availability for sampling.

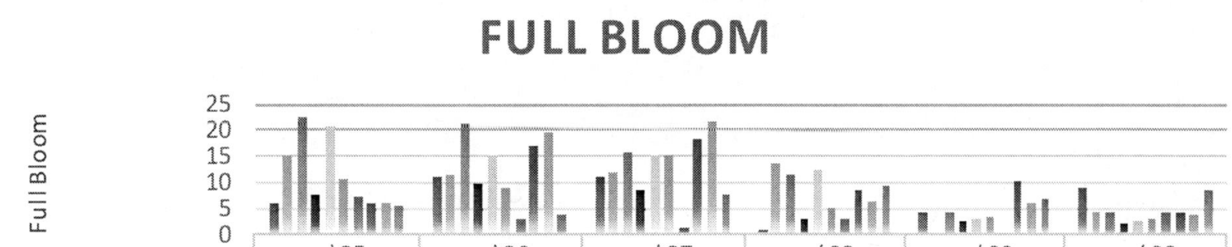

	wk25	wk26	wk27	wk28	wk29	wk30
BD Lavender Lace	6	11	11	0.6666667	4	8.6666667
BD Crimson Red	15.3333333	11.3333333	12	13.3333333	0	4
Plum Magic	22.5	21	15.5	11.5	4	4
BD Pure White	7.6666667	9.6666667	8.3333333	3	2.3333333	2
Red Rooster	20.6666667	15.3333333	15.3333333	12.3333333	2.6666667	2.3333333
BD Red Hot	10.6666667	8.6666667	15	5	3.3333333	3
BD Shell Pink	7	2.6666667	1	2.6666667	0.3333333	4
Delta Moonlight	6	17	18	8.3333333	10	4
Delta Eclipse	6	19.6666667	21.6666667	6.3333333	6	3.6666667
Delta Breeze	5.3333333	3.6666667	7.6666667	9.3333333	6.6666667	8.3333333

Cultivars

Figure 3. Number of full Blooms (flower clusters) found on ten selected crapemyrtle cultivars over a 6-week time period from week 25 to week 30 of 2018 (week of June 18 to week of July 23rd, 2018)

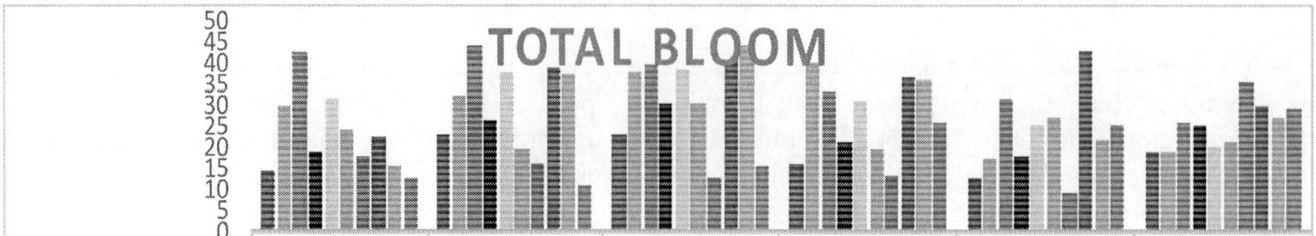

	wk25	wk26	wk27	wk28	wk29	wk30
BD Lavender Lace	14	23	23	16	12.6666667	18.6666667
BD Crimson Red	30	32.3333333	37.6666667	40	17	19
Plum Magic	42.5	44.5	39.5	33.5	31.5	26
BD Pure White	18.6666667	26.3333333	30.6666667	21.3333333	18	25.3333333
Red Rooster	31.6666667	37.6666667	38.3333333	31	25.3333333	20
BD Red Hot	24	19.3333333	30.3333333	19.3333333	26.6666667	21.3333333
BD Shell Pink	18	16	12.6666667	13	9	35.3333333
Delta Moonlight	22.3333333	39	41	37	43	29.6666667
Delta Eclipse	15.3333333	37.3333333	44	36	22	26.6666667
Delta Breeze	12.6666667	10.6666667	15.6666667	25.6666667	25.3333333	29.3333333

Figure 3. Total number of blooms (flower clusters) provided by ten selected crapemyrtle cultivars over a 6-week time period from week 25 to week 30 of 2018 (week of June 18 to week of July 23rd, 2018).

DISCUSSIONS

Based on a six-week observation, we know there is a big difference among cultivars in terms of bloom time, number of full bloom (when the pollen is most available) and total number of blooms. This significant difference in flowering phenology indicate that cultivar selection is critical in future neonicotinoid pathway research. Also, many of these popular new cultivars will be planted in large numbers by homeowners and landscape contractors. Knowing their flowering phenology will greatly assist in the development of an IPM program that will avoid acute impact of insecticides and be friendlier to beneficial insects and pollinators.

Many challenges were encountered during the pollen collection trials. The cigarette cotton filter caught quite an amount of pollen in it, and the cotton was later covered with parafilm and tube was remade only to use duct-tape around the pipe on the vacuum and the clear tube chamber. Comparing the three trials, collecting flowers the day before pollen collection may cause mold or loss of pollen to insects (i.e., thrips) that were collected and stored in the same container as flowers. Although all six 'White Natchez' trees used in the trials were of the same age and size, we observed difference in the overall amount of pollen in full blooms among them. Prolonged rainy weather and everyday thunderstorms also significantly decreased pollen availability.

Using the compact vacuum and flowers collected on the same day, we were able to collect 0.7-gram pollen over a two hours of work time. Apparently, the compact vacuum is still the best way to collect pollen once the timing of the day and weeks of pollen availability are determined. However, it would be very difficult to collect 4 grams of pollen as currently required by contracted laboratories (personal communication). Newer methodology for pollen analysis requiring smaller amount of pollen is available (David et al., 2015 and 2016).

Acknowledgement

We thank LSU EXPLORE Internship Program for the financial support to this summer intern project, and staffs at the LSU AgCenter Hammond Research for technical support. This project is part of the USDA NIFA Specialty Crop Research Initiative project award No. 2017-51181-26831, "Systemic Strategies to Manage Crapemyrtle Bark Scale, a New Exotic Pest".

Literature Cited

David, A., Botias, C., Abdul-Sada, A. Nicholls, E., Rotheray, E.I., Hill, E.M and D. Goulson D. (2016). Widespread contamination of wildflower and bee-collected pollen with complex mixtures of neonicotinoids and fungicides commonly applied to crops. Environ. Internat. *88*: 169-178.

David, A., Botias, C., Abdul-Sada, A., Goulson, and Hill, E.M. (2015.) Sensitive determination of mixtures of neonicotinoid and fungicide residues in pollen and single bumblebees using a scaled down QuEChERS method for exposure assessment. Ann Bioanalyt. Chem. *407*:8151-8162.

Stoner, K.A. and Eitzer. B.D. (2012). Movement of soil applied imidacloprid and thiamethoxam into nectar and pollen of squash (*Cucurbita pepo*). PLoS ONE e39114.

Wang, Z., Chen, Y., Gu, M., Vafaie, E., Merchant, M. and Diaz, R. (2016). Crapemyrtle bark scale: A new threat for crapemyrtles, a popular landscape plant in the U.S. Insects. *7*(4):78.
doi: 10.3390/insects7040078.

Combined Proceedings IPPS 68:408-414. 2018.

Performance Plus Plants from the Trial Gardens at UGA

John M. Ruter
Department of Horticulture, The University of Georgia, Athens, GA 30602 USA

ruter@uga.edu

Keywords: Plant breeding companies, teaching, breeders, growers, public, plant evaluations, perennials, annuals.

INTRODUCTION

The Trial Gardens at UGA were started in 1982 and Dr. John Ruter has served as Director of the Trial Gardens since 2013. In 2017, Brandon Coker joined as manager of the Trial Gardens. The mission of the Garden includes teaching, research, and introduction of new plants to the industry. The Trial Garden is an essential testing site for heat and humidity tolerance for many of the world's breeding companies. During most summers some 56 days reach ≥32°C (≥90°F), while the average rainfall during the growing season (May-September) is 49cm (19.2 in.). In 2018 we had 74 days above 32°C (90°F) and 75 cm (29.7 in.) of rainfall; 47 cm (18.6 in) occurred during May and June.

Teaching in the garden focuses on two classes taught by Dr. Ruter, HORT 3500 taught during the fall semester which focuses on annuals, vines, and fall-blooming perennials, and HORT 3510, taught in the spring which focuses on bulbs, spring ephemerals, and early blooming perennials. Both classes are taught as half-semester courses. The Trial Gardens are also utilized by classes from Agricultural Communications, Entomology, Landscape Architecture, Plant Pathology, and other departments from the Arts & Sciences.

As for research and trialing, we work with many of the major breeding companies from around the world. In 2018 we evaluated over 350 annual taxa in ground beds, hanging baskets, and containers. Trials also include numerous perennials and ~125 landscape roses. Overall there are approximately 1200 different taxa growing on less than 0.3 ha (.75 ac). Several plants have been introduced to the trade via the Trail Gardens over the past 20 years (ugatrial.hort.uga.edu). This tradition continues as Dr. Ruter releases plants from his breeding program at UGA.

TRIAL ESTABLISHMENT AND DATA COLLECTION

Seed of small seeded crops such as begonia is received and planted during the month of January. Cuttings and rooted plugs are received in February and March. Propagules are shifted into 8.3 cm (3.25 in.) containers and grown in the greenhouse using recommended protocols. Substrate is a custom mix from Old Castle Lawn & Garden. Hanging baskets made from recycled newspaper (Henry Molded Products, OR) are also planted and grown-out during this time. In-ground trial beds are rototilled in March-April and appropriate soil amendments are added based on soil tests. Outdoor beds are covered with pine straw mulch before planting. Irrigation is accomplished with overhead sprinklers and hand-watering of containers and hanging baskets. Pelargoniums and cold-tolerant annuals such as petunias and Calibrachoa's are placed out into the garden at the end of April, with all other plants going out before the 15th of May. Plants are liquid-fed during the growing season using several different fertilizer products.

Data collection begins around the end of May and is taken every two weeks until the end of September. All data is taken by Brandon Coker, Trial Garden manager. Having as single evaluator is essential for consistency of data collection. Data is entered into Excel and is downloaded to our Trial Garden web site (ugatrial.hort.uga.edu). Plants are rated on a scale of 1-5 (5 being best) on overall vigor, appearance, flower production, and disease and insect resistance. All this data can be found on the web site and is graphically tracked for each variety being evaluated. Graphing is important as it allows for viewing of performance over time and shows how the plant performed from spring until fall. Data is also sent to the National Trials database at www.planttrials.org.

During the first week of June each year breeders and growers are invited to attend our Industry Open House. A week or two later we hold a public open house. These events allow industry professionals as well as the gardening public to see a variety plants all growing at the same location. During each event participants are asked to select five outstanding plants in the garden. This data is collected and shared on the web site and through email communications.

Periodically throughout the growing season ~ 10 plants are selected as "Plants of Distinction" for their performance in the garden. These plants are posted on the web site and are also emailed out to all interested parties. At the end of the season the "Classic City Awards" are given to the best 10-12 plants that had excellent performance all summer. The best cultivars for each genus are also listed under the "Best of the Best" link on the web site.

CLASSIS CITY AWARD WINNERS FOR 2018

***Begonia* 'Canary Wings' - Ball Ingenuity** (Figure 1). It's refreshing to relax in the shade with a nice glass of sweet tea here in Georgia…then you spot Begonia 'Canary Wings' from Ball Ingenuity and all the sudden you have the excitement and vigor to go forth and conquer the world! Well the plant world that is. We have been watching this novel shade begonia in the garden since day one. To our satisfaction shady areas become bright and vibrant with Canary Wing's chartreuse leaves and bright red flowers. It is an exceptional plant for both containers and in-ground plantings.

Figure 1. *Begonia* 'Canary Wings' - Ball Ingenuity.

***Calibrachoa* Superbells® 'Holy Smokes!' - Proven Winners** (Figure 2). 'Holy Smokes!' is what you say when you lay your eyes on this beauty from Proven Winners Superbells® *Calibrachoa* collection. Summer longevity is a must for the Southeast as we have a growing season that can be stretched out over 25 weeks. So, for any plant to maintain interest over such a period of time it's got to be a stunner and 'Holy Smokes!' did it like a champ. White petals with what seem to be watercolor-like purple/blue swatches and yellow centers. It maintained a mounding habit throughout spring and summer that made it without doubt one of the showiest flowering plants in the garden.

Figure 2. *Calibrachoa* Superbells® 'Holy Smokes!' - Proven Winners

***Calibrachoa* Lia™ 'White' – Danziger** (Figure 3). Hanging baskets bring brilliant color to eye level and are a staple form of growing plants around the country. Calibrachoa Lia™ 'White' from Danziger sets the bar for other *Calibrachoa's* with its nearly perfect shape as it has been in full bloom for months. Medium-sized pure white petals with yellow throats are anything but just another white flower in the garden. Anybody wanting to have constant color needs Lia™ 'White' added to their hanging baskets or mixed planter's as white is an essential color to make all other colors pop.

Figure 3. Calibrachoa Lia™ 'White' – Danziger.

***Echinacea* Sombrero® 'Tres Amigos' - Darwin Perennials** (Figure 4). Show-stoppers are a must for every garden but they can be hard to find if you don't know where to look, but look no more. *Echinacea* Sombrero® 'Tres Amigos' is going to be your new favorite perennial. Flowers transition thru three distinct colors that emerge from peachy-coral to rose and finish with a hint of burgundy. You would think these kinds of colors are only found in books and movies, but they are real, and this selection is a must for any perennial garden. The plants perform well in full sun and can take drier soil conditions. Leave the flowers for the goldfinches to enjoy late in the season.

Figure 4. *Echinacea* Sombrero® 'Tres Amigos' - Darwin Perennials.

***Euphorbia* Crystal White™ - Green Fuse Botanicals** (Figure 5). *Euphorbia* is a well-known landscape plant, especially in hot parts of the country where it thrives. Crystal White™ from Green Fuse Botanicals has a desirable shape that's about half the size of other popular varieties on the market. Only about a foot tall - it packs a punch with flower production and has very tight branching. The plant was blooming in April when it was planted and was still in full bloom going into September – very nice!

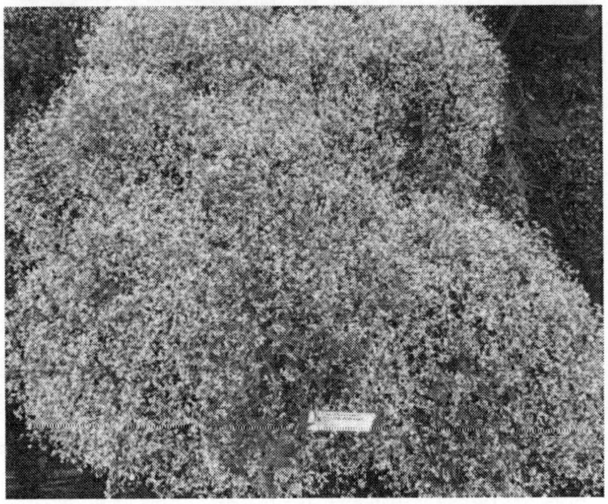

Figure 5. Euphorbia Crystal White™ - Green Fuse Botanicals.

***Gaillardia* SpinTop™ 'Red Starburst' - Dummen Orange** (Figure 6). *Gaillardia* SpinTop™ 'Red Starburst' from Dummen Orange just makes people smile. The flowers on this plant exude color with their red centers that burst into a red/orange and finish with bright yellow tips. This plant would make an excellent border plant or container planting. This selection needs to be beside your front door so you can get a little burst of happiness whenever you walk by!

Figure 6. *Gaillardia* SpinTop™ 'Red Starburst' - Dummen Orange.

Impatiens New Guinea Harmony® Radiance™ 'Hot Pink' – Danziger (Figure 7). Hot Pink is a vast understatement; something more like 'Hot Hot Hot Pink' is better suited to Danziger's Impatiens New Guinea Harmony® Radiance™ 'Hot Pink'! We are growing this plant in a location getting no more than 4 hours of sun in the morning and it is thriving.

Three uninterrupted months of blooming is what separates the good plants from the stellar plants. In the south, where so many of our mature landscapes need color under mature trees, this plant will light up those dark corners with exceptional color.

Figure 7. Impatiens New Guinea Harmony® Radiance™ 'Hot Pink' – Danziger.

***Pelargonium* Calliope® 'Large Rose Mega Splash' - Syngenta Flowers** (Figure 8.)
Geraniums tend to do either great or not so good here in the south. Calliope® 'Large Rose Mega Splash' from Syngenta Flowers tops the list in the great category. Super showy pink petals with rose starbursts from the center of the flowers are sure to please in containers, hanging baskets, or in- ground plantings. Production of flowers has been non-stop. This cultivar has been the best of all our Geraniums this year and it has maintained excellent disease resistance even with a wetter than usual summer in Athens, Georgia.

Figure 8. *Pelargonium* Calliope® 'Large Rose Mega Splash' - Syngenta Flowers.

***Petunia* Dekko™ 'Star Coral' - Syngenta Flowers** (Figure 9). Picking the best Petunia each year is a challenge because Petunias produce vibrant colors and make bold statements in containers here in the South. The Dekko™ series from Syngenta Flowers had several stellar performers this year and we have chosen 'Star Coral' to top the list simply because it has bloomed the longest and had the best overall shape. The hot pink coral- colored blooms have a touch of white in the petals that gives them a dramatic coloration. This cultivar would lend itself well to just about any other range of colors as companion plants, not to mention how well it performed in a standalone planter.

Figure 9. *Petunia* Dekko™ 'Star Coral' - Syngenta Flowers.

***Portulaca* Hot Shots™ 'Tangerine Glow' - Green Fuse Botanicals** (Figure 10). The winner column this year at the UGA Trial Gardens is packed is superior plants each excelling in a particular way. Portulaca Hot Shots™ 'Tangerine Glow' is in a league of its own in terms of overall interest. Our planting of 'Tangerine Glow' took over an area 6' long by 3' wide and is just simply beautiful. The added interest of a flower that opens in the morning and closes in the afternoon is just too cool. Bright tangerine petals with scorching yellow centers bloom abundantly like a carpet of wonderfulness that deserves to be in every garden with full sun.

Figure 10. *Portulaca* Hot Shots™ 'Tangerine Glow' - Green Fuse Botanicals.

***Salvia* Skyscraper™ 'Pink' – Selecta** (Figure 11). We all know and love salvia and who would not? They attract bees, butterflies, hummingbirds, and are just downright beautiful with a plethora of colors and sizes to choose from. Selecta struck a vein with *Salvia* Skyscraper™ 'Pink' and the plant speaks for itself. We are in the business of novel plants and we are all used to blue, purple, and red salvias, but pink? The flowers burst forth with a pale magenta calyx followed by a bright pink corolla that emerges to tower over the lush dark green foliage on this unique cultivar.

Figure 11. *Salvia* Skyscraper™ 'Pink' – Selecta.

***Solenostemon* ColorBlaze® 'Torchlight' Proven Winners** (Figure 12). The ColorBlaze® Series has been nothing short of spectacular this year so the top pick from this category had to be nearly perfect. 'Torchlight' has superior qualities in every way, from its lush storybook- like green with red and hot pink veined leaves to its seemingly indestructible growth habit. This plant is also very responsive to trimming. Every time we pruned the plants the new leaves came back with increased color. When 2019 comes around get this plant first and use it in multiple plantings.

Figure 12. *Solenostemon* ColorBlaze® 'Torchlight' - Proven Winners.

BEST OF THE BEST – 2018

- Angelonia Angelface® 'Wedgewood Blue' - Proven Winners
- Begonia 18OS06 Trial Entry - All America Selections
- Begonia Tophat™ 'Pink' - Syngenta
- Caladium 'Bottle Rocket' - Classic Caladiums
- Capsicum 'Onyx Red' - American Takii (2018 AAS National Winner)
- Capsicum 'Purple Flash' - PanAmerican Seed
- Coreopsis 'Super Star' - Darwin Perennials
- Gomphrena Truffula™ 'Pink' - Proven Winners
- Hibiscus Hollywood™ 'Jolly Polly' - J. Berry Nursery
- Hibiscus moscheutos 2015-38 'Variegated' - Dr. John Ruter - UGA
- Impatiens New Guinea Sun Harmony™ 'Blushing Orchid' – Danziger

- Leucanthemum 'Birdy' - Dümmen Orange
- Ocimum 'Amazel' - Proven Winners
- Pelargonium Calliope® 'Large Lavender' - Syngenta Flowers
- Pelargonium 'Super Moon Red' - Selecta
- Pentas BeeBright™ 'Pink' - Syngenta Flowers
- Salvia 'Big Blue' - PanAmerican Seed
- Salvia Rockin™ 'Deep Purple' - Proven Winners
- Solenostemon ColorBlaze® 'Golden Dreams' - Proven Winners
- Verbena Firehouse™ 'Pink' - Ball FloraPlant

Combined Proceedings IPPS 68:415-419. 2018.

Propagation of Shumaka™ Crape Myrtle

Jenny B. Ryals, Patricia R. Knight, Scott Langlois, Eugene Blythe, J. Skylar Baldwin, Christine Coker, Gary Bachman, Jim DelPrince, and Anthony T. Bowden
Coastal Research and Extension Center, South Mississippi Branch Experiment Station
P.O. Box 193, Poplarville, Mississippi 39470, USA

j.ryals@msstate.edu

Keywords: Hardwood cuttings, wounding, auxins, Hortus IBA, Dip'N Grow®, *Lagerstroemia*.

Abstract

Shumaka™ crape myrtle is a hybrid resulting from the cross of *Lagerstroemia* 'Arapaho' and an unknown pollen donor. Industry recommendations are needed for propagating this new Mississippi State University crape myrtle selection. Hence, the objective of this research was to determine optimal commercial auxin formulation and concentration, with and without basal wounding, for hardwood cuttings of Shumaka™. Hardwood cuttings quick-dipped in Hortus IBA Water Soluble Salts™ (Hortus IBA) at 5000 ppm had higher root quality ratings compared to cuttings that were wounded and received no auxin. However, cuttings quick-dipped in Hortus IBA did have better root quality compared to cuttings treated with Dip'N Grow® (DNG); root quality and number increased with increasing concentrations of Hortus IBA. Cutting quality (rooted, transplant-ready cuttings) also increased when Hortus IBA concentrations increased. Results suggest that hardwood cuttings of Shumaka™ crape myrtle will root without wounding or use of auxin. However, a basal quick-dip in Hortus IBA 5000 ppm results in a higher quality liner.

INTRODUCTION

Crape myrtle, *Lagerstroemia*, is considered a staple in many southern landscapes (Wilson et al., 2007). A number of crape myrtle cultivars flower for more than 100 days, providing an aesthetically appealing landscape element with a color palette that is constantly expanding (Byers, 1983; Knight et al., 2006). Mississippi State University has been active in developing new crape myrtle selections, including Shumaka™. Shumaka™ is a hybrid resulting from crossing *Lagerstroemia* 'Arapaho' (Pooler, 2006) and an unknown pollen donor.

Shumaka™ has a unique very light pink flower color and large growth habit. Three-year-old plants in a research setting have grown more than 6 m (20 ft) and flower from early June through late August. The bark is smooth to exfoliating, with an outer bark that is grayish brown in color.

When releasing a new crape myrtle, like Shumaka™, it is important for nurseries to know the best propagation and production methods to scale-up plant numbers to meet market demands. Propagation of crape myrtle via softwood or hardwood cuttings is widely described as easy (Byers, 1983; Dirr and Heuser, 1987). Wade and Woodward (2001) reported that propagation of a crapemyrtle is easiest when using semi-hardwood cuttings from new growth. Byers (1983) propagated 20 cm (8-in.) hardwood cuttings taken after frost and stored overwinter. Dirr and Heuser (1987) reported that hardwood cuttings propagated in early February rooted better (43%) than cuttings propagated in early January or early March; bottom heat and peat:perlite or bark media was used. Propagation scheduling and preference, at a nursery, will determine whether it is more beneficial to propagate hardwood cuttings, soft or semi-hardwood cuttings.

Easy-to-root species like crape myrtle can certainly be scheduled around crops that have more specific propagation requirements. To assess the best way to propagate Shumaka™, two studies were conducted. Study one evaluated the optimal commercial auxin formulation and concentration and basal wounding for hardwood cutting propagation, while study two evaluated the optimal commercial auxin formulation and concentration - and impact of cutting stem position for soft or semi-hardwood cutting propagation of Shumaka™. However, this paper will only discuss results from the hardwood cutting study.

MATERIALS AND METHODS

Medial hardwood cuttings with a length of 12.7 cm (5 in.) were taken from Shumaka™ stock plants and stuck to a depth of 2.5 cm (1 in.) on 1 Feb. 2017. Cuttings were placed in 100% perlite in 7.6 cm (3-in.) containers. The 2x3x3 factorial experiment included two basal wounding treatments (wounded or non-wounded), three commercial auxin formulations [Hortus IBA (Hortus IBA Water Soluble Salts™), Dip'N Grow®, or Hortus IBA + KNAA (Hortus IBA Water Soluble Salts™ + NAA potassium salt)], and three auxin concentrations (0, 1000, or 5000 ppm IBA). DNG and Hortus IBA + KNAA formulations contained NAA at one-half the rate of IBA. KNAA is a research-only product but was added to Hortus IBA in selected treatments for comparison with DNG. A randomized complete block experimental design with five single cutting replications was utilized. Data were collected 60 days after sticking cuttings in perlite and included: rooting percentage, growth index (new shoots), cutting quality (0-5, with 0 = dead and 5 = transplant-ready cutting), total root number, average root length (of three longest roots), and root quality (0-5, with 0=no roots and 5=healthy, vigorous root system). Data were analyzed using linear mixed models and generalized linear mixed models with the GLIMMIX procedure of SAS (ver. 9.4; SAS Institute Inc., Cary, NC).

RESULTS

Rooting percentage, root number, average length of three longest roots, and growth indices were similar among treatments (Table 1). Cuttings that were not wounded and dipped in DNG 5000 ppm had higher root quality ratings compared to wounded control cuttings, wounded cuttings dipped in Hortus IBA 1000 ppm, non-wounded cuttings dipped in DNG 1000 or 5000 ppm, or non-wounded cuttings dipped

Table 1. Influence of basal wounding treatment, auxin concentration, auxin source on rooting percentage, root number, average length of three longest roots, root quality, cutting quality, and growth of Shumaka™ crapemyrtle.

Treatment[z] (ppm)	Rooting (%)	Roots (no.)	Mean length of 3 longest roots (inches)	Root quality rating[y]	Cutting quality rating[x]	Growth index[w]
Wounded control	40a[v]	4.0a	4.0a	1.9c	2.1ab	5.4a
Non-wounded control	60a	4.1a	4.3a	2.4abc	2.2ab	5.5a
Wounded Hortus IBA 1000	80a	3.4a	3.2a	2.2bc	1.8b	6.3a
Non-wounded Hortus IBA 1000	100a	1.8a	3.2a	2.3abc	2.1ab	5.4a
Wounded Hortus IBA 5000	100a	4.3a	4.0a	2.5ab	2.3ab	5.9a
Non-wounded Hortus IBA 5000	80a	7.4a	4.6a	2.8a	2.6a	6.2a
Wounded Dip'N Grow 1000	60a	1.9a	4.9a	2.3abc	2.2ab	4.7a
Non-wounded Dip'N Grow 1000	40a	1.7a	3.5a	2.1bc	1.9ab	6.1a
Wounded Dip'N Grow 5000	80a	4.5a	4.4a	2.4abc	2.2ab	5.7a
Non-wounded Dip'N Grow 5000	80a	2.0a	3.6a	2.2bc	2.0ab	5.1a
Wounded Hortus IBA 1000 K-NAA 500	100a	2.7a	4.3a	2.3abc	2.2ab	5.6a
Non-wounded Hortus IBA 1000 KNAA 500	100a	2.2a	2.8a	2.2bc	2.0ab	5.2a
Wounded Hortus IBA 5000 K-NAA 2500	100a	3.5a	5.2a	2.4abc	2.3ab	5.9a
Non-wounded Hortus IBA 5000 K-NAA 2500	100a	4.5a	4.0a	2.4abc	2.2ab	5.1a

[z]Dip'N Grow and Hortus IBA + KNAA formulations contained NAA at one-half the rate of IBA.
[y]Root quality (0-5, with 0=no roots and 5=healthy, vigorous root system).
[x]Cutting quality (0-5, with 0=dead and 5=transplant ready cutting).
[w]Growth index=(width1+width2+height)/3.
[v]Means followed by the same letter are similar according to Holm-Simulated method for simultaneous comparisons ($\alpha = 0.05$).

in Hortus IBA (1000 ppm) + KNAA (500 pm). Cuttings that were not wounded and dipped in DNG 5000 ppm had higher cutting quality ratings compared to cuttings that were wounded and dipped in Hortus IBA 1000 ppm. All other cuttings had similar quality regardless of treatment.

For a more thorough examination of treatment factors, selected treatment combinations were compared using the Shaffer-Simulated method for simultaneous comparisons. Treatment comparisons were as follows: wounded vs. non-wounded, Hortus IBA vs. no auxin, DNG vs. no auxin, Hortus IBA + KNAA vs. no auxin, Hortus IBA vs. DNG, DNG vs. Hortus IBA + KNAA, Hortus IBA vs. Hortus IBA + KNAA, Hortus IBA at 5000 vs. 1000 ppm IBA, DNG at 5000 vs. 1000 ppm IBA, and Hortus IBA + KNAA at 5000 ppm IBA vs. 1000 ppm IBA.

Rooting percentages, average length of three longest roots, and growth indices were similar regardless of treatment comparison (Table 2). Wounding or use of DNG or Hortus IBA + KNAA had no influence on cutting data.

Cuttings that were dipped in Hortus IBA, regardless of concentration, had a better root quality compared to cuttings dipped in DNG. Root quality, root number, and cutting quality all increased when Hortus IBA concentrations increased.

Table 2. Direct comparisons of selected treatment combinations on rooting percentage, root number, average length of three longest roots, root quality, cutting quality, and growth of Shumaka™ crape myrtle.

Comparison	Rooting (%)	Roots (no.)	Mean length of 3 longest roots (inches)	Root quality rating[z]	Cutting quality rating[y]	Growth index[x]
Wounded vs. non-wounded	NS[w]	NS	NS	NS	NS	NS
Hortus IBA vs. no auxin	NS	NS	NS	*	NS	NS
Dip'N Grow vs. no auxin	NS	NS	NS	NS	NS	NS
Hortus IBA + NAA vs. no auxin	NS	NS	NS	NS	NS	NS
Hortus IBA vs. Dip'N Grow	NS	NS	NS	*	NS	NS
Dip'N Grow vs. Hortus IBA + NAA	NS	NS	NS	NS	NS	NS
Hortus IBA vs. Hortus IBA + NAA	NS	NS	NS	NS	NS	NS
Hortus IBA 5000 vs. Hortus IBA 1000	NS	*	NS	**	**	NS
Dip'N Grow 5000 vs. Dip'N Grow 1000	NS	NS	NS	NS	NS	NS
Hortus IBA + NAA 5000 vs. Hortus IBA + NAA 1000	NS	NS	NS	NS	NS	NS

[z]Root quality (0-5, with 0=no roots and 5=healthy, vigorous root system).
[y]Cutting quality (0-5, with 0=dead and 5=transplant ready cutting).
[x]Growth index=(width1+width2+height)/3.
[w]NS=Not significant or significant at $\alpha = 0.01$ (**) or 0.05 (*) using the Shaffer-Simulated method for simultaneous comparisons.

DISCUSSION

Rooting percentages ranged from 40% to 100%, within the range reported by Dirr and Heuser (1987) for hardwood cuttings and similar to those reported by Dirr (1990) for summer-propagated 'Natchez' crape myrtle using 5000 ppm IBA or 95% ethanol, the solvent used for IBA. Blythe et al. (2003) reported greater than 90% rooting when using 1000 ppm DNG for 'Natchez' crape myrtle. Blythe et al. (2003) reported that 'Natchez' crape myrtle cuttings receiving DNG 1000 ppm as a basal dip had more roots compared to cuttings receiving no auxin, but 'Natchez' cuttings receiving K-IBA as a basal dip had similar root numbers compared to cuttings receiving no auxin. These results are similar to those reported by Dirr and

Heuser (1987) indicating use of auxin increased root number and improved quality of 'Natchez', 'Tuscarora', and 'Muskogee' crape myrtle cuttings. Differences in rooting percentages may be due to differences in cultivars evaluated or cultural conditions of the parent material (Davies et al., 2018).

Propagation methods that provide the grower with a quality product while allowing for maximum efficiency - are critical to the success of that plant in the market. Overall, the results from this study suggest that hardwood cuttings of Shumaka™ crape myrtle will root without wounding or use of auxin. However, a basal quick-dip in Hortus IBA 5000 ppm does result in a better liner.

Literature Cited

Blythe, E.K., Sibley, J.L., Tilt, K.M., and Ruter, J.M. (2003). Foliar application of auxin for rooting stem cuttings of selected ornamental crops. J. Environ. Hort. *21*:131-136.

Byers, D. (1983). Selection and propagation of crape myrtle. Comb. Proc. Int. Plant Prop. Soc. *33*:542-545.

Davies, F.T. Jr., Geneve, R.L., Wilson, S.B. (2018). Hartmann & Kester's Plant Propagation: Principles and Practices, 9th ed. Pearson. NY, NY.

Dirr, M.A. (1990). Effects of P-ITB and IBA on the rooting response of 19 landscape taxa. J. Environ. Hort. *8*:83-85.

Dirr, M.A. and Heuser, C.W. Jr. (1987). The Reference Manual of Woody Plant Propagation. Varsity Press, Inc. Athens, Georgia.

Knight, P.R., Anderson, J.M., Lee, D., Lee, M., DeJean, L., and Murchison, D. (2006). Crape myrtle evaluations in South Mississippi. HortScience *41*:512.

Pooler, M.R. (2006). 'Arapaho' and 'Cheyenne' *Lagerstroemia.* HortScience *41*:855-856.

Wade, G.L. and Woodward, L.W. (2001). Crape Myrtle Culture. Georgia Coop. Ext. Serv. Leaflet 331.

Wilson, J., Henn, A., Layton, B. (2018). Crapemyrtle: Flower of the South. Publication P2007, Mississippi State University Extension Service, Mississippi State, MS.

Combined Proceedings IPPS 68:420-425. 2018.

Whorled Sunflower (*Helianthus verticillatus*): A Potential Landscape Plant

R. N. Trigiano[1], S.B. Wilson[2], and C.N. Steppe[2]
[1]University of Tennessee, Department of Entomology and Plant Pathology, Knoxville, Tennessee, 37996-4560, USA; [2]University of Florida, Department of Environmental Horticulture, Gainesville, Florida, 32611-0675, USA.

rtrigian@utk.edu

Keywords: Clonal propagation, seed propagation, cuttings, tissue culture, endangered species, plant breeding.

INTRODUCTION

Helianthus verticillatus or whorled sunflower is a recently designated endangered species found only on a few prairie remnants in Tennessee, Georgia, and Alabama. It is a perennial species that spreads via rhizomes (Mathews et al., 2002) and produces copious amounts of small, black or gray seeds if compatible genotypes are present. This sunflower species is self-incompatible (Ellis and McCauley, 2009).

Plants emerge in late winter in Tennessee as a basal rosette (Figure 1) and reach 3-4 meters by August (Figure 2). It flowers profusely in August and September, and through the middle of October or until a killing frost. The plants bear up 20, 4-6 cm diameter yellow ray flowers with brown to black discs at the 20-30 cm of the termini of the stalks. The flowers attract over 30 species of potential pollinators, which include mostly native bees (Figure 3) and flies, and a variety of moths and butterflies. Inflorescences are

Figure 1. *Helianthus verticillatus* emerging in the spring.

covered with pollination bags when brown and shriveled, and seeds collected in November after the flowers are thoroughly dried (Fig. 4). This sunflower also appears to be drought tolerant.

The specific aims of this research were to determine germination conditions for seeds and to develop clonal propagation methods that utilize tissue culture and rooted cuttings.

Figure 2. Rapid vegetative growth (early August) and prolific flowering on tall stalks of *Helianthus verticillatus* in September.

Figure 3. In the center of the flower is a native bee (*Bombus* sp.), a potential pollinator of *Helianthus verticillatus*.

Figure 4. Pollination bags covering multiple heads of *Helianthus verticillatus*. Seeds were harvested in November 2017.

MATERIALS AND METHODS

Seed viability and germination.

Seeds were collected from a 200-stem stand growing in Maryville, Tennessee in November 2017. Seeds were separated from somatic tissues, stored in glass vials at room temperature, and shipped via postal service to the University of Florida in early February 2018. Pre-germination and seed viability were assessed with 100 seeds using a tetrazolium staining method (Peters, 2005) over 18-24 hours. Seeds with embryos that evenly stained dark pink- to -red were considered viable (Figure 5).

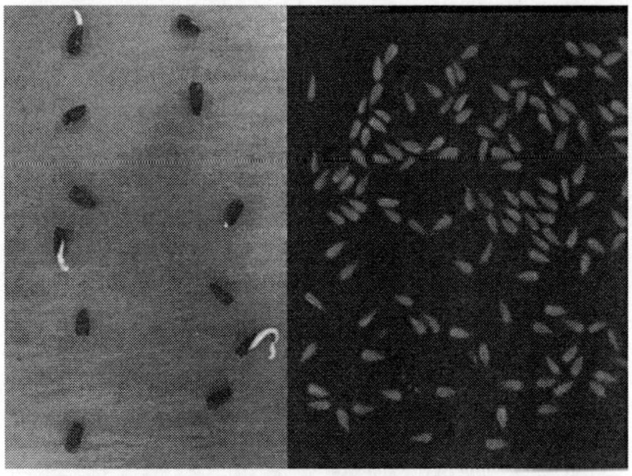

Figure 5. Seed germination (left) and X-ray analysis for viability (right) in *Helianthus verticillatus.*

Germination of seeds was evaluated using four replications of 100 seeds (AOSA, 2016) each at the following four temperature regiments: 22/11^{0}C; 27/17^{0}C; 29/19^{0}C; and 33/24^{0}C with a 12- hour photoperiod. Tests were conducted over 28 days. Post-germination tetrazolium tests were performed on ungerminated seeds after 28 days and percent dormancy, total viability, and germination of viable seeds determined.

Clonal Propagation - Leaf culture.

Young, not fully expanded leaves of less than 0.5-meter tall plants were harvested in April and surface disinfested in 25% NaOCl for 10 min and then rinsed thoroughly with sterile distilled water. Dissected leaves were placed on Murashige and Skoog (MS) (1962) medium amended with a range of concentrations of various cytokinins with and without auxin and cultured at 22^{0}C with a 12-hour photoperiod. Explants were transferred to MS medium without growth regulators after two weeks and incubated under the same conditions for the next four weeks. Explants were assessed for callus, shoot, and root formation after six weeks.

Clonal Propagation - Nodes/Axillary buds.

Nodes, each with two or three axillary buds, were harvested from less than one-meter plants. Leaves were removed and the nodal segments surface disinfested in 25% NaOCl for 15 min and then rinsed in sterile distilled water. Nodes were transferred to 60 mm diameter plastic test tubes containing MS medium supplemented with a range of concentrations of various cytokinins. Cultures were incubated at 22^{0}C with a 12-hour photoperiod. Shoots were harvested after four weeks and rooting attempted on MS medium amended with auxin.

Clonal Propagation – Cuttings.

The terminal 15 cm of young plants were harvested in late May 2017 and either treated with IAA or water for 10 min. Shoots were then placed in Promix B. In a second experiment, the upper 45 cm of plants were harvested from plants in late June 2017. Three node cuttings were placed in Promix B after a 10 min treatment with either IAA or water. Cuttings were rooted in the greenhouse with mist. Rooting was determined four weeks after treatment.

RESULTS AND DISCUSSION

Seed viability and germination.

Greater than 98% of the seeds were viable (Figure 6). Seeds did not lose viability within 7 months of dry ambient storage.

Germination of seeds occurred in all treatments, but only about 77% of the seeds in the high-temperature treatment germinated within 28 days. In the other three treatments over 96% of the seeds germinated.

Fifty percent of the seed in the coldest treatment germinated within 7-8 days, whereas 50% of the seed in the three warmer treatments germinated within 1-2 days. Seeds are easy to germinate and will be critical in breeding efforts that combine resistance to powdery mildew and flower morphology.

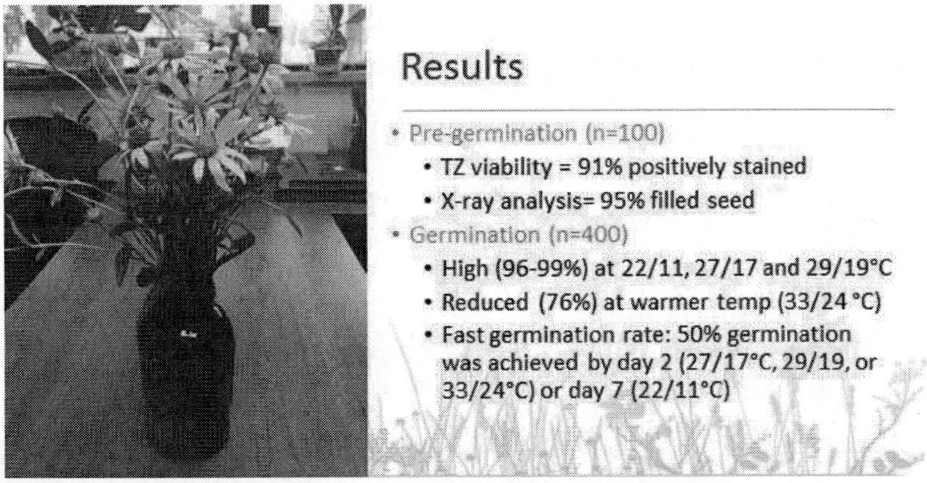

Figure 6. Results of seed germination studies of *Helianthus verticillatus.*

Tissue culture. More than 50% of cultures became contaminated within ten days of culture initiation. Very few shoots were formed from explants cultured on medium containing cytokinin (Figure 7). However, leaves cultured on MS without growth regulators formed numerous roots and some callus. More than 75% of the nodes cultured became contaminated with bacteria and fungi after two weeks and were discarded.

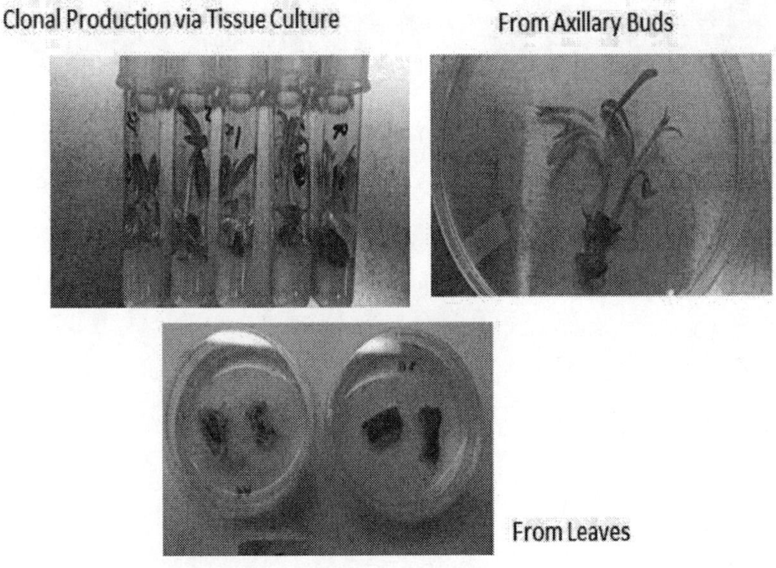

Figure 7. Clonal propagation by tissue culture using explants from axillary buds and leaves of *Helianthus verticillatus.*

On the remaining 25% of the cultures, axillary buds elongated on medium with or without cytokinins, but excised shoots failed to produce roots. Additional shoots did not form on nodes transferred to fresh medium. Preliminary results suggest that tissue culture propagation methods do not seem to be well suited to this sunflower, but additional studies are necessary. Besides controlling contamination, shoot proliferation appears to be very problematic.

Cuttings. Almost all (98%) of the cuttings harvested in late May formed robust root systems after four weeks when treated with auxin or water (Figure 8). These plants grew well and flowered normally in September. Only about 20% of the three node cuttings harvested in late June produced roots in response to either water of auxin treatment. Only one surviving whole plant flowered in September. Propagation of plants by terminal cuttings harvested in late May is very efficient. Rooted cuttings typically produce as many as four additional shoots via rhizomes during the growing season.

All of these are clonal, which is essential for determining compatibility and controlled (not open pollination) breeding for desirable traits.

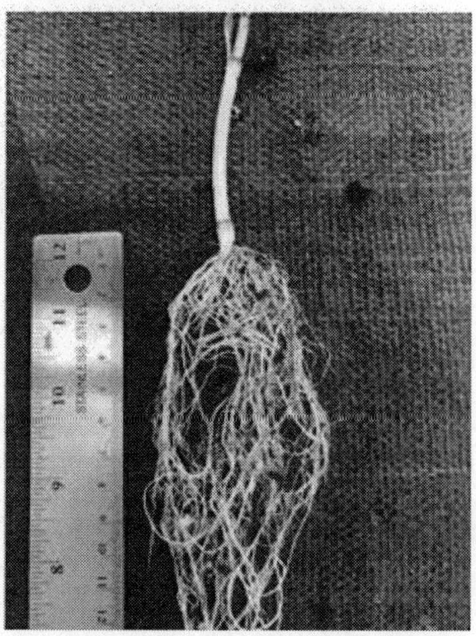

Figure 8. Clonal propagation by cuttings of *Helianthus verticillatus.*

Literature Cited

Association of Official Seed Analysts (AOSA). (2016). Rules for testing seeds. *16*:1-113.

Ellis J.R. and D.E. McCauley. (2009). Phenotypic differentiation in fitness related traits between populations of an extremely rare sunflower: conservation management of isolated populations. Biol. Conservation *142*: 1836-1843.

Matthews J.F., Allison J.R., Ware R.T., Sr., and Nordman C. (2002). *Helianthus verticillatus* Small (Asteraceae) rediscovered and redescribed. *Castanea 67*: 13-24.

Murashige T. and Skoog F. (1962). A revised medium for rapid growth and bioassays with tobacco tissue cultures. Physiol. Plant. *15*:473-497.

Peters, J. (2005). Tetrazolium testing handbook. Association of Official Seed Analysts, Las Cruces, NM.

Combined Proceedings IPPS 68:426-428. 2018.

Pest Management in Propagation

Suzanne Wainwright-Evans
Buglady Consulting, 4660 Trestle Lane, Slatington , Pennsylvania 18080, USA

buglady@bugladyconsulting.com

Keywords: Stock plants, scouting, biocontrol, integrated pest management (IPM), two-spotted spider mites, whitefly, aphids, western flower thrips, inspect cutting material, insecticide soaps, oils, microbials (bacteria), *Beauveria bassiana*, predatory mites, *Neoseiulus cucumeris*.

INTRODUCTION

When it comes to pest management the best place to start your program is from the beginning, and this means in propagation. If your young plants are not clean of insects and disease, it is going to be an uphill battle.

To start, if you have your own stock plants, keep them clean. This can be challenging because these plants are often allowed to grow large and dense in order to produce many cuttings. Getting good spray coverage on plants with dense growth can be very difficult. Often pest populations seemed to be under control, but their numbers are frequently just knocked back. In these situations, systemic insecticides can be useful but remember they may not control all pests present - and in some cases, create more problems.

If your target pests are two-spotted spider mites, whitefly, aphids or western flower thrips, biocontrol has proven to be very effective for these pests. These beneficials can find their way into very dense plants, getting where contact sprays sometimes cannot get. Always make sure you have correctly identified the pest or pathogen before you start any pest management program.

SCOUT CUTTINGS PRIOR TO STICKING

Once cuttings have been taken (or if you buy them in), a thorough inspection should be done to check for potential problems. Sometimes you can see large adult insects, but the tiny eggs and immature insects/ mites can easily slip by - hitch-hiking their way in. This is because of their small size and some eggs, like those of the western flower thrips, are laid inside of the plant tissue so they cannot be seen. Also, when cuttings are shipped in, they often arrive in large volumes, and frequently it's a race against the clock to get them stuck. This rush may not leave time for adequate inspections.

One approach growing in popularly is to dip the cuttings before sticking them (Figure 1). Researchers in Canada have been looking at immersing cuttings in treatments like insecticide soaps, oils and microbials (like *Beauveria bassiana*). The work was done at Vineland Research and Innovation Centre and the dip trials proved to be very successful.

Figure 1. Immersing cuttings in treatments containing insecticide soaps, oils and microbials (like *Beauveria bassiana*) is an excellent IPM for pest control.

To learn more, you can look on their website, http://www.vinelandresearch.com, where they cover rates, pests and how to. They even have videos, showing how to do the process.

This does not mean you should immediately run out and start dunking your cuttings to control pests. Testing must always be done to make sure there are no phototoxicity issues from these treatments and that the treatment works to control the target pest.

Even if you start with a clean cutting, that does not ensure that pests will not arrive on their own. One of the easiest things to do to help monitor whether a problem has arrived it to use sticky cards. Yellow cards will attract most flying insects including fungus gnats, shore flies, thrips, winged aphids, whiteflies and others (Figures 2 and 3). Sticky cards will not help you monitor for mites, since mites do not fly.

Figure 2. Sticky cards are essential part of managing pests in a greenhouse. In propagation they can help monitor fungus gnat populations which are a major pest problem for young plant producers. They will also trap pests like thrips, whiteflies and other flying insects.

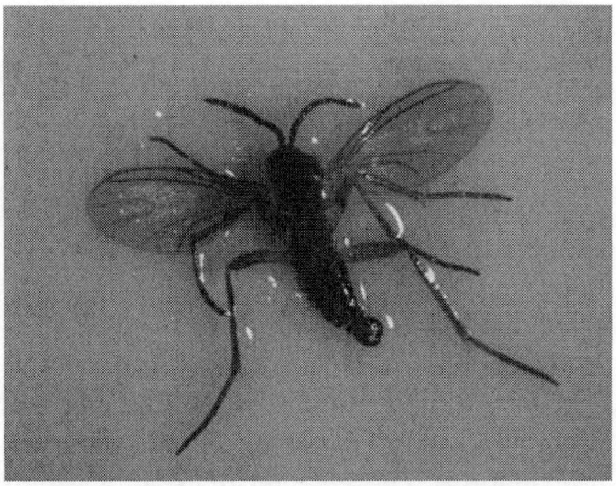

Figure 3. Adult fungus gnats can be easily identified by their mosquito shaped body and the "Y" shaped vain on their wings.

STARTING A BIOLOGICAL CONTROL PROGRAM IN PROPAGATION

Once the cuttings have been stuck – is the optimum time to start a biological control program. For some reason growers often wait to treat for pests until they are out of the propagation stage, which just gives pests time to multiply. Beneficial insects, mites and nematodes work well in propagation, looking in every nook and cranny for pests to feed on. Pesticide applications in propagation can be challenging because you need get good spray coverage without causing phototoxicity. Sprays can get washed off in mist and with systemic drenches. And, you need the cuttings' roots to develop before the pesticide can be taken up.

Predatory mites can be applied on the stuck cuttings and will get right to work feeding on pests. One example is *Neoseiulus cucumeris*, a predatory mite that forges on lot of different pest species – such as western flower thrips, broad mite, and spider mites. It is very economical to use easy to get from insectaries, such as Beneficial Insectary, Bioline, Biobest, or Koppert.

Slow release sachets can be used in propagation (Figure 4).

Figure 4. Slow release sachets can be used in propagation. These water-resistant systems slowly release predatory mites for many weeks; they feed on pests like thrips, broad mites, cyclamen mites, spider mites and whitefly.

These water-resistant systems slowly release predatory mites for many weeks; they feed on pests, including thrips, broad mites, cyclamen mites, spider mites and whitefly.

One control method that thrives in propagation under mist are the beneficial nematodes. They are increasingly being used by many commercial growers to very effectively control fungus gnats (which feed on roots of cuttings) and shore flies (which feeds on algae and bacteria); the nematodes love the propagation environment. They are also compatible with most pesticides (talk to your supplier) - and are very affordable to use. Another advantage is that you do not have to worry about developing resistance issues with them, which occurs with pesticide usage.

SUMMARY

Controlling pests early is key to a successful pest management program. Starting with clean propagation/ liner material will help make the rest of the growing process a lot easier. Using biological control in this early stage will not only stop the pests from starting, but also will reduce the amount of pesticide sprays used - reducing chances of resistance issues. Also, if young plants are free of non-biocontrol friendly pesticide residue - once growers pots up their young plants - they can immediately start using beneficials without worrying about pesticide residues.

Product Insectaries

Beneficial insectary
https://www.insectary.com/

Bioline https://www.biolineagrosciences.com/

Biobest https://www.biobestgroup.com/

Koppert https://www.koppertus.com/

Combined Proceedings IPPS 68:429-431. 2018.

Unraveling the Rose Rosette Puzzle

Alan Windham, Mark Windham, Frank Hale, Katherine Solo and Sara Collins
Entomology and Plant Pathology, University of Tennessee, 5201 Marchant Drive
Nashville, Tennessee 37211, USA

awindha1@utk.edu

Keywords: Eriophyid mite, *Phyllocoptes fructiphilus,* best management practices (BMPs), virus, *Emaravirus* sp.

INTRODUCTION

Rose rosette is currently the major plant disease of the rose in the world. A virus disease vectored by an eriophyid mite, *Phyllocoptes fructiphilus*, has caused millions of dollars of losses just this year to major rose growers (Windham, 2018). The Combating Rose Rosette team formed with funding from an USDA Small Crops Research Initiative (SCRI) Grant is looking at all aspects of the disease. The Tennessee group is tasked to develop best management practices (BMP) and evaluate rose species for resistance and study the biology of the mite vector.

Rose rosette symptoms are variable but common symptoms are: abnormal reddening of canes, excessive thorns on canes, rosettes (witches' broom), thickened canes and death (Figure 1). Symptoms may vary by species or cultivar and by season. The most reliable symptom is the rosette.

Early best management practices (BMPs) studies found that pruning out rosettes from an infected plant did not save that plant (Windham, 2016). Infected roses should be removed promptly to stop spread of the disease. Recent studies have shown that symptomatic rose tissue has 40-to-80 times the number of mites as non-symptomatic tissue. This provides incentive to remove infected plants. We also found that breaking up mass plantings of roses with taller non-host plants slowed the spread of rose rosette.

Also, in earlier studies we looked at miticides as a means to protect roses. While common products such as horticultural oil, carbaryl, and Avid were not successful, some products did protect roses on a very short spray interval of one week (Windham, 2017). Current studies are looking at these products at 2, 4- and 6-week intervals.

Figure 1. Symptoms of rose rosette disease may include: a witch's broom (rosette)(top-right), thickened cane, excessive thorns (left), burgundy new growth (left, bottom right).

Resistance studies have looked at various rose species and species. Most rose cultivars, numbered crosses and rose species are susceptible to rose rosette. This fall we plan to release a list of roses that have survived 4 years in our test plots. Recent roses added to our resistance trials include 1600 diploid and polyploidy rose crosses from Dr. Dave Byrnes laboratory at Texas A&M University. We are also trialing the Brindabella series of roses from Suntory in Australia.

A recent graduate of our program studied the biology of *Phyllocoptes fructiphilus* and conducted an extensive survey of the Deep South (Solo, 2017). Stories of a "southern line" of rose rosette had circulated in the rose world. Our student found indeed that was true. If you draw a line from Vicksburg, MS to Tuscaloosa, AL, to Birmingham, AL to Macon, GA - she found little rose rosette below the line. She found that mite numbers were highest above this line, but also found that even though the disease could not be found, mites could be found south of this line. She also found one eriophyid mite species in several locations, south of the line that has only been reported on roses in Israel. In another study, this student found that mites have a tendency to stay on rose tissue even after it's removed from the plant. Early BMP's suggested bagging infected plants during removal. This may not be needed as mites remained on tissue removed from roses up to 48 hours.

The Combating Rose Rosette team is comprised of rose breeders, horticulturalists, plant pathologists, entomologists and diagnosticians.

Work continues on a quick test for rose rosette, biological studies of the mites associated with rose rosette, evaluation for resistance to the virus or mite, mechanisms of resistance to the virus and the complexity of the virus.

Literature Cited

Solo, K., Collins, S. B., Bauchan, J., Ochoa, R., Hale, F.A., Windham, A. S., Williams-Woodward, J., Jacobi, J., Henn, A., and Windham, M.T. (2017). Where might that mite be? A survey of the southeast region for rose rosette virus and its eriophyid mite vector. The Fire Fly: Proc. Ann. Meet. Tenn. Entomol. Soc.

Windham, M. T., Windham, A. S., Henn, A. (2018). A Guide to Rose Diseases and their Management. The American Rose Society.

Windham, M. T., Windham, A. S., and Hale, F. A. (2017). Controlling rose rosette disease with cultural and chemical methods. VII Int. Symp. Rose Res. and Cultiv., Session 7, Poster 8. Angers, France, July 2-7.

Windham, M. T., Hale, F. A., and Windham, A. S. (2016). Managing rose rosette in the landscape-ideas based on experimental data. American Rose, Nov-Dec: 34-36